T0322080

Matrix Calculus, Kronecker Product and Tensor Product

A Practical Approach to Linear Algebra, Multilinear Algebra and Tensor Calculus with Software Implementations

Third Edition

Matrix Calculus, Kronecker Product and Tensor Product

A Practical Approach to Linear Algebra, Multilinear Algebra and Tensor Calculus with Software Implementations

Third Edition

Yorick Hardy
University of the Witwatersrand, South Africa

Willi–Hans Steeb
University of Johannesburg, South Africa

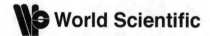
World Scientific

NEW JERSEY • LONDON • SINGAPORE • BEIJING • SHANGHAI • HONG KONG • TAIPEI • CHENNAI • TOKYO

Published by

World Scientific Publishing Co. Pte. Ltd.

5 Toh Tuck Link, Singapore 596224

USA office: 27 Warren Street, Suite 401-402, Hackensack, NJ 07601

UK office: 57 Shelton Street, Covent Garden, London WC2H 9HE

Library of Congress Cataloging-in-Publication Data

Names: Hardy, Yorick, 1976– author. | Steeb, W.-H., author. | Steeb, W.-H.
　　Matrix calculus and Kronecker product.

Title: Matrix calculus, Kronecker product, and tensor product : a practical approach to
　　linear algebra, multilinear algebra, and tensor calculus, with software implementations / by
　　Yorick Hardy (University of the Witwatersrand, South Africa),
　　Willi-Hans Steeb (University of Johannesburg, South Africa).

Other titles: Matrix calculus and Kronecker product

Description: 3rd edition. | New Jersey : World Scientific, 2019. | Previous edition:
　　Matrix calculus and Kronecker product : a practical approach to linear and
　　multilinear algebra / Willi-Hans Steeb, Yorick Hardy (2011). |
　　Includes bibliographical references and index.

Identifiers: LCCN 2019003179 | ISBN 9789811202513 (hardcover : alk. paper)

Subjects: LCSH: Matrices. | Kronecker products.

Classification: LCC QA188 .S662 2019 | DDC 512.9/434--dc23

LC record available at https://lccn.loc.gov/2019003179

British Library Cataloguing-in-Publication Data

A catalogue record for this book is available from the British Library.

For any available supplementary material, please visit
https://www.worldscientific.com/worldscibooks/10.1142/11338#t=suppl

Printed in Singapore

Preface

The Kronecker product of matrices plays an important role in mathematics and in applications found in theoretical physics. Such applications are signal processing where the Fourier and Hadamard matrices play the central role. In group theory and matrix representation theory, the Kronecker product also comes into play. In statistical mechanics, we apply the Kronecker product in the calculation of the partition function and free energy of spin and Fermi systems. Furthermore the spectral theorem for finite dimensional hermitian matrices can be formulated using the Kronecker product. The so-called quantum groups rely heavily on the Kronecker product. Most books on linear algebra and matrix theory investigate the Kronecker product only superficially. This book gives a comprehensive introduction to the Kronecker product of matrices with a large number of applications.

In chapter 1 we give a comprehensive introduction into matrix algebra. The basic definitions and notations are given in section 1.1 and the basic operations are given in section 1.2. The Gram-Schmidt orthonormalization technique is very important in matrix algebra. This technique is described in section 1.3. Matrices are closely associated with linear equations. This connection is described in section 1.4. In section 1.5 unbiased bases are constructed. The trace and determinant of square matrices are introduced and their properties are discussed in section 1.6. The eigenvalue problem plays a central role in physics. Section 1.7 is devoted to this problem. Section 1.8 presents the Cayley-Hamilton theorem which is very useful for computing functions of matrices such as the exponential function. Projection matrices and projection operators are important in Hilbert space theory and quantum mechanics. They are also used in group theoretical reduction in finite group theory. Section 1.9 discusses these matrices. In signal processing

Unitary, Fourier and Hadamard matrices play a central role for the fast Fourier transform and fast Hadamard transform, respectively. Section 1.10 is devoted to these matrices. Transformations of matrices are described in section 1.11. The invariance of the trace and determinant are also discussed. The Cayley transform is defined in section 1.12 and a number of examples are given. Finite groups can be represented as permutation matrices. These matrices are investigated in section 1.13. The spectral theorem for matrices is provided in section 1.14. Various useful matrix decompositions are introduced in section 1.15. The pseudo inverse for all matrices is defined in section 1.16. The vec operator describes an important connection between matrices and vectors. This operator is also important in connection with the Kronecker product. Section 1.17 introduces this operator. The different vector and matrix norms are defined in section 1.18. The relationships between the different norms are explained. We also describe the connection with the eigenvalues of the matrices. Furthermore we discuss the approximation of a matrix by a lower rank matrix. Sequences of vectors and matrices are introduced in section 1.19 and in particular the exponential function is discussed. The commutator is the basic operation of Lie algebras and is central in quantum mechanics. Commutators and anti-commutators are described in section 1.20. Groups and matrices are studied in section 1.21. A number of their properties are given. Section 1.22 introduces Lie algebras. The exponential function of a square matrix is useful in many applications, for example, Lie groups, Lie transformation groups and for the solution of systems of ordinary differential equations. In section 1.23 methods for calculating functions of matrices are described. One classifies square matrices into normal and nonnormal matrices. These concepts are discussed in section 1.24.

Sections 2.1, 2.2 and 2.3 in chapter 2 give an introduction to the Kronecker product. In particular, the connection with matrix multiplication is discussed. In section 2.4 permutation matrices are discussed. Section 2.5 is devoted to the trace and determinant of a matrix and their relation to the Kronecker product. The eigenvalue problem and the Kronecker product is studied in section 2.6. We calculate the eigenvalues and eigenvectors of Kronecker products of matrices. We consider projection matrices and the Kronecker product in section 2.7. Fourier and Hadamard matrices are important in spectral analysis, such as fast Fourier transforms. These matrices are introduced in section 2.8 and their connection with the Kronecker product is described. The direct sum and the Kronecker sum are studied in sections 2.9 and 2.10. Some matrix decompositions which are specific to Kronecker product structures are described in section 2.11. Section 2.12 is devoted to the vec operator and its connection with the Kronecker product. Groups and the Kronecker product are investigated in sections 2.13 and 2.14. In particular the matrix representation of groups is described. In section 2.15, the relation between the Kronecker product, the commutator and the anticommutator is investigated. The inversion of partitioned matrices is discussed in section 2.16. Approximation of matrices by a nearest Kronecker product is the topic of section 2.17. In section 2.18, the Gâteaux derivative is introduced and applications for maps involving matrices are given.

In chapter 3, we study applications in statistical mechanics, quantum mechanics, Lax representation and signal processing for the Kronecker product. The trace and partial trace are defined in section 3.1 and applications are provided. Next, we introduce Pauli spin matrices and give some applications in section 3.2. Spin coherent states playing a central role in quantum mechanics are studied in section 3.3. The Pauli group, Clifford groups and Bell group are discussed in section 3.4. Applications in quantum theory are given in section 3.5. We assume that the Hamilton operator is given by a Hermitian matrix. We investigate the time evolution of the wave function (Schrödinger equation) and the time evolution of a matrix (Heisenberg equation of motion). The eigenvalue problem of the two-point Heisenberg model is solved in detail. For a number of Hamilton operators involving the Kronecker product the partition function is evaluated in section 3.6. The one dimensional Ising model is solved in section 3.7. Fermi systems are studied in section 3.8. We then study the dimer problem, which is a combinatorial problem in section 3.9. The two dimensional Ising model is solved in section 3.10. In section 3.11 the one dimensional Heisenberg model is discussed applying the famous Yang-Baxter relation. An introduction to Hopf algebras is provided in section 3.12. Quantum groups are discussed in section 3.13. Section 3.14 describes the connection of the Kronecker product with the Lax representation for ordinary differential equations. Signal processing and the Kronecker product is discussed in section 3.15. Section 3.16 describes Clebsch-Gordan series. Section 3.17 considers the connection between the Kronecker product and Braid-like relations. The Kronecker product can be used to find fast transforms. Section 3.18 examines the fast Fourier transform in terms of the Kronecker product. Entanglement and the Kronecker product is discussed in section 3.19. The hyperdeterminant is an extension of the determinant introduced by Cayley. In section 3.20 its definition is provided and a number of applications are given. The tensor eigenvalue problem is studied in section 3.21. Carleman and Bell matrices are introduced in section 3.22 and the link with analytic function is established.

The tensor product can be considered as an extension of the Kronecker product to infinite dimensions. Chapter 4 gives an introduction into the tensor product and some applications. The Hilbert space is introduced in section 4.1 and the tensor product in section 4.2. Sections 4.3 and 4.4 give two applications. In the first one, we consider a spin-orbit system and the second one, a Bose-spin system. For the interpretation of quantum mechanics (system and the measuring apparatus), the tensor product and Kronecker product are of central importance. We describe this connection in section 4.5. Section 4.6 introduces the universal enveloping algebra. Tensor fields are introduced in section 4.7 and applications are provided for metric tensor fields.

Chapter 5 presents a number of computer algebra implementations in SymbolicC++ and Maxima on examples and concepts from the previous chapters.

In most sections, a large number of examples and problems serve to illustrate the mathematical tools. The end of a proof is indicated by □. The end of an example is indicated by ♣.

If you have any comments about the book, please send them to us at:

email: yorickhardy@gmail.com
email: steebwilli@gmail.com

Webpage: http://issc.uj.ac.za

The International School for Scientific Computing (ISSC) provides certificate courses for this subject. Please contact the second author if you want to do this course or other courses of the ISSC.

Contents

Symbol Index

$\emptyset$	empty set
$A \cup B$	union of the sets A and B
$A \cap B$	intersection of the sets A and B
$\mathbb{I}$	countable index set
$\mathbb{N}$	the set of positive integers: natural numbers
$\mathbb{Z}$	the set of integers
$\mathbb{Q}$	the set of rational numbers
$\mathbb{R}$	the set of real numbers
$\mathbb{R}^+$	nonnegative real numbers
$\mathbb{C}$	the set of complex numbers
$\mathbb{F}$	field
$\mathcal{A}$	algebra
$\mathbb{R}^n$	the n dimensional real linear space
$\mathbb{C}^n$	the n dimensional complex linear space
i	$:= \sqrt{-1}$
z	complex number
$\Re(z)$	real part of z
$\Im(z)$	imaginary part of z
$\mathbf{u}, \mathbf{v}$	column vectors in $\mathbb{C}^n$
$\mathbf{e}_{1,n}, \ldots, \mathbf{e}_{n,n}$	standard basis in $\mathbb{C}^n$
$\mathbf{0}$	zero vector in $\mathbb{C}^n$
$\otimes$	Kronecker product (direct product, tensor product)
$\oplus$	direct sum
$\oplus_K$	Kronecker sum
$\| \cdot \|$	norm
$\langle , \rangle$	scalar product (inner product) in $\mathbb{C}^n$
tr	trace

det	determinant
$r(A)$	rank of the matrix A
I	unit matrix (identity matrix)
I_n	$n \times n$ unit matrix (identity matrix)
0_n	$n \times n$ zero matrix
$0_{m \times n}$	$m \times n$ zero matrix
A^T	transpose of the matrix A
σ	permutation
P	permutation matrix
U	unitary matrix
Π	projection matrix
$[\,,\,]$	commutator
$[\,,\,]_+$	anti-commutator
δ_{jk}	Kronecker delta 1 if $j = k$ and 0 if $j \neq k$
λ	eigenvalue
G	group
$GL(n, \mathbb{F})$	group of all $n \times n$ invertible matrices over the field $\mathbb{F}$
L	Lie algebra
σ	spin
$\sigma_1, \sigma_2, \sigma_3$	Pauli spin matrices
S_1, S_2, S_3	spin matrices $[S_1, S_2] = iS_3$, $[S_2, S_3] = iS_1$, $[S_3, S_1] = iS_2$
ϵ	real parameter
$c^\dagger, c$	Fermi creation and annihilation operators
$b^\dagger, b$	Bose creation and annihilation operators
T	transfer matrix
$\hat{H}$	Hamilton operator
ω	frequency
t	time
N	Number of lattice sites
J	exchange constant
$Z(\beta)$	partition function $\beta = 1/(k_B T)$

Chapter 1

Matrix Calculus

1.1 Definitions and Notation

We assume that the reader is familiar with some basic terms in linear algebra such as vector spaces, linearly dependent vectors, matrix addition and matrix multiplication (Horn and Johnson [32], Laub [41], Axler [2]).

Throughout we consider matrices over the field of complex numbers $\mathbb{C}$ or real number $\mathbb{R}$. Let $z \in \mathbb{C}$ with $z = x + iy$ and $x, y \in \mathbb{R}$. Then $\bar{z} = x - iy$. In some cases we restrict the underlying field to the real numbers $\mathbb{R}$. The matrices are denoted by A, B, C, D, X, Y. The matrix elements (entries) of the matrix A are denoted by a_{jk}. For the column vectors we write $\mathbf{u}$, $\mathbf{v}$, $\mathbf{w}$. The zero column vector is denoted by $\mathbf{0}$. Let A be a matrix. Then A^T denotes the transpose and $\bar{A}$ is the complex conjugate matrix. We call A^* the adjoint matrix, where $A^* := \bar{A}^T$. A special role is played by the $n \times n$ matrices, i.e. the square matrices. In this case we also say the matrix is of order n. I_n denotes the $n \times n$ unit matrix (also called identity matrix). The $n \times n$ zero matrix is denoted by 0_n.

Let V be a vector space of finite dimension n, over the field $\mathbb{R}$ of real numbers, or the field $\mathbb{C}$ of complex numbers. If there is no need to distinguish between the two, we speak of the field $\mathbb{F}$ of *scalars*. A *basis* of V is a set $\{\mathbf{e}_1, \mathbf{e}_2, \ldots, \mathbf{e}_n\}$ of n linearly independent vectors of V, denoted by $(\mathbf{e}_j)_{j=1}^n$. Every vector $\mathbf{v} \in V$ then has the unique representation

$$\mathbf{v} = \sum_{j=1}^n v_j \mathbf{e}_j$$

the scalars v_j, which we will sometimes denote by $(\mathbf{v})_j$, being the *components* of the vector $\mathbf{v}$ relative to the basis $(\mathbf{e})_j$. As long as a basis is fixed unambiguously, it is always possible to identify V with $\mathbb{F}^n$. In matrix notation, the vector $\mathbf{v}$ will always be represented by the *column vector*

$$\mathbf{v} = \begin{pmatrix} v_1 \\ v_2 \\ \vdots \\ v_n \end{pmatrix}.$$

while $\mathbf{v}^T$ and $\mathbf{v}^*$ will denote the following *row vectors*

$$\mathbf{v}^T = (v_1, v_2, \ldots, v_n), \qquad \mathbf{v}^* = (\bar{v}_1, \bar{v}_2, \ldots, \bar{v}_n)$$

where $\bar{\alpha}$ is the complex conjugate of α. The row vector $\mathbf{v}^T$ is the *transpose* of the column vector $\mathbf{v}$, and the row vector $\mathbf{v}^*$ is the *conjugate transpose* of the column vector $\mathbf{v}$.

Definition 1.1. Let $\mathbb{C}^n$ be the familiar n dimensional vector space. Let $\mathbf{u}, \mathbf{v} \in \mathbb{C}^n$. Then the *scalar* (or *inner*) *product* is defined as

$$\langle \mathbf{u}, \mathbf{v} \rangle := \sum_{j=1}^{n} \bar{u}_j v_j.$$

Obviously $\langle \mathbf{u}, \mathbf{v} \rangle = \overline{\langle \mathbf{v}, \mathbf{u} \rangle}$ and $\langle \mathbf{u}_1 + \mathbf{u}_2, \mathbf{v} \rangle = \langle \mathbf{u}_1, \mathbf{v} \rangle + \langle \mathbf{u}_2, \mathbf{v} \rangle$.

Since $\mathbf{u}$ and $\mathbf{v}$ are considered as column vectors, the scalar product can be written in matrix notation as $\langle \mathbf{u}, \mathbf{v} \rangle \equiv \mathbf{u}^* \mathbf{v}$.

Definition 1.2. Two vectors $\mathbf{u}, \mathbf{v} \in \mathbb{C}^n$ are called *orthogonal* if $\langle \mathbf{u}, \mathbf{v} \rangle = 0$.

Example 1.1. Let $\mathbf{u} = \begin{pmatrix} 1 & 1 \end{pmatrix}^T$, $\mathbf{v} = \begin{pmatrix} 1 & -1 \end{pmatrix}^T$. Then $\langle \mathbf{u}, \mathbf{v} \rangle = 0$. ♣

The scalar product induces a *norm* of $\mathbf{u}$ defined by

$$\|\mathbf{u}\| := \sqrt{\langle \mathbf{u}, \mathbf{u} \rangle}$$

with $\|\mathbf{u}\| \geq 0$.

Definition 1.3. A vector $\mathbf{u} \in \mathbb{C}^n$ is called *normalized* if $\langle \mathbf{u}, \mathbf{u} \rangle = 1$.

Example 1.2. The vectors

$$\mathbf{u} = \frac{1}{\sqrt{2}} \begin{pmatrix} 1 \\ 1 \end{pmatrix}, \qquad \mathbf{v} = \frac{1}{\sqrt{2}} \begin{pmatrix} 1 \\ -1 \end{pmatrix}$$

are normalized and form an orthonormal basis in the vector space $\mathbb{R}^2$. ♣

Let V and W be two vector spaces over the same field, equipped with bases $(\mathbf{e}_j)_{j=1}^n$ and $(\mathbf{f}_i)_{i=1}^m$, respectively. Relative to these bases, a linear transformation

$$\mathcal{A} : V \to W$$

is represented by the matrix having m rows and n columns

$$A = \begin{pmatrix} a_{11} & a_{12} & \cdots & a_{1n} \\ a_{21} & a_{22} & \cdots & a_{2n} \\ \vdots & \vdots & \ddots & \vdots \\ a_{m1} & a_{m2} & \cdots & a_{mn} \end{pmatrix}.$$

The elements a_{ij} of the matrix A are defined uniquely by the relations

$$\mathcal{A}\mathbf{e}_j = \sum_{i=1}^m a_{ij}\mathbf{f}_i, \qquad j = 1, 2, \ldots, n.$$

Equivalently, the jth column vector

$$\begin{pmatrix} a_{1j} \\ a_{2j} \\ \vdots \\ a_{mj} \end{pmatrix}$$

of the matrix A represents the vector $\mathcal{A}\mathbf{e}_j$ relative to the basis $(\mathbf{f}_i)_{i=1}^m$. We call

$$(a_{i1}\, a_{i2}\, \ldots\, a_{in})$$

the ith row vector of the matrix A. A matrix with m rows and n columns is called a matrix of type (m, n), and the vector space over the field $\mathbb{F}$ consisting of matrices of type (m, n) with elements in $\mathbb{F}$ is denoted by $\mathcal{A}_{m,n}$. A column vector is then a matrix of type $(m, 1)$ and a row vector a matrix of type $(1, n)$. A matrix is called real or complex according to whether its elements are in the field $\mathbb{R}$ or the field $\mathbb{C}$. A matrix A with elements a_{ij} is written as $A = (a_{ij})$ the first index i always designating the row and the second, j, the column.

Definition 1.4. A matrix with all its elements 0 is called the *zero matrix* or *null matrix*.

Definition 1.5. Given a matrix $A \in \mathcal{A}_{m,n}(\mathbb{C})$, the matrix $A^* \in \mathcal{A}_{n,m}(\mathbb{C})$ denotes the *adjoint* of the matrix A and is defined uniquely by the relations

$$\langle A\mathbf{u}, \mathbf{v} \rangle_m = \langle \mathbf{u}, A^*\mathbf{v} \rangle_n \text{ for every } \mathbf{u} \in \mathbb{C}^n, \ \mathbf{v} \in \mathbb{C}^m$$

which imply that $(A^*)_{ij} = \bar{a}_{ji}$.

Definition 1.6. Given a matrix $A = \mathcal{A}_{m,n}(\mathbb{R})$, the matrix $A^T \in \mathcal{A}_{n,m}(\mathbb{R})$ denotes the *transpose* of a matrix A and is defined uniquely by the relations

$$\langle A\mathbf{u}, \mathbf{v} \rangle_m = \langle \mathbf{u}, A^T \mathbf{v} \rangle_n \quad \text{for every} \ \ \mathbf{u} \in \mathbb{R}^n, \ \mathbf{v} \in \mathbb{R}^m$$

which imply that $(A^T)_{ij} = a_{ji}$.

To the composition of linear transformations there corresponds the multiplication of matrices.

Definition 1.7. If $A = (a_{ij})$ is a matrix of type (m, l) and $B = (b_{kj})$ of type (l, n), their *matrix product* AB is the matrix of type (m, n) defined by

$$(AB)_{ij} = \sum_{k=1}^{l} a_{ik} b_{kj}.$$

We have $(AB)^T = B^T A^T$, $(AB)^* = B^* A^*$.

Note that $AB \neq BA$, in general, where A and B are $n \times n$ matrices. If $AB = BA$ we say that the matrices A and B commute.

Definition 1.8. A matrix of type (n, n) is said to be *square*, or a matrix of *order* n if it is desired to make explicit the integer n; it is convenient to speak of a matrix as rectangular if it is not necessarily square.

Definition 1.9. If $A = (a_{ij})$ is a square matrix, the elements a_{ii} are called *diagonal elements*, and the elements $a_{ij}, i \neq j$, are called *off-diagonal elements*.

Definition 1.10. The $n \times n$ *identity matrix* (also called $n \times n$ *unit matrix*) is the square matrix $I := (\delta_{ij})$, where δ_{ij} is the *Kronecker delta* with $\delta_{ii} = 1$ for $i = 1, \ldots, n$ and 0 otherwise.

Definition 1.11. A square matrix A is *invertible* if there exists a matrix (which is unique, if it does exist), written as A^{-1} and called the *inverse* of the matrix A, which satisfies $AA^{-1} = A^{-1}A = I$. Otherwise, the matrix is said to be *singular*.

If A and B are invertible $n \times n$ matrices, then

$$(AB)^{-1} = B^{-1}A^{-1}, \qquad (A^T)^{-1} = (A^{-1})^T, \qquad (A^*)^{-1} = (A^{-1})^*.$$

Definition 1.12. A square matrix A is *symmetric* if A is real and $A = A^T$.

The sum of two symmetric matrices is again a symmetric matrix.

Definition 1.13. A square matrix A is *skew-symmetric* if A is real and $A = -A^T$.

The sum of two skew-symmetric matrices is again a skew-symmetric matrix.

Every square matrix A over $\mathbb{R}$ can be written as sum of a symmetric matrix S and a skew-symmetric matrix T, i.e. $A = T + S$. Thus

$$S = \frac{1}{2}(A + A^T), \qquad T = \frac{1}{2}(A - A^T).$$

Definition 1.14. A square matrix A over $\mathbb{C}$ is *Hermitian* if $A = A^*$.

The sum of two Hermitian matrices is again a Hermitian matrix.

Definition 1.15. A square matrix A over $\mathbb{C}$ is *skew-Hermitian* if $A = -A^*$.

The sum of two skew-Hermitian matrices is again a skew-Hermitian matrix. Let A be a skew-Hermitian matrix. Then iA is a hermitian matrix.

Definition 1.16. A Hermitian $n \times n$ matrix A is *positive semidefinite* if

$$\mathbf{v}^* A \mathbf{v} \geq 0$$

for all nonzero $\mathbf{v} \in \mathbb{C}^n$.

Example 1.3. The 2×2 matrices

$$A = \begin{pmatrix} 1 & 1 \\ 1 & 1 \end{pmatrix}, \quad B = \begin{pmatrix} 1 & -1 \\ -1 & 1 \end{pmatrix}$$

are positive semidefinite. The eigenvalues of both matrices are 0, 2 and $AB = BA = 0_2$. ♣

Let B be an arbitrary $m \times n$ matrix over $\mathbb{C}$. Then the $n \times n$ matrix B^*B is positive semidefinite.

Definition 1.17. A square matrix A is *orthogonal* if A is real and

$$AA^T = A^T A = I.$$

Thus for an orthogonal matrix A we have $A^{-1} = A^T$. The product of two orthogonal matrices is again an orthogonal matrix. The inverse of an orthogonal matrix is again an orthogonal matrix. The orthogonal matrices form a group under matrix multiplication.

Definition 1.18. A square matrix A is *unitary* if $AA^* = A^*A = I$.

Thus for a unitary matrix we have $A^* = A^{-1}$. The matrix product of two unitary matrices is again a unitary matrix and the inverse of a unitary matrix is again a unitary matrix. The unitary matrices form a group under matrix multiplication.

Example 1.4. Consider the 2×2 Pauli spin matrix

$$\sigma_2 = \begin{pmatrix} 0 & -i \\ i & 0 \end{pmatrix}.$$

The matrix σ_2 is Hermitian and unitary. We have $\sigma_2^* = \sigma_2$ and $\sigma_2^* = \sigma_2^{-1}$. Furthermore $\sigma_2^2 = I_2$. The eigenvalues are $+1$ and -1. ♣

Definition 1.19. A square matrix is *normal* if $AA^* = A^*A$.

Example 1.5. Consider the 2×2 matrices

$$A = \begin{pmatrix} 0 & i \\ -i & 0 \end{pmatrix}, \quad B = \begin{pmatrix} 0 & i \\ 0 & 0 \end{pmatrix}.$$

The matrix A is normal. The matrix B is not a normal matrix. We call such a matrix nonnormal. Note that B^*B and $B + B^*$ are normal. ♣

Normal matrices include diagonal, real symmetric, real skew-symmetric, orthogonal, Hermitian, skew-Hermitian, and unitary matrices.

Definition 1.20. A matrix $A = (a_{ij})$ is *diagonal* if $a_{ij} = 0$ for $i \neq j$ and is written as

$$A = \mathrm{diag}(a_{ii}) = \mathrm{diag}(a_{11}, a_{22}, \ldots, a_{nn}).$$

The matrix product of two $n \times n$ diagonal matrices is again a diagonal matrix.

Definition 1.21. Let $A = (a_{ij})$ be an $m \times n$ matrix over a field $\mathbb{F}$. The columns of A generate a subspace of $\mathbb{F}^m$, whose dimension is called the *column rank* of A. The rows generate a subspace of $\mathbb{F}^n$ whose dimension is called the *row rank* of A. In other words: the column rank of A is the maximum number of linearly independent columns, and the row rank is the maximum number of linearly independent rows. The row rank and the column rank of A are equal to the same number r. Thus r is simply called the *rank* of the matrix A.

Example 1.6. The rank of the 2×3 matrix

$$A = \begin{pmatrix} 1 & 2 & 3 \\ 4 & 5 & 6 \end{pmatrix}$$

and the 3×2 matrix A^T is $r(A) = r(A^T) = 2$. ♣

The rank of the matrix product of two matrices cannot exceed the rank of either factors.

Definition 1.22. The *kernel* (or *null space*) of an $m \times n$ matrix A is the subspace of vectors $\mathbf{x}$ in $\mathbb{C}^n$ for which $A\mathbf{x} = \mathbf{0}$. The dimension of this subspace is the *nullity* of A.

Example 1.7. Consider the matrix

$$A = \begin{pmatrix} 1 & 1 \\ 1 & 1 \end{pmatrix}.$$

Then from the linear equation $A\mathbf{x} = \mathbf{0}$ we obtain $x_1 + x_2 = 0$. The null space of A is the set of solutions to this equation, i.e. a line through the origin of $\mathbb{R}^2$. The nullity of A is equal to 1. ♣

Definition 1.23. Let A, B be $n \times n$ matrices. Then the *commutator* of A and B is defined by $[A, B] := AB - BA$. Obviously we have $[A, B] = -[B, A]$ and if C is another $n \times n$ matrix

$$[A, B + C] = [A, B] + [A, C].$$

Consider the vector space of all $n \times n$ matrices over $\mathbb{C}$. A basis is given by the *elementary matrices*

$$(E_{ij}), \qquad i, j = 1, 2, \ldots, n$$

where (E_{ij}) is the matrix having 1 in the (i, j) position and 0 elsewhere. Since

$$(E_{ij})(E_{kl}) = \delta_{jk}(E_{il})$$

it follows that the commutator is given by

$$[(E_{ij}), (E_{kl})] = \delta_{jk}(E_{il}) - \delta_{li}(E_{kj}).$$

Thus the coefficients are all ± 1 or 0.

Let A, B be $n \times n$ matrices. Then the *anticommutator* of A and B is defined by $[A, B]_+ := AB + BA$.

Let

$$\sigma_1 = \begin{pmatrix} 0 & 1 \\ 1 & 0 \end{pmatrix}, \quad \sigma_2 = \begin{pmatrix} 0 & -i \\ i & 0 \end{pmatrix}.$$

Then $[\sigma_1, \sigma_2]_+ = 0_2$.

Exercises. (1) Let A, B be $n \times n$ upper triangular matrices. Can we conclude that $AB = BA$?

(2) Let A be an arbitrary $n \times n$ matrix. Let B be a diagonal matrix. Is $AB = BA$?

(3) Let A be a normal matrix and U be a unitary matrix. Show that U^*AU is a normal matrix. Let B be a nonnormal matrix. Is U^*BU a nonnormal matrix?

(4) Show that the following operations, called elementary transformations, on a matrix do not change its rank:
(i) The interchange of the i-th and j-th rows.
(ii) The interchange of the i-th and j-th columns.

(5) Let A and B be two $n \times n$ matrices. Is it possible to have $AB + BA = 0_n$?

(6) Let A_k, $1 \leq k \leq m$, be matrices of order n satisfying $\sum_{k=1}^{m} A_k = I$. Show that the following conditions are equivalent

  (i) $A_k = (A_k)^2$, $1 \leq k \leq m$
  (ii) $A_k A_l = 0$ for $k \neq l$, $1 \leq k, l \leq m$
  (iii) $\displaystyle\sum_{k=1}^{m} r(A_k) = n$

where $r(A)$ denotes the rank of the matrix A.

(7) Prove that if A is of order $m \times n$, B is of order $n \times p$ and C is of order $p \times q$, then (*associative law*) $A(BC) = (AB)C$.

(8) Let S be an invertible $n \times n$ matrix. Let A be an arbitrary $n \times n$ matrix and $\tilde{A} = SAS^{-1}$. Show that $\tilde{A}^2 = SA^2S^{-1}$.

(9) Is the $n \times n$ matrix $M = I_n + \mathbf{v}\mathbf{v}^*$ invertible, where $\mathbf{v}$ is a normalized (column) vector in $\mathbb{C}^n$?

(10) Can one find 2×2 matrices A, B such that $[A, B] = I_2$, where I_2 is the 2×2 identity matrix?

1.2 Matrix Operations

Let $\mathbb{F}$ be a field, for example the set of real numbers $\mathbb{R}$ or the set of complex numbers $\mathbb{C}$. Let m, n be two integers ≥ 1. An array A of numbers in $\mathbb{F}$

$$\begin{pmatrix} a_{11} & a_{12} & a_{13} & \cdots & a_{1n} \\ a_{21} & a_{22} & a_{23} & \cdots & a_{2n} \\ \vdots & \vdots & \vdots & \ddots & \vdots \\ a_{m1} & a_{m2} & a_{m3} & \cdots & a_{mn} \end{pmatrix} = (a_{ij})$$

is called an $m \times n$ *matrix* with entry a_{ij} in the ith row and jth column. A *row vector* is a $1 \times n$ matrix. A *column vector* is an $n \times 1$ matrix. We have a *zero matrix*, in which $a_{ij} = 0$ for all i, j.

Let $A = (a_{ij})$ and $B = (b_{ij})$ be two $m \times n$ matrices. We define $A + B$ to be the $m \times n$ matrix whose entry in the i-th row and j-th column is $a_{ij} + b_{ij}$. The $m \times n$ matrices over a field $\mathbb{F}$ form a vector space.

As described above matrix multiplication is only defined between two matrices if the number of columns of the first matrix is the same as the number of rows of the second matrix. If A is an $m \times n$ matrix and B is an $n \times p$ matrix, then the matrix product AB is an $m \times p$ matrix defined by

$$(AB)_{ij} = \sum_{r=1}^{n} a_{ir} b_{rj}$$

for each pair i and j, where $(AB)_{ij}$ denotes the (i, j)th entry in AB.

Example 1.8. Let $x \in \mathbb{R}$. Consider the column vectors in $\mathbb{R}^3$

$$\mathbf{v}(x) = \begin{pmatrix} 1 \\ x \\ x^2 \end{pmatrix}, \quad \mathbf{w}(x) = \begin{pmatrix} x^2 \\ -2x \\ 1 \end{pmatrix}.$$

Then $\mathbf{v}^T(x)\mathbf{w}(x) = 0$, $d\mathbf{w}(x)/dx = (2x, -2, 0)^T$ and $\mathbf{v}^T(x)d\mathbf{w}(x)/dx = 0$. Let $M(x) = \mathbf{w}(x)\mathbf{v}^T(x)$. Then

$$M(x) = \mathbf{w}(x)\mathbf{v}^T(x) = \begin{pmatrix} x^2 & x^3 & x^4 \\ -2x & -2x^2 & -2x^3 \\ 1 & x & x^2 \end{pmatrix}, \quad \frac{dM}{dx} = \begin{pmatrix} 2x & 3x^2 & 4x^3 \\ -2 & -4x & -6x^2 \\ 0 & 1 & 2x \end{pmatrix}.$$

Obviously $M(x)$ is nilpotent with $M^2(x) = 0_3$. Furthermore

$$M(x)\frac{dM}{dx} = \frac{dM}{dx}M(x) = 0_3.$$

♣

Let $A = (a_{ij})$ and $B = (b_{ij})$ be two $m \times n$ matrices with entries in some field. Then their *entrywise product* (also called *Hadamard product*) of A and B, that is the $m \times n$ matrix $A \bullet B$ whose (i, j)th entry is $a_{ij} b_{ij}$.

Example 1.9. Let

$$A = \begin{pmatrix} 1 & i \\ -1 & i \end{pmatrix}, \quad B = \begin{pmatrix} 2 & -1 \\ 1 & 0 \end{pmatrix}.$$

Then

$$A + B = \begin{pmatrix} 3 & i-1 \\ 0 & i \end{pmatrix}, \quad A \bullet B = \begin{pmatrix} 2 & -i \\ -1 & 0 \end{pmatrix}, \quad AB = \begin{pmatrix} i+2 & -1 \\ i-2 & 1 \end{pmatrix}. \quad \clubsuit$$

Exercises. (1) Let A, B be $n \times n$ matrices over $\mathbb{C}$. We define

$$A \Diamond B := AB - A \bullet B$$

where $\bullet$ is the entrywise multiplication. Is $A \Diamond B = B \Diamond A$? Is $(A \Diamond B) \Diamond C = A \Diamond (B \Diamond C)$?

(2) Consider the three *rotation matrices* in $\mathbb{R}^3$

$$R_{12}(\theta) = \begin{pmatrix} \cos(\theta) & -\sin(\theta) & 0 \\ \sin(\theta) & \cos(\theta) & 0 \\ 0 & 0 & 1 \end{pmatrix}, \quad R_{23}(\phi) = \begin{pmatrix} 1 & 0 & 0 \\ 0 & \cos(\phi) & -\sin(\phi) \\ 0 & \sin(\phi) & \cos(\phi) \end{pmatrix},$$

$$R_{13}(\psi) = \begin{pmatrix} \cos(\psi) & 0 & -\sin(\psi) \\ 0 & 1 & 0 \\ \sin(\psi) & 0 & \cos(\psi) \end{pmatrix}.$$

Let $R(\theta, \phi, \psi) = R_{12}(\theta) R_{23}(\phi) R_{13}(\psi)$. Find θ, ϕ, ψ such that

$$R(\theta, \phi, \psi) = \frac{1}{\sqrt{2}} \begin{pmatrix} 1 \\ 0 \\ 1 \end{pmatrix} = \begin{pmatrix} 0 \\ 1 \\ 0 \end{pmatrix}.$$

(3) Let $n \geq 2$ and E_{ij} $(i, j = 1, \ldots, n)$ be the *elementary matrices*. One defines

$$V_{ij}(1) = E_{ij}$$

$$V_{ij}(2) = \sum_{i_1=1}^{n} E_{ii_1} E_{i_1 j}$$

$$\vdots$$

$$V_{ij}(q) = \sum_{i_1, i_2, \ldots, i_{q-1}=1}^{n} E_{ii_1} E_{i_1 i_2} \ldots E_{i_{q-1} j}.$$

Show that

$$[E_{ij}, V_{k\ell}(q)] = \delta_{jk} V_{i\ell}(q) - \delta_{i\ell} V_{kj}(q).$$

1.3 Gram-Schmidt Orthonormalization

The *Gram-Schmidt algorithm* is as follows: Let $\mathbf{v}_1, \mathbf{v}_2, \ldots, \mathbf{v}_n$ be a basis in the Hilbert space $\mathbb{C}^n$. We define

$$\mathbf{w}_1 := \mathbf{v}_1$$

$$\mathbf{w}_2 := \mathbf{v}_2 - \frac{\langle \mathbf{w}_1, \mathbf{v}_2 \rangle}{\langle \mathbf{w}_1, \mathbf{w}_1 \rangle} \mathbf{w}_1$$

$$\mathbf{w}_3 := \mathbf{v}_3 - \frac{\langle \mathbf{w}_2, \mathbf{v}_3 \rangle}{\langle \mathbf{w}_2, \mathbf{w}_2 \rangle} \mathbf{w}_2 - \frac{\langle \mathbf{w}_1, \mathbf{v}_3 \rangle}{\langle \mathbf{w}_1, \mathbf{w}_1 \rangle} \mathbf{w}_1$$

$$\vdots$$

$$\mathbf{w}_n := \mathbf{v}_n - \frac{\langle \mathbf{w}_{n-1}, \mathbf{v}_n \rangle}{\langle \mathbf{w}_{n-1}, \mathbf{w}_{n-1} \rangle} \mathbf{w}_{n-1} - \cdots - \frac{\langle \mathbf{w}_1, \mathbf{v}_n \rangle}{\langle \mathbf{w}_1, \mathbf{w}_1 \rangle} \mathbf{w}_1.$$

Then the vectors $\mathbf{w}_1, \mathbf{w}_2, \ldots, \mathbf{w}_n$ form an orthogonal basis in $\mathbb{C}^n$. Normalizing these vectors yields an orthonormal basis in $\mathbb{C}^n$.

Example 1.10. Let

$$\mathbf{v}_1 = \begin{pmatrix} 1 \\ 0 \\ 0 \\ 0 \end{pmatrix}, \quad \mathbf{v}_2 = \begin{pmatrix} 1 \\ 1 \\ 0 \\ 0 \end{pmatrix}, \quad \mathbf{v}_3 = \begin{pmatrix} 1 \\ 1 \\ 1 \\ 0 \end{pmatrix}, \quad \mathbf{v}_4 = \begin{pmatrix} 1 \\ 1 \\ 1 \\ 1 \end{pmatrix}.$$

We find the standard basis

$$\mathbf{w}_1 = \begin{pmatrix} 1 \\ 0 \\ 0 \\ 0 \end{pmatrix}, \quad \mathbf{w}_2 = \begin{pmatrix} 0 \\ 1 \\ 0 \\ 0 \end{pmatrix}, \quad \mathbf{w}_3 = \begin{pmatrix} 0 \\ 0 \\ 1 \\ 0 \end{pmatrix}, \quad \mathbf{w}_4 = \begin{pmatrix} 0 \\ 0 \\ 0 \\ 1 \end{pmatrix}.$$

These vectors are already normalized. ♣

Example 1.11. Let

$$\mathbf{v}_1 = \begin{pmatrix} 1 \\ 1 \\ 1 \\ 1 \end{pmatrix}, \quad \mathbf{v}_2 = \begin{pmatrix} 1 \\ 1 \\ 1 \\ 0 \end{pmatrix}, \quad \mathbf{v}_3 = \begin{pmatrix} 1 \\ 1 \\ 0 \\ 0 \end{pmatrix}, \quad \mathbf{v}_4 = \begin{pmatrix} 1 \\ 0 \\ 0 \\ 0 \end{pmatrix}.$$

We find

$$\mathbf{w}_1 = \begin{pmatrix} 1 \\ 1 \\ 1 \\ 1 \end{pmatrix}, \quad \mathbf{w}_2 = \frac{1}{4} \begin{pmatrix} 1 \\ 1 \\ 1 \\ -3 \end{pmatrix}, \quad \mathbf{w}_3 = \frac{1}{3} \begin{pmatrix} 1 \\ 1 \\ -2 \\ 0 \end{pmatrix}, \quad \mathbf{w}_4 = \frac{1}{2} \begin{pmatrix} 1 \\ -1 \\ 0 \\ 0 \end{pmatrix}$$

which after normalization gives the orthonormal basis

$$\left\{ \frac{1}{2}(1,1,1,1)^T, \frac{1}{\sqrt{12}}(1,1,1,-3)^T, \frac{1}{\sqrt{6}}(1,1,-2,0)^T, \frac{1}{\sqrt{2}}(1,-1,0,0)^T \right\}. \clubsuit$$

Exercise. (1) Let $x \in \mathbb{R}$. Show that the vectors in $\mathbb{R}^4$

$$\begin{pmatrix} 1 \\ x \\ x^2 \\ x^3 \end{pmatrix}, \quad \begin{pmatrix} 0 \\ 1 \\ 2x \\ 3x^2 \end{pmatrix}, \quad \begin{pmatrix} 0 \\ 0 \\ 2 \\ 6x \end{pmatrix}, \quad \begin{pmatrix} 0 \\ 0 \\ 0 \\ 6 \end{pmatrix}$$

are linearly independent. Apply the Gram-Schmidt orthonormalization algorithm to this set of vectors.

(2) Let $x \in \mathbb{R}$. Apply the Gram-Schmidt orthonormalization algorithm to the 2×2 matrices

$$\begin{pmatrix} 1 & x \\ x^2 & x^3 \end{pmatrix}, \quad \begin{pmatrix} 0 & 1 \\ 2x & 3x^2 \end{pmatrix}, \quad \begin{pmatrix} 0 & 0 \\ 2 & 6x \end{pmatrix}, \quad \begin{pmatrix} 0 & 0 \\ 0 & 6 \end{pmatrix}$$

with the scalar product $\langle A, B \rangle = \operatorname{tr}(AB^T)$.

(3) Show that the four vectors in $\mathbb{R}^4$ with $\theta = 2\pi/5$

$$\mathbf{v}_1 = \begin{pmatrix} \cos(\theta) \\ \sin(\theta) \\ \cos(2\theta) \\ \sin(2\theta) \end{pmatrix}, \quad \mathbf{v}_2 = \begin{pmatrix} \cos(2\theta) \\ \sin(2\theta) \\ \cos(4\theta) \\ \sin(4\theta) \end{pmatrix},$$

$$\mathbf{v}_3 = \begin{pmatrix} \cos(3\theta) \\ \sin(3\theta) \\ \cos(6\theta) \\ \sin(6\theta) \end{pmatrix}, \quad \mathbf{v}_4 = \begin{pmatrix} \cos(4\theta) \\ \sin(4\theta) \\ \cos(8\theta) \\ \sin(8\theta) \end{pmatrix}$$

are linearly independent. Show that $\det(\mathbf{v}_1 \, \mathbf{v}_2 \, \mathbf{v}_3 \, \mathbf{v}_4) = 5\sqrt{5}/4$.

(4) Let $x \in \mathbb{R}$. Are the four vectors in $\mathbb{R}^4$

$$\mathbf{v}_0 = \begin{pmatrix} 1 \\ x - 1/2 \\ x^2/2 - x/2 \\ x^3/6 - x^2/4 + 1/24 \end{pmatrix}, \quad \mathbf{v}_1 = \begin{pmatrix} 0 \\ 1 \\ x - 1/2 \\ x^2/2 - x/2 \end{pmatrix},$$

$$\mathbf{v}_2 = \begin{pmatrix} 0 \\ 0 \\ 1 \\ x - 1/2 \end{pmatrix}, \quad \mathbf{v}_3 = \begin{pmatrix} 0 \\ 0 \\ 0 \\ 1 \end{pmatrix}$$

linearly independent? If so apply the Gram-Schmidt algorithm.

1.4   Linear Equations

Let A be an $m \times n$ matrix over a field $\mathbb{F}$. Let $b_1, \ldots, b_m$ be elements of the field $\mathbb{F}$. The system of equations

$$a_{11}x_1 + a_{12}x_2 + \cdots + a_{1n}x_n = b_1$$
$$a_{21}x_1 + a_{22}x_2 + \cdots + a_{2n}x_n = b_2$$
$$\vdots$$
$$a_{m1}x_1 + a_{m2}x_2 + \cdots + a_{mn}x_n = b_m$$

is called a system of linear equations. We also write $A\mathbf{x} = \mathbf{b}$, where $\mathbf{x}$ and $\mathbf{b}$ are considered as column vectors. The system is said to be homogeneous if all the numbers $b_1, \ldots, b_m$ are equal to 0. The number n is called the number of unknowns, and m is called the number of equations. The system of homogeneous equations also admits the trivial solution

$$x_1 = x_2 = \cdots = x_n = 0.$$

A system of homogeneous equations of m linear equations in n unknowns with $n > m$ admits a nontrivial solution. An under determined linear system is either inconsistent or has infinitely many solutions.

An important special case is $m = n$. Then for the system of linear equations $A\mathbf{x} = \mathbf{b}$ we investigate the cases A^{-1} exists and A^{-1} does not exist. If A^{-1} exists we can write the solution as

$$\mathbf{x} = A^{-1}\mathbf{b}.$$

If $m > n$, then we have an overdetermined system and it can happen that no solution exists. One solves these problems in the least-square sense.

Example 1.12. Consider the system of three linear equations

$$3x_1 - x_2 = -1, \quad x_2 - x_3 = 0, \quad x_1 + x_2 + x_3 = 1.$$

These equations have the matrix representation

$$\begin{pmatrix} 3 & -1 & 0 \\ 0 & 1 & -1 \\ 1 & 1 & 1 \end{pmatrix} \begin{pmatrix} x_1 \\ x_2 \\ x_3 \end{pmatrix} = \begin{pmatrix} -1 \\ 0 \\ 1 \end{pmatrix}.$$

The matrix on the left-hand side is invertible and the solution is given by

$$\begin{pmatrix} x_1 \\ x_2 \\ x_3 \end{pmatrix} = \begin{pmatrix} 2/7 & 1/7 & 1/7 \\ -1/7 & 3/7 & 3/7 \\ -1/7 & -4/7 & 3/7 \end{pmatrix} \begin{pmatrix} -1 \\ 0 \\ 1 \end{pmatrix} = \begin{pmatrix} -1/7 \\ 4/7 \\ 4/7 \end{pmatrix}.$$

♣

Exercises. (1) The sum $1^2 + 2^2 + 3^2 + \cdots + n^2$ can be written as

$$1^2 + 2^2 + 3^2 + \cdots + n^2 = an^3 + bn^2 + cn$$

where the unknown coefficients a, b, c can be determined from the system of three linear equations obtained from $n = 1$, $n = 2$, $n = 3$. Find this system of linear equations. Apply Gauss elimination that finds the solution given by $a = 1/3$, $b = 1/2$, $c = 1/6$.

(2) Consider the system of three linear equations

$$\begin{pmatrix} 1 & 1 & 1 \\ 1 & 2 & 4 \\ 1 & 4 & 10 \end{pmatrix} \begin{pmatrix} x_1 \\ x_2 \\ x_3 \end{pmatrix} = \begin{pmatrix} 1 \\ \epsilon \\ \epsilon^2 \end{pmatrix}$$

where $\epsilon \in \mathbb{R}$. Find the condition on ϵ such that the linear system admits a solutions. Note that the matrix is not invertible, i.e. the determinant is equal to 0.

(3) Find all solutions of the matrix equation

$$\begin{pmatrix} x_1 & x_2 \\ x_3 & x_4 \end{pmatrix} \begin{pmatrix} 0 & 0 \\ 1 & 0 \end{pmatrix} \begin{pmatrix} x_4 & -x_2 \\ -x_3 & x_1 \end{pmatrix} = \begin{pmatrix} 0 & 0 \\ 1 & 0 \end{pmatrix}$$

with the constraint $x_1 x_4 - x_2 x_3 = 1$.

(4) Let $\alpha \in \mathbb{R}$ be fixed. Find all solutions of the equation

$$\begin{pmatrix} 1 & 1 \\ 1 & 1 \end{pmatrix} \begin{pmatrix} x_1 \\ x_2 \end{pmatrix} = \begin{pmatrix} \cos(\alpha) \\ \sin(\alpha) \end{pmatrix}.$$

(5) Let $n > 2$. Consider the system of equations

$$a_k x_1 + b_k x_2 = c_k, \quad k = 1, 2, \ldots, n.$$

Hence we have more equations then the number of unknowns x_1, x_2. Given a point $(\alpha, \beta) \in \mathbb{E}^2$ and the line $a_j x_1 + b_j x_2 = c_j$. The *Euclidean distance* from this point to the line is given by

$$\frac{|a_j \alpha + b_j \beta - c_j|}{\sqrt{a_j^2 + b_j^2}}.$$

Consider the function $f : \mathbb{R}^2 \to \mathbb{R}$

$$f(x_1, x_2) = \sum_{j=1}^{n} \frac{(a_j x_1 + b_j x_2 - c_j)^2}{a_j^2 + b_j^2}.$$

We want to find the point (x_1^*, x_2^*) that minimizes this function. Show that this leads to the system of equations

$$\left(\sum_{j=1}^{n} \frac{a_j^2}{a_j^2 + b_j^2}\right) x_1^* + \left(\sum_{j=1}^{n} \frac{a_j b_j}{a_j^2 + b_j^2}\right) x_2^* = \sum_{j=1}^{n} \frac{a_j c_j}{a_j^2 + b_j^2}$$

$$\left(\sum_{j=1}^{n} \frac{a_j b_j}{a_j^2 + b_j^2}\right) x_1^* + \left(\sum_{j=1}^{n} \frac{b_j^2}{a_j^2 + b_j^2}\right) x_2^* = \sum_{j=1}^{n} \frac{b_j c_j}{a_j^2 + b_j^2}.$$

(6) Let I_3 be the 3×3 identity matrix and

$$X = \begin{pmatrix} 2 & -6 & 6 \\ 3 & -7 & 6 \\ 3 & -6 & 5 \end{pmatrix}.$$

Consider the matrix equation

$$c_0 I_3 + c_1 X + c_2 X^2 = 0_3.$$

Show that the solution is given by $c_0 = -2c_2$, $c_1 = -c_2$.

(7) Solve the system of linear equations

$$\sqrt{2}u = x_1 - x_2, \quad \sqrt{6}v = x_1 + x_2 - 2x_3, \quad 3R = x_1 + x_2 + x_3$$

with respect to x_1, x_2, x_3. In matrix form we have

$$\begin{pmatrix} 1 & -1 & 0 \\ 1 & 1 & -2 \\ 1 & 1 & 1 \end{pmatrix} \begin{pmatrix} x_1 \\ x_2 \\ x_3 \end{pmatrix} = \begin{pmatrix} \sqrt{2}u \\ \sqrt{6}v \\ 3R \end{pmatrix}.$$

1.5 Mutually Unbiased Bases

A complex inner product vector space (also called pre-Hilbert space) is a complex vector space together with an inner product. An inner product vector space which is complete with respect to the norm induced by the inner product is called a Hilbert space. Two finite dimensional Hilbert spaces are considered. The Hilbert space $\mathbb{C}^d$ with the scalar product

$$\langle \mathbf{v}_1, \mathbf{v}_2 \rangle = \mathbf{v}_1^* \mathbf{v}_2$$

and the Hilbert space of $m \times m$ matrices X, Y over $\mathbb{C}$ with the scalar product

$$\langle X, Y \rangle := \mathrm{tr}(XY^*).$$

Let $\mathcal{H}_1$ and $\mathcal{H}_2$ be two finite dimensional Hilbert spaces with

$$\dim(\mathcal{H}_1) = \dim(\mathcal{H}_2) = d.$$

Let $\mathcal{B}_{1,1} = \{|j_1\rangle\}$ and $\mathcal{B}_{1,2} = \{|j_2\rangle\}$ $(j_1, j_2 = 1, \ldots, d$ be two orthonormal bases in the Hilbert space $\mathcal{H}_1$. They are called *mutually unbiased* iff

$$|\langle j_1 | j_2 \rangle|^2 = \frac{1}{d} \quad \text{for all} \quad j_1, j_2 = 1, \ldots, d.$$

In other words two orthonormal basis in the Hilbert space $\mathbb{C}^d$

$$\mathcal{A} = \{\, \mathbf{e}_1, \ldots, \mathbf{e}_d \,\}, \quad \mathcal{B} = \{\, \mathbf{f}_1, \ldots, \mathbf{f}_d \,\}$$

are called mutually unbiased if for every $1 \leq j, k \leq d$

$$|\langle \mathbf{e}_j, \mathbf{f}_k \rangle| = \frac{1}{\sqrt{d}}.$$

Example 1.13. Consider the standard basis in $\mathbb{R}^2$

$$\mathcal{B}_1 = \left\{ \begin{pmatrix} 1 \\ 0 \end{pmatrix}, \begin{pmatrix} 0 \\ 1 \end{pmatrix} \right\}.$$

Applying the *Hadamard matrix* to the standard basis

$$\frac{1}{\sqrt{2}} \begin{pmatrix} 1 & 1 \\ 1 & -1 \end{pmatrix} \begin{pmatrix} 1 \\ 0 \end{pmatrix} = \frac{1}{\sqrt{2}} \begin{pmatrix} 1 \\ 1 \end{pmatrix}, \quad \frac{1}{\sqrt{2}} \begin{pmatrix} 1 & 1 \\ 1 & -1 \end{pmatrix} \begin{pmatrix} 0 \\ 1 \end{pmatrix} = \frac{1}{\sqrt{2}} \begin{pmatrix} 1 \\ -1 \end{pmatrix}$$

provides a mutually unbiased basis.   ♣

Example 1.14. Consider the two normalized vectors

$$\mathbf{v}(\alpha) = \begin{pmatrix} \cos(\alpha) \\ \sin(\alpha) \end{pmatrix}, \quad \mathbf{u}(\beta) = \begin{pmatrix} \cos(\beta) \\ \sin(\beta) \end{pmatrix}$$

in $\mathbb{R}^2$. Then the condition on $\alpha, \beta \in \mathbb{R}$ such that

$$|\mathbf{v}^*(\alpha)\mathbf{u}(\beta)|^2 = \frac{1}{2}$$

is given by $\alpha - \beta \in \{\, \pi/4, \ 3\pi/4, \ 5\pi/4, \ 7\pi/4 \,\} \bmod 2\pi$.   ♣

Example 1.15. Consider the Hilbert space $M_d(\mathbb{C})$ of $d \times d$ matrices with scalar product $\langle A, B \rangle := \text{tr}(AB^*)$, $A, B \in M_d(\mathbb{C})$. Consider an orthogonal basis of d^2 $d \times d$ hermitian matrices $B_1, B_2, \ldots, B_{d^2}$, i.e.

$$\langle B_j, B_k \rangle = \text{tr}(B_j B_k) = d\delta_{jk}$$

since $B_k^* = B_k$ for a hermitian matrix. Let M be a $d \times d$ hermitian matrix. Let

$$m_j = \text{tr}(B_j M), \qquad j = 1, \ldots, d^2.$$

Given m_j and B_j $(j = 1, \ldots, d^2)$. Then M is given by

$$M = \frac{1}{d} \sum_{j=1}^{d^2} m_j B_j$$

since

$$MB_k = \frac{1}{d} \sum_{j=1}^{d^2} m_j B_j B_k.$$

Thus

$$\mathrm{tr}\left(\frac{1}{d} \sum_{j=1}^{d^2} m_j B_j B_k \right) = \frac{1}{d} \sum_{j=1}^{d^2} m_j \mathrm{tr}(B_j B_k) = \sum_{j=1}^{d^2} m_j \delta_{jk} = m_k.$$

♣

Exercises. (1) Consider the two normalized vectors

$$\mathbf{v}(\alpha, \phi) = \begin{pmatrix} e^{i\phi} \cos(\alpha) \\ \sin(\alpha) \end{pmatrix}, \quad \mathbf{u}(\beta, \psi) = \begin{pmatrix} e^{i\psi} \cos(\beta) \\ \sin(\beta) \end{pmatrix}$$

in $\mathbb{C}^2$. Find the condition on $\alpha, \beta, \phi, \psi \in \mathbb{R}$ such that

$$|\mathbf{v}^*(\alpha, \phi)\mathbf{u}(\beta, \psi)|^2 = \frac{1}{2}.$$

Note that

$$\mathbf{v}^*(\alpha, \phi)\mathbf{u}(\beta, \psi) = e^{i(\psi - \phi)} \cos(\alpha) \cos(\beta) + \sin(\alpha) \sin(\beta)$$

and

$$|\mathbf{v}^*(\alpha, \phi)\mathbf{u}(\beta, \psi)|^2 = \cos^2(\alpha) + 2\cos(\psi - \phi)\cos(\alpha)\cos(\beta) + \sin^2(\beta).$$

(2) Consider the two normalized vectors in $\mathbb{R}^3$

$$\mathbf{v}_1(\phi_1, \theta_1) = \begin{pmatrix} \cos(\phi_1)\sin(\theta_1) \\ \sin(\phi_1)\sin(\theta_1) \\ \cos(\theta_1) \end{pmatrix}, \quad \mathbf{v}_2(\phi_2, \theta_2) = \begin{pmatrix} \cos(\phi_2)\sin(\theta_2) \\ \sin(\phi_2)\sin(\theta_2) \\ \cos(\theta_2) \end{pmatrix}.$$

Find the conditions on ϕ_1, θ_1, ϕ_2, θ_2 such that $|\mathbf{v}_1^*(\phi_1, \theta_1)\mathbf{v}_2(\phi_2, \theta_2)| = \frac{1}{\sqrt{3}}$.

(3) Consider the Hilbert space $M(2, \mathbb{C})$ of the 2×2 matrices over the complex numbers with the scalar product $\langle A, B \rangle = \mathrm{tr}(AB^*)$ and $A, B \in M(2, \mathbb{C})$. Two mutually unbiased bases are given by

$$\begin{pmatrix} 1 & 0 \\ 0 & 0 \end{pmatrix}, \quad \begin{pmatrix} 0 & 1 \\ 0 & 0 \end{pmatrix}, \quad \begin{pmatrix} 0 & 0 \\ 1 & 0 \end{pmatrix}, \quad \begin{pmatrix} 0 & 0 \\ 0 & 1 \end{pmatrix}$$

and

$$\frac{1}{2}\begin{pmatrix} 1 & 1 \\ 1 & 1 \end{pmatrix}, \quad \frac{1}{2}\begin{pmatrix} 1 & -1 \\ 1 & -1 \end{pmatrix}, \quad \frac{1}{2}\begin{pmatrix} 1 & 1 \\ -1 & -1 \end{pmatrix}, \quad \frac{1}{2}\begin{pmatrix} 1 & -1 \\ -1 & 1 \end{pmatrix}.$$

Construct mutually unbiased bases applying the Kronecker product for the Hilbert space $M(4, \mathbb{C})$.

(4) Consider the Hilbert space $\mathbb{C}^5$ and the standard basis $\mathbf{e}_1, \ldots, \mathbf{e}_5$. Let $\omega = \exp(2\pi i/5)$. Consider the 5×5 unitary matrix

$$U = \frac{1}{\sqrt{5}} \begin{pmatrix} 1 & 1 & 1 & 1 & 1 \\ \omega^4 & 1 & \omega & \omega^2 & \omega^3 \\ \omega & \omega^3 & 1 & \omega^2 & \omega^4 \\ \omega & \omega^4 & \omega^2 & 1 & \omega^3 \\ \omega^4 & \omega^3 & \omega^2 & \omega & 1 \end{pmatrix}.$$

Show that $\mathbf{e}_1, \ldots, \mathbf{e}_5$ and $U\mathbf{e}_1, \ldots, U\mathbf{e}_5$ are mutually unbiased bases.

1.6   Trace and Determinant

In this section we introduce the trace and determinant of an $n \times n$ matrix and summarize their properties.

Definition 1.24. The *trace* of a square matrix $A = (a_{jk})$ of order n is defined as the sum of its diagonal elements

$$\text{tr}(A) := \sum_{j=1}^{n} a_{jj}.$$

Example 1.16. Let

$$A = \begin{pmatrix} 1 & 2 \\ 0 & -1 \end{pmatrix}.$$

Then $\text{tr}(A) = 0$ and $\text{tr}(A^2) = 2$. The eigenvalues λ_1 and λ_2 of A can be found from the two equations $\lambda_1 + \lambda_2 = \text{tr}(A)$, $\lambda_1^2 + \lambda_2^2 = \text{tr}(A^2)$.   ♣

The properties of the trace are as follows. The trace of the $n \times n$ unit matrix is obviously n. Let $a, b \in \mathbb{C}$ and A, B, C be three $n \times n$ matrices.

Then

$$\mathrm{tr}(aA + bB) = a\mathrm{tr}(A) + b\mathrm{tr}(B)$$
$$\mathrm{tr}(A^T) = \mathrm{tr}(A)$$
$$\mathrm{tr}(AB) = \mathrm{tr}(BA)$$
$$\mathrm{tr}(A) = \mathrm{tr}(S^{-1}AS) \qquad S \text{ nonsingular } n \times n \text{ matrix}$$
$$\mathrm{tr}(A^*A) = \mathrm{tr}(AA^*)$$
$$\mathrm{tr}(ABC) = \mathrm{tr}(CAB) = \mathrm{tr}(BCA).$$

Thus the trace is a linear functional. From the third property we find that

$$\mathrm{tr}([A, B]) = \mathrm{tr}(AB - BA) = \mathrm{tr}(AB) - \mathrm{tr}(BA) = 0$$

where $[\,,\,]$ denotes the commutator. The last property is called the *cyclic invariance* of the trace. Notice, however, that

$$\mathrm{tr}(ABC) \neq \mathrm{tr}(BAC)$$

in general. An example is given by the following three matrices

$$A = \begin{pmatrix} 1 & 0 \\ 0 & 0 \end{pmatrix}, \qquad B = \begin{pmatrix} 0 & 1 \\ 0 & 0 \end{pmatrix}, \qquad C = \begin{pmatrix} 0 & 0 \\ 1 & 0 \end{pmatrix}.$$

We have $\mathrm{tr}(ABC) = 1$ but $\mathrm{tr}(BAC) = 0$.

If λ_j, $j = 1, 2, \ldots, n$ are the eigenvalues of the $n \times n$ matrix A, then

$$\mathrm{tr}(A) = \sum_{j=1}^{n} \lambda_j, \quad \mathrm{tr}(A^2) = \sum_{j=1}^{n} \lambda_j^2, \quad \ldots, \quad \mathrm{tr}(A^n) = \sum_{j=1}^{n} \lambda_j^n.$$

More generally, if p designates a polynomial of degree r

$$p(x) = \sum_{j=0}^{r} a_j x^j$$

then

$$\mathrm{tr}(p(A)) = \sum_{k=1}^{n} p(\lambda_k).$$

Moreover we find

$$\mathrm{tr}(AA^*) = \mathrm{tr}(A^*A) = \sum_{j,k=1}^{n} |a_{jk}|^2 \geq 0.$$

Thus $\sqrt{\mathrm{tr}(AA^*)}$ is a norm of A.

Example 1.17. Let

$$A = \begin{pmatrix} 0 & -i \\ i & 0 \end{pmatrix}.$$

Then $AA^* = I_2$, where I_2 is the 2×2 identity matrix. Thus $\|A\| = \sqrt{2}$. ♣

Let $\mathbf{x}$, $\mathbf{y}$ be column vectors in $\mathbb{R}^n$. Then $\mathbf{x}^T\mathbf{y} = \mathrm{tr}(\mathbf{x}\mathbf{y}^T) = \mathrm{tr}(\mathbf{y}\mathbf{x}^T)$. Let A an $n \times n$ matrix over $\mathbb{R}$. Then we have $\mathbf{x}^T A\mathbf{y} = \mathrm{tr}(A\mathbf{y}\mathbf{x}^T)$.

Next we introduce the definition of the *determinant* of an $n \times n$ matrix. Then we give the properties of the determinant.

Definition 1.25. The *determinant* of an $n \times n$ matrix A is a scalar quantity denoted by $\det(A)$ and is given by

$$\det(A) := \sum_{j_1, j_2, \ldots, j_n} p(j_1, j_2, \ldots, j_n) a_{1j_1} a_{2j_2} \cdots a_{nj_n}$$

where $p(j_1, j_2, \ldots, j_n)$ is a permutation equal to ± 1 and the summation extends over $n!$ permutations $j_1, j_2, \ldots, j_n$ of the integers $1, 2, \ldots, n$. For an $n \times n$ matrix there exist $n!$ permutations. Therefore

$$p(j_1, j_2, \ldots, j_n) = \mathrm{sign} \prod_{1 \leq s < r \leq n} (j_r - j_s).$$

Example 1.18. For a matrix of order $(3,3)$ we find

$$p(1,2,3) = 1, \quad p(1,3,2) = -1, p(3,1,2) = 1$$
$$p(3,2,1) = -1, p(2,3,1) = 1, \quad p(2,1,3) = -1.$$

Then the determinant for a 3×3 matrix is given by

$$\det \begin{pmatrix} a_{11} & a_{12} & a_{13} \\ a_{21} & a_{22} & a_{23} \\ a_{31} & a_{32} & a_{33} \end{pmatrix} = a_{11}a_{22}a_{33} - a_{11}a_{23}a_{32} + a_{13}a_{21}a_{32}$$
$$- a_{13}a_{22}a_{31} + a_{12}a_{23}a_{31} - a_{12}a_{21}a_{33}. \quad \clubsuit$$

Definition 1.26. We call a square matrix A a *nonsingular* matrix if

$$\det(A) \neq 0$$

whereas if $\det(A) = 0$ the matrix A is called a *singular* matrix.

If $\det(A) \neq 0$, then A^{-1} exists. Conversely, if A^{-1} exists, then $\det(A) \neq 0$.

Example 1.19. The matrix

$$\begin{pmatrix} 1 & 1 \\ 1 & 0 \end{pmatrix}$$

is nonsingular since its determinant is -1, and the matrix

$$\begin{pmatrix} 0 & 1 \\ 0 & 0 \end{pmatrix}$$

is singular since its determinant is 0. $\quad \clubsuit$

Next we list some properties of determinants.

1. Let A be an $n \times n$ matrix and A^T the transpose. Then

$$\det(A) = \det(A^T).$$

Remark. Let

$$A = \begin{pmatrix} 0 & -i \\ 1 & 0 \end{pmatrix} \Rightarrow A^T = \begin{pmatrix} 0 & 1 \\ -i & 0 \end{pmatrix}, \quad A^* \equiv \bar{A}^T = \begin{pmatrix} 0 & 1 \\ i & 0 \end{pmatrix}.$$

Obviously $\det(A) \neq \det(A^*)$.

2. Let A be an $n \times n$ matrix and $\alpha \in \mathbb{R}$. Then

$$\det(\alpha A) = \alpha^n \det(A).$$

3. Let A be an $n \times n$ matrix. If two adjacent columns are equal, i.e. $A_j = A_{j+1}$ for some $j = 1, 2, \ldots, n-1$, then $\det(A) = 0$.

4. Let A be an $n \times n$ matrix. If any vector in A is a zero vector then $\det(A) = 0$.

5. Let A be an $n \times n$ matrix. Let j be some integer, $1 \leq j < n$. If the j-th and $(j+1)$-th columns are interchanged, then the determinant changes by a sign.

6. Let $A_1, \ldots, A_n$ be the column vectors of an $n \times n$ matrix A. If they are linearly dependent, then $\det(A) = 0$.

7. Let A and B be $n \times n$ matrices. Then $\det(AB) = \det(A) \det(B)$.

8. Let A be an $n \times n$ diagonal matrix. Then $\det(A) = a_{1,1} a_{2,2} \cdots a_{n,n}$.

9. Let A be an $n \times n$ counter diagonal matrix. Then

$$\det(A) = (-1)^n a_{1,n} a_{2,n-1} a_{3,n-2} \cdots a_{n,1}.$$

10. Let $a_{j,k}(t)$ be differentiable functions. Then $(d/dt) \det(A(t)) = $ sum of the determinants where each of them is obtained by differentiating the rows of A with respect to t one at a time, then taking its determinant.

Proof. Since

$$\det(A(t)) = \sum_{j_1, \ldots, j_n} p(j_1, \ldots, j_n) a_{1j_1}(t) \cdots a_{nj_n}(t)$$

we find

$$
\frac{d}{dt}\det(A(t)) = \sum_{j_1,\ldots,j_n} p(j_1,\ldots,j_n)\frac{da_{1j_1}(t)}{dt}a_{2j_2}(t)\cdots a_{nj_n}(t)
$$

$$
+ \sum_{j_1,\ldots,j_n} p(j_1,\ldots,j_n)a_{1j_1}(t)\frac{da_{2j_2}(t)}{dt}\cdots a_{nj_n}(t)
$$

$$
+\cdots+ \sum_{j_1,\ldots,j_n} p(j_1,\ldots,j_n)a_{1j_1}(t)\cdots a_{n-1j_{n-1}}(t)\frac{da_{nj_n}(t)}{dt}. \qquad \square
$$

Example 1.20. We have

$$
\frac{d}{dt}\det\begin{pmatrix} e^t & \cos(t) \\ 1 & \sin(t^2) \end{pmatrix} = \det\begin{pmatrix} e^t & -\sin(t) \\ 1 & \sin(t^2) \end{pmatrix} + \det\begin{pmatrix} e^t & \cos(t) \\ 0 & 2t\cos(t^2) \end{pmatrix}.
$$
♣

11. Let A be an invertible $n \times n$ symmetric matrix over $\mathbb{R}$. Then

$$
\mathbf{v}^T A^{-1}\mathbf{v} = \frac{\det(A + \mathbf{v}\mathbf{v}^T)}{\det(A)} - 1
$$

for every vector $\mathbf{v} \in \mathbb{R}^n$.

Example 1.21. Let

$$
A = \begin{pmatrix} 0 & 1 \\ 1 & 0 \end{pmatrix}, \qquad \mathbf{v} = \begin{pmatrix} 1 \\ 1 \end{pmatrix}.
$$

Then $A^{-1} = A$ and therefore $\mathbf{v}^T A^{-1}\mathbf{v} = 2$. Since

$$
\mathbf{v}\mathbf{v}^T = \begin{pmatrix} 1 & 1 \\ 1 & 1 \end{pmatrix}
$$

and $\det(A) = -1$ we obtain

$$
\frac{\det(A + \mathbf{v}\mathbf{v}^T)}{\det(A)} - 1 = 2.
$$
♣

12. The determinant of a diagonal matrix or triangular matrix is the product of its diagonal elements.

13. Let A be an $n \times n$ matrix. Then $\det(\exp(A)) \equiv \exp(\operatorname{tr}(A))$.

14. Let A be an $n \times n$ matrix and $\lambda_1, \lambda_2, \ldots, \lambda_n$ be the eigenvalues of A. Then $\det(A) = \lambda_1\lambda_2\cdots\lambda_n$ and $\operatorname{tr}(A) = \lambda_1 + \lambda_2 + \cdots + \lambda_n$.

15. Let A be a Hermitian matrix. Then $\det(A)$ is a real number.

16. Let U be a unitary matrix. Then $\det(U) = e^{i\phi}$ for some $\phi \in \mathbb{R}$. Thus $|\det(U)| = 1$.

17. Let A, B, C be $n \times n$ matrices and let 0_n be the $n \times n$ zero matrix. Then

$$\det \begin{pmatrix} A & 0_n \\ C & B \end{pmatrix} = \det(A)\det(B).$$

18. The determinant of the $n \times n$ matrix

$$A_n := \begin{pmatrix} b_1 & a_2 & 0 & \dots & 0 & 0 \\ -1 & b_2 & a_3 & \dots & 0 & 0 \\ \vdots & \vdots & \vdots & \ddots & \vdots & \vdots \\ 0 & 0 & 0 & \dots & b_{n-1} & a_n \\ 0 & 0 & 0 & \dots & -1 & b_n \end{pmatrix}, \qquad n = 1, 2, \dots$$

satisfies the *recursion relation*

$$\det(A_n) = b_n \det(A_{n-1}) + a_n \det(A_{n-2}), \quad \det(A_0) = 1, \quad \det(A_1) = b_1$$

where $n = 2, 3, \dots$.

19. Let A be a 2×2 matrix. Then $\det(I_2 + A) \equiv 1 + \operatorname{tr}(A) + \det(A)$.

20. Let A be an invertible $n \times n$ matrix, i.e. $\det(A) \neq 0$. Then the inverse of A can be calculated as

$$(A)^{-1}_{kj} = \frac{\partial}{\partial(A)_{jk}} \ln(\det(A)).$$

21. The equation of a hyperplane passing through the points

$$\mathbf{x}_1, \mathbf{x}_2, \dots, \mathbf{x}_n \in \mathbb{R}^n$$

can be given in the form

$$\det \begin{pmatrix} 1 & 1 & 1 & \cdots & 1 \\ \mathbf{x} & \mathbf{x}_1 & \mathbf{x}_2 & \cdots & \mathbf{x}_n \end{pmatrix} = 0.$$

22. The volume of an *n-simplex* in n-dimensional Euclidean space $\mathbb{E}^n$ with vertices $(\mathbf{v}_0\, \mathbf{v}_1\, \dots\, \mathbf{v}_n)$ is given by

$$\left| \frac{1}{n!} \det(\mathbf{v}_1 - \mathbf{v}_0\, \mathbf{v}_2 - \mathbf{v}_0\, \dots\, \mathbf{v}_n - \mathbf{v}_0) \right|$$

i.e. the $n \times n$ matrix $(\mathbf{v}_1 - \mathbf{v}_0 \ \mathbf{v}_2 - \mathbf{v}_0 \ \cdots \ \mathbf{v}_n - \mathbf{v}_0)$ contains the differences of the vectors $\mathbf{v}_j - \mathbf{v}_0$ $(j = 1, \ldots, n)$.

Exercises. (1) Let X and Y be $n \times n$ matrices over $\mathbb{R}$. Show that

$$\langle X, Y \rangle := \operatorname{tr}(XY^T)$$

defines a scalar product, i.e. prove that $\langle X, X \rangle \geq 0$, $\langle X, Y \rangle = \langle Y, X \rangle$, $\langle cX, Y \rangle = c \langle X, Y \rangle$ $(c \in \mathbb{R})$, $\langle X + Y, Z \rangle = \langle X, Z \rangle + \langle Y, Z \rangle$.

(2) Let A and B be $n \times n$ matrices. Show that $\operatorname{tr}([A, B]) = 0$.

(3) Use (2) to show that the relation $[A, B] = \lambda I$, $(\lambda \in \mathbb{C})$ for finite dimensional matrices can only be satisfied if $\lambda = 0$. For certain infinite dimensional matrices A and B we can find a nonzero λ.

(4) Let A and B be $n \times n$ matrices. Suppose that AB is nonsingular. Show that A and B are nonsingular matrices.

(5) Let

$$A = \begin{pmatrix} A_{11} & A_{12} \\ A_{21} & A_{22} \end{pmatrix}$$

be a square matrix partitioned into blocks. Assuming the submatrix A_{11} to be invertible, show that

$$\det(A) = \det(A_{11}) \det(A_{22} - A_{21} A_{11}^{-1} A_{12}).$$

(6) An $n \times n$ matrix A for which $A^k = 0_n$, where k is a positive integer, is called *nilpotent*. Let A be a nilpotent matrix. Show that $\det(A) = 0$.

(7) Let A be an $n \times n$ skew-symmetric matrix over $\mathbb{R}$, i.e. $A = -A^T$. Show that if n is odd then $\det(A) = 0$. Hint. Apply $\det(A) = (-1)^n \det(A)$.

(8) Let A be an $n \times n$ matrix with $A^2 = I_n$. Calculate $\det(A)$. Let B be an $n \times n$ matrix with $B^2 = -I_n$. Calculate $\det(B)$.

(9) Let A, B be $n \times n$ positive definite matrices and $n \in \mathbb{N}$. Show that

$$\operatorname{tr}(A^n B^n) \geq \operatorname{tr}((AB)^n).$$

(10) Consider the smooth map $\mathbf{f} : \mathbb{R}^3 \to \mathbb{R}^3$

$$f_1(\mathbf{x}) = x_1 + e^{x_2}, \quad f_2(\mathbf{x}) = x_2 + e^{x_3}, \quad f_3(\mathbf{x}) = x_3 + e^{x_1}.$$

Show that the *functional determinant* is given by

$$\det \begin{pmatrix} \partial f_1/\partial x_1 & \partial f_1/\partial x_2 & \partial f_1/\partial x_3 \\ \partial f_2/\partial x_1 & \partial f_2/\partial x_2 & \partial f_2/\partial x_3 \\ \partial f_3/\partial x_1 & \partial f_3/\partial x_2 & \partial f_3/\partial x_3 \end{pmatrix} = \det \begin{pmatrix} 1 & e^{x_2} & 0 \\ 0 & 1 & e^{x_3} \\ e^{x_1} & 0 & 1 \end{pmatrix} = 1 + e^{x_1 + x_2 + x_3}.$$

Thus the functional determinant is nonzero. Why can the inverse of the map **f** not be constructed?

(11) Let A be an $n \times n$ matrix and 0_n be the $n \times n$ zero matrix. Consider

$$B = \begin{pmatrix} 0_n & A \\ A & 0_n \end{pmatrix}.$$

Show that $\det(B - \lambda I_{2n}) = \det(\lambda I_n - A) \det(\lambda I_n + A) = 0$.

(12) Find the determinants of

$$A_2 = \frac{1}{2!} \det \begin{pmatrix} 1 & 0 & 0 \\ 1 & x_{1,1} & x_{1,2} \\ 1 & x_{2,1} & x_{2,2} \end{pmatrix}, \quad A_3 = \frac{1}{3!} \det \begin{pmatrix} 1 & 0 & 0 & 0 \\ 1 & x_{1,1} & x_{1,2} & x_{1,3} \\ 1 & x_{2,1} & x_{2,2} & x_{2,3} \\ 1 & x_{3,1} & x_{3,2} & x_{3,3} \end{pmatrix}.$$

What does the determinant describe?

(13) Let $\tau \in \mathbb{R}$. Find the maxima and minima of the polynomials

$$f_2(\tau) = \det \begin{pmatrix} 1 & \tau \\ \tau & 1 \end{pmatrix}, \quad f_3(\tau) = \det \begin{pmatrix} 1 & \tau & \tau^2 \\ \tau & \tau^2 & 1 \\ \tau^2 & \tau & 1 \end{pmatrix}, \quad f_4(\tau) = \det \begin{pmatrix} 1 & \tau & \tau^2 & \tau^3 \\ \tau & \tau^2 & \tau^3 & 1 \\ \tau^2 & \tau^3 & 1 & \tau \\ \tau^3 & 1 & \tau & \tau^2 \end{pmatrix}.$$

Extend to higher dimensions.

(14) Show that the equation for a *parabola* with horizontal axis passing through the points (x_1, y_1), (x_2, y_2), (x_3, y_3) is given by

$$\det \begin{pmatrix} y^2 & x & y & 1 \\ y_1^2 & x_1 & y_1 & 1 \\ y_2^2 & x_2 & y_2 & 1 \\ y_3^2 & x_3 & y_3 & 1 \end{pmatrix} = 0.$$

(15) Let A be an 2×2 matrix. What is the condition on A such that

$$\det(I_2 + A) = \det(I_2) + \det(A) \equiv 1 + \det(A)?$$

(16) Let $x \in \mathbb{R}$. Show that the determinants of the 3×3 matrix and 4×4 matrix

$$A_3(x) = \begin{pmatrix} 1 & 1 & 1 \\ 1 & 2 & 1 \\ 1 & 3-x & 2 \end{pmatrix}, \quad A_4(x) = \begin{pmatrix} 1 & 1 & 1 & 1 \\ 1 & 2 & 1 & 1 \\ 1 & 3-x & 2 & 1 \\ 1 & 4 & 3 & 2 \end{pmatrix}$$

do not depend on x. Extend to N dimensions $(N \geq 3)$

$$A_N(x) = \begin{pmatrix} 1 & 1 & 1 & 1 & \cdots & 1 & 1 \\ 1 & 2 & 1 & 1 & \cdots & 1 & 1 \\ 1 & 3-x & 2 & 1 & \cdots & 1 & 1 \\ 1 & 4 & 3 & 2 & \cdots & 1 & 1 \\ & & \cdots & & \ddots & \\ 1 & N & N-1 & N-2 & \cdots & 3 & 2 \end{pmatrix}.$$

(17) Let $n \geq 1$ and $x \in \mathbb{R}$. The *Chebyshev polynomials* $T_n(x)$ are given as the determinant of the $n \times n$ symmetric tridiagonal matrix

$$T_n(x) = \begin{pmatrix} x & -1 & 0 & 0 & \cdots & 0 \\ -1 & 2x & -1 & 0 & \cdots & 0 \\ 0 & -1 & 2x & -1 & \cdots & 0 \\ \vdots & \vdots & \ddots & \ddots & \ddots & \vdots \\ 0 & 0 & \cdots & -1 & 2x & -1 \\ 0 & 0 & \cdots & 0 & -1 & 2x \end{pmatrix}.$$

Show that the generating function $f(y)$ is given by

$$f(y) = \frac{1}{2}\left(\frac{1-y^2}{1-2xy+y^2} + 1 \right) = \sum_{n=0}^{\infty} T_n(x)y^n \quad \text{with} \quad T_n(x) = \frac{f^{(n)}(0)}{n!}.$$

(18) The sign of a permutation p of n elements is

$$\text{sgn}(p) = \prod_{i=1}^{n-1}\left(\prod_{j=i+1}^{n} \frac{p(j)-p(i)}{j-i} \right).$$

p is called "even" if $\text{sgn}(p) = +1$ and "odd" if $\text{sgn}(p) = -1$. Given the two permutation p and q. Show that

$$\text{sgn}(pq) = \text{sgn}(p)\text{sgn}(q).$$

(19) (i) The *Pfaffian* of a $2n \times 2n$ antisymmetric matrix $A = (a_{jk})$ is defined as

$$pf(A) = \frac{1}{n!2^n} \sum_{\pi \in S_{2n}} (-1)^\pi a_{\pi(1),\pi(2)} \cdots a_{\pi(2n-1),\pi(2n)}.$$

Note that $(pf(A))^2 = \det(A)$. Find the Pfaffian of the 4×4 matrix

$$A = \begin{pmatrix} 0 & 1 & 0 & 0 \\ -1 & 0 & 1 & 0 \\ 0 & -1 & 0 & 1 \\ 0 & 0 & -1 & 0 \end{pmatrix}.$$

(ii) The *Hafnian* of a $2n \times 2n$ symmetric matrix $A = (a_{jk})$ is defined by

$$\text{haf}(A) = \frac{1}{n! 2^n} \sum_{\pi \in S_{2n}} a_{\pi(1),\pi(2)} \cdots a_{\pi(2n-1),\pi(2n)}.$$

Find the Hafnian for the 4×4 matrix

$$A = \begin{pmatrix} 0 & 1 & 0 & 0 \\ 1 & 0 & 1 & 0 \\ 0 & 1 & 0 & 1 \\ 0 & 0 & 1 & 0 \end{pmatrix}.$$

(20) Let $\mathbf{v}_1, \mathbf{v}_2, \mathbf{v}_3 \in \mathbb{R}^3$. Show that $\mathbf{v}_1 \cdot (\mathbf{v}_2 \times \mathbf{v}_3) \equiv (\mathbf{v}_1 \times \mathbf{v}_2) \cdot \mathbf{v}_3$ is in absolute value equal to the volume of a *parallelepiped* with sides $\mathbf{v}_1$, $\mathbf{v}_2$, $\mathbf{v}_3$. Find $\det(\mathbf{v}_1\ \mathbf{v}_2\ \mathbf{v}_3)$.

(21) Let A be a 2×2 matrix. Show that

$$A^2 - \text{tr}(A)A + \frac{1}{2}((\text{tr}(A))^2 - \text{tr}(A^2))I_2 = 0_2.$$

(22) (i) Find all 2×2 matrices A such that $A^2 = \det(A)$.
(ii) Find all 2×2 matrices B such that $B^2 = \frac{1}{2}\text{tr}(B)B$.

(23) Let A be an $n \times n$ matrix. The *determinant* and *permanent* are defined as

$$\det(A) := \sum_{\pi \in S_n} (-1)^\pi \prod_{j=1}^n a_{j,\pi(j)}, \qquad \text{per}(A) := \sum_{\pi \in S_n} \prod_{j=1}^n a_{j,\pi(j)}.$$

(i) Let A be a positive semi-definite hermitian $n \times n$ matrix. Show that

$$\det(A) \leq \prod_{j=1}^n a_{j,j}.$$

What are the conditions on A for equality?
(ii) Let A be a positive semi-definite hermitian $n \times n$ matrix. Show that

$$\text{per}(A) \geq \prod_{j=1}^n a_{j,j}.$$

What are the conditions on A for equality?

1.7   Eigenvalue Problem

The eigenvalue problem plays a central role in theoretical and mathematical physics (Steeb [60]). We give a short introduction into the eigenvalue calculation for finite dimensional matrices. In section 2.6 we study the eigenvalue problem for Kronecker products of matrices.

Definition 1.27. A complex number λ is said to be an *eigenvalue* (or *characteristic value*) of an $n \times n$ matrix A, if there is at least one nonzero vector $\mathbf{v} \in \mathbb{C}^n$ satisfying the *eigenvalue equation*

$$A\mathbf{v} = \lambda\mathbf{v}, \qquad \mathbf{v} \neq \mathbf{0}.$$

Each nonzero vector $\mathbf{v} \in \mathbb{C}^n$ satisfying the eigenvalue equation is called an *eigenvector* (or *characteristic vector*) of A with eigenvalue λ.

The eigenvalue equation can be written as

$$(A - \lambda I)\mathbf{v} = \mathbf{0}$$

where I is the $n \times n$ unit matrix and $\mathbf{0}$ is the zero vector. This system of n linear simultaneous equations in $\mathbf{v}$ has a nontrivial solution for the vector $\mathbf{v}$ only if the matrix $(A - \lambda I)$ is singular, i.e.

$$\det(A - \lambda I) = 0.$$

The expansion of the determinant gives a polynomial in λ of degree equal to n, which is called the characteristic polynomial of the matrix A. The n roots of the equation $\det(A - \lambda I) = 0$, called the *characteristic equation*, are the eigenvalues of A.

Definition 1.28. Let λ be an eigenvalue of an $n \times n$ matrix A. The vector $\mathbf{u}$ is a *generalized eigenvector* of A corresponding to λ if

$$(A - \lambda I)^n \mathbf{u} = \mathbf{0}.$$

The eigenvectors of a matrix are also generalized eigenvectors of the matrix.

Theorem 1.1. *Every $n \times n$ matrix A has at least one eigenvalue and corresponding eigenvector.*

Proof. We follow the proof in Axler [2]. Suppose $\mathbf{v} \in \mathbb{C}^n \setminus \{\mathbf{0}\}$. Then $\{\mathbf{v}, A\mathbf{v}, \dots, A^n\mathbf{v}\}$ must be a linearly dependent set of vectors, i.e. there exist $c_0, \dots, c_n \in \mathbb{C}$ such that

$$c_0\mathbf{v} + c_1 A\mathbf{v} + \cdots + c_n A^n\mathbf{v} = \mathbf{0}.$$

Let $m \in \{0, 1, \ldots, n\}$ be the largest index satisfying $c_m \neq 0$. Consider the following polynomial in x and its factorization over $\mathbb{C}$

$$c_0 + c_1 x + \cdots + c_m x^m = c_m (x - x_0)(x - x_1) \cdots (x - x_m)$$

for some $x_0, \ldots, x_m \in \mathbb{C}$ (i.e. the roots of the polynomial). Then $c_0 \mathbf{v} + c_1 A\mathbf{v} + \cdots + c_n A^n \mathbf{v} = c_m (A - x_0 I_n)(A - x_1 I_n) \cdots (A - x_m I_n)\mathbf{v} = \mathbf{0}$. It follows that there is a largest $j \in \{0, 1, \ldots, m\}$ satisfying

$$(A - x_j I_n)\left[(A - x_{j+1} I_n) \cdots (A - x_m I_n)\mathbf{v}\right] = \mathbf{0}.$$

This is a solution to the eigenvalue equation, the eigenvalue is x_j and the corresponding eigenvector is

$$(A - x_{j+1} I_n) \cdots (A - x_m I_n)\mathbf{v}. \qquad \square$$

Definition 1.29. The *spectrum* of the matrix A is the subset

$$\mathrm{sp}(A) := \bigcup_{i=1}^{n} \{\, \lambda_i(A) \,\}$$

of the complex plane. The *spectral radius* of the matrix A is the nonnegative number defined by

$$\varrho(A) := \max\{\, |\lambda_j(A)| \; : \; 1 \le j \le n \,\}.$$

If $\lambda \in \mathrm{sp}(A)$, the vector subspace $\{\, \mathbf{v} \in V \; : \; A\mathbf{v} = \lambda \mathbf{v} \,\}$ (of dimension at least 1) is called the *eigenspace* corresponding to the eigenvalue λ.

Example 1.22. Consider the 2×2 hermitian and unitary matrix

$$A = \begin{pmatrix} 0 & -i \\ i & 0 \end{pmatrix}.$$

Then $\det(A - \lambda I_2) \equiv \lambda^2 - 1 = 0$. Therefore the eigenvalues are given by $\lambda_1 = 1$, $\lambda_2 = -1$. To find the eigenvector of the eigenvalue $\lambda_1 = 1$ we have to solve

$$\begin{pmatrix} 0 & -i \\ i & 0 \end{pmatrix} \begin{pmatrix} u_1 \\ u_2 \end{pmatrix} = 1 \begin{pmatrix} u_1 \\ u_2 \end{pmatrix}.$$

Therefore $u_2 = iu_1$ and the eigenvector of $\lambda_1 = 1$ is given by

$$\mathbf{u}_1 = \begin{pmatrix} 1 \\ i \end{pmatrix}.$$

For $\lambda_2 = -1$ we have

$$\begin{pmatrix} 0 & -i \\ i & 0 \end{pmatrix} \begin{pmatrix} u_1 \\ u_2 \end{pmatrix} = -1 \begin{pmatrix} u_1 \\ u_2 \end{pmatrix}$$

and hence the second eigenvector is

$$\mathbf{u}_2 = \begin{pmatrix} 1 \\ -i \end{pmatrix}.$$

We see that $\langle \mathbf{u}_1, \mathbf{u}_2 \rangle \equiv \mathbf{u}_2^* \mathbf{u}_1 = 0$. Both eigenvectors are not normalized. ♣

Example 1.23. The $n \times n$ *permutation matrix*

$$P = \begin{pmatrix} 0 & 1 & 0 & 0 & \ldots & 0 \\ 0 & 0 & 1 & 0 & \ldots & 0 \\ 0 & 0 & 0 & 1 & \ldots & 0 \\ \vdots & \vdots & \vdots & \vdots & \ddots & \vdots \\ 0 & 0 & 0 & 0 & \ldots & 1 \\ 1 & 0 & 0 & 0 & \ldots & 0 \end{pmatrix}$$

satisfies $P^n = I_n$ and admits the eigenvalues

$$\exp(2\pi i k/n), \quad k = 0, 1, \ldots, n-1. \qquad \clubsuit$$

A special role in theoretical physics is played by the Hermitian matrices. In this case we have the following theorem.

Theorem 1.2. *Let A be a Hermitian matrix, i.e. $A^* = A$, where $A^* \equiv \bar{A}^T$. The eigenvalues of A are real, and two eigenvectors corresponding to two different eigenvalues are mutually orthogonal.*

Proof. The eigenvalue equation is $A\mathbf{u} = \lambda\mathbf{u}$, where $\mathbf{u} \neq \mathbf{0}$. Now we have the identity

$$(A\mathbf{u})^*\mathbf{u} \equiv \mathbf{u}^*A^*\mathbf{u} \equiv \mathbf{u}^*(A^*\mathbf{u}) \equiv \mathbf{u}^*(A\mathbf{u})$$

since A is Hermitian, i.e. $A = A^*$. Inserting the eigenvalue equation into this equation yields

$$(\lambda\mathbf{u})^*\mathbf{u} = \mathbf{u}^*(\lambda\mathbf{u})B \quad \text{or} \quad \bar{\lambda}(\mathbf{u}^*\mathbf{u}) = \lambda(\mathbf{u}^*\mathbf{u}).$$

Since $\mathbf{u}^*\mathbf{u} \neq 0$, we have $\bar{\lambda} = \lambda$ and therefore λ must be real. Let

$$A\mathbf{u}_1 = \lambda_1\mathbf{u}_1, \qquad A\mathbf{u}_2 = \lambda_2\mathbf{u}_2.$$

Now

$$\lambda_1\langle\mathbf{u}_1, \mathbf{u}_2\rangle = \langle\lambda_1\mathbf{u}_1, \mathbf{u}_2\rangle = \langle A\mathbf{u}_1, \mathbf{u}_2\rangle = \langle\mathbf{u}_1, A\mathbf{u}_2\rangle = \langle\mathbf{u}_1, \lambda_2\mathbf{u}_2\rangle = \lambda_2\langle\mathbf{u}_1, \mathbf{u}_2\rangle.$$

Since $\lambda_1 \neq \lambda_2$, we find that $\langle\mathbf{u}_1, \mathbf{u}_2\rangle = 0$. $\qquad\square$

Theorem 1.3. *The eigenvalues λ_j of a unitary matrix U satisfy $|\lambda_j| = 1$.*

Proof. Since U is a unitary matrix we have $U^* = U^{-1}$, where U^{-1} is the inverse of U. Let $U\mathbf{u} = \lambda\mathbf{u}$ be the eigenvalue equation. It follows that

$$(U\mathbf{u})^* = (\lambda\mathbf{u})^* \quad \text{or} \quad \mathbf{u}^*U^* = \bar{\lambda}\mathbf{u}^*.$$

Thus we obtain $\mathbf{u}^*U^*U\mathbf{u} = \bar{\lambda}\lambda\mathbf{u}^*\mathbf{u}$. Owing to $U^*U = I$ we obtain

$$\mathbf{u}^*\mathbf{u} = \bar{\lambda}\lambda\mathbf{u}^*\mathbf{u}.$$

Since $\mathbf{u}^*\mathbf{u} \neq 0$ we have $\bar{\lambda}\lambda = 1$. Hence the eigenvalue λ can be written as $\lambda = \exp(i\alpha)$, $(\alpha \in \mathbb{R})$. Consequently $|\lambda| = 1$. $\qquad\square$

Theorem 1.4. *If* $\mathbf{x}$ *is an eigenvalue of the* $n \times n$ *normal matrix* A *corresponding to the eigenvalue* λ, *then* $\mathbf{x}$ *is also an eigenvalue of* A^* *corresponding to the eigenvalue* $\overline{\lambda}$.

Proof.   Since $(A - \lambda I_n)\mathbf{x} = \mathbf{0}$ and $AA^* = A^*A$ we find

$$
\begin{aligned}
((A^* - \overline{\lambda}I_n)\mathbf{x}, (A^* - \overline{\lambda}I_n)\mathbf{x}) &= \left[(A^* - \overline{\lambda}I_n)\mathbf{x}\right]^* \left[(A^* - \overline{\lambda}I_n)\mathbf{x}\right] \\
&= \mathbf{x}^*(A - \lambda I_n)(A^* - \overline{\lambda}I_n)\mathbf{x} \\
&= \mathbf{x}^*(A^* - \overline{\lambda}I_n)(A - \lambda I_n)\mathbf{x} \\
&= \mathbf{x}^*(A^* - \overline{\lambda}I_n)\mathbf{0} = 0.
\end{aligned}
$$

Consequently $(A^* - \overline{\lambda}I_n)\mathbf{x} = \mathbf{0}$. □

Theorem 1.5. *The eigenvalues of a skew-Hermitian matrix* $(A^* = -A)$ *can only be 0 or (purely) imaginary.*

The proof is left as an exercise to the reader.

Example 1.24. The skew-Hermitian matrix

$$
A = \begin{pmatrix} 0 & i \\ i & 0 \end{pmatrix} \equiv i\sigma_1
$$

has the eigenvalues $\pm i$.   ♣

Now consider the general case. Let λ be an eigenvalue of A with the corresponding eigenvector $\mathbf{x}$, and let μ be an eigenvalue of A^* with the corresponding eigenvector $\mathbf{y}$. Then

$$
\mathbf{x}^* A^* \mathbf{y} = (\mathbf{x}^* A^*)\mathbf{y} = (A\mathbf{x})^*\mathbf{y} = (\lambda\mathbf{x})^*\mathbf{y} = \overline{\lambda}\mathbf{x}^*\mathbf{y}
$$

and $\mathbf{x}^* A^* \mathbf{y} = \mathbf{x}^*(A^*\mathbf{y}) = \mu\mathbf{x}^*\mathbf{y}$. It follows that $\mathbf{x}^*\mathbf{y} = 0$ or $\overline{\lambda} = \mu$.

Example 1.25. Consider

$$
A = \begin{pmatrix} 0 & 1 \\ 0 & 0 \end{pmatrix} \quad \Rightarrow \quad A^* = \begin{pmatrix} 0 & 0 \\ 1 & 0 \end{pmatrix}.
$$

Both eigenvalues of A and A^* are zero. The eigenspaces corresponding to the eigenvalue 0 of A and A^* are

$$
\left\{ \begin{pmatrix} t \\ 0 \end{pmatrix} : t \in \mathbb{C}, t \neq 0 \right\} \quad \text{and} \quad \left\{ \begin{pmatrix} 0 \\ t \end{pmatrix} : t \in \mathbb{C}, t \neq 0 \right\}
$$

respectively. Obviously both conditions above are true. The eigenvalues of the hermitian and positive definite matrices A^*A and AA^* are given by 0 and 1.   ♣

Exercises. (1) Let A be an $n \times n$ matrix over $\mathbb{C}$. Show that the eigenvectors of A corresponding to distinct eigenvalues are linearly independent.

(2) Show that

$$\operatorname{tr}(A) = \sum_{j=1}^{n} \lambda_j(A), \qquad \det(A) = \prod_{j=1}^{n} \lambda_j(A).$$

(3) Let U be a Hermitian and unitary matrix. What can be said about the eigenvalues of U?

(4) Let A be an invertible matrix whose elements, as well as those of A^{-1}, are all nonnegative. Show that there exists a permutation matrix P and a matrix $D = \operatorname{diag}(d_j)$, with d_j positive, such that $A = PD$ (the converse is obvious).

(5) Let A and B be two square matrices of the same order. Show that the matrices AB and BA have the same characteristic polynomial.

(6) Let $a, b, c \in \mathbb{R}$. Find the eigenvalues and eigenvectors of the 4×4 matrix

$$A = \begin{pmatrix} a & b & 0 & 0 \\ c & a & b & 0 \\ 0 & c & a & b \\ 0 & 0 & c & a \end{pmatrix}.$$

(7) Let $a_1, a_2, \ldots, a_n \in \mathbb{R}$. Show that the eigenvalues of the matrix

$$A = \begin{pmatrix} a_1 & a_2 & a_3 & \cdots & a_{n-1} & a_n \\ a_n & a_1 & a_2 & \cdots & a_{n-2} & a_{n-1} \\ a_{n-1} & a_n & a_1 & \cdots & a_{n-3} & a_{n-2} \\ \vdots & \vdots & \vdots & \ddots & \vdots & \vdots \\ a_3 & a_4 & a_5 & \cdots & a_1 & a_2 \\ a_2 & a_3 & a_4 & \cdots & a_n & a_1 \end{pmatrix}$$

called a *circulant matrix*, are of the form

$$\lambda_{l+1} = a_1 + a_2 \xi_l + a_3 \xi_l^2 + \cdots + a_n \xi_l^{n-1}, \qquad l = 0, 1, \ldots, n-1$$

where $\xi_l := e^{2i\pi l/n}$.

(8) Let $\phi \in \mathbb{R}$. Consider the 4×4 matrix

$$A(\phi) = \begin{pmatrix} 0 & 1 & 0 & 0 \\ 0 & 0 & 1 & 0 \\ 0 & 0 & 0 & 1 \\ e^{i\phi} & 0 & 0 & 0 \end{pmatrix}.$$

Show that the matrix is unitary. Find the eigenvalues and normalized eigenvectors. Then write down the spectral decomposition of $A(\phi)$. Of course first one has to check whether the spectral decomposition can be applied. Can the matrix be written as Kronecker product of two 2×2 matrices?

(9) Consider the function $\mathbf{f} : \mathbb{R}^3 \to \mathbb{R}^3$

$$f_1(\mathbf{x}) = x_2 x_3, \quad f_2(\mathbf{x}) = x_1 x_3, \quad f_3(\mathbf{x}) = x_1 x_2.$$

Find the 3×3 matrix $A(\mathbf{x})$

$$a_{jk}(\mathbf{x}) = (\partial f_j / \partial x_k - \partial f_k / \partial x_j), \quad j, k = 1, 2, 3$$

and the eigenvalues.

(10) Let $\mathbf{v}$ be a normalized (column vector) in the Hilbert space $\mathbb{C}^n$. Consider the $n \times n$ matrix

$$A = I_n - 2\mathbf{v}\mathbf{v}^*.$$

Find the eigenvalues and normalized eigenvectors of A. First calculate A^2. Is the matrix hermitian? Is the matrix unitary? Note that

$$A\mathbf{v} = (I_n - 2\mathbf{v}\mathbf{v}^*)\mathbf{v} = \mathbf{v} - 2\mathbf{v}\mathbf{v}^*\mathbf{v} = \mathbf{v} - 2\mathbf{v} = -\mathbf{v}.$$

(11) (i) Find the eigenvalues of the $n \times n$ matrix over $\mathbb{R}$

$$\begin{pmatrix} a_1 & a_2 & \cdots & a_{n-1} & a_n \\ b_1 & 0 & \cdots & 0 & 0 \\ 0 & b_2 & \cdots & 0 & 0 \\ \vdots & \vdots & \ddots & \vdots & \vdots \\ 0 & 0 & \cdots & b_{n-1} & 0 \end{pmatrix}.$$

(ii) Find the eigenvalues of the $n \times n$ matrix over $\mathbb{R}$

$$\begin{pmatrix} 0 & 1 & 0 & \cdots & 0 \\ 0 & 0 & 1 & \cdots & 0 \\ \vdots & \vdots & \vdots & \ddots & \vdots \\ 0 & 0 & 0 & \cdots & 1 \\ c_1 & c_2 & c_3 & \cdots & c_n \end{pmatrix}.$$

(12) Show that the function $f : \mathbb{R}^2 \to \mathbb{R}$

$$f(x_1, x_2) = 2x_1 + 6x_2 - 2x_1^2 - 3x_2^2 + 4x_1 x_2$$

can be written in matrix form as

$$f(x_1, x_2) = (2\ 6) \begin{pmatrix} x_1 \\ x_2 \end{pmatrix} + \frac{1}{2} (x_1\ x_2) \begin{pmatrix} -4 & 4 \\ 4 & -6 \end{pmatrix} \begin{pmatrix} x_1 \\ x_2 \end{pmatrix}$$

with the symmetric matrix

$$H = \begin{pmatrix} -4 & 4 \\ 4 & -6 \end{pmatrix}.$$

Show that both eigenvalues of H are negative and thus the matrix is negative definite. Hence conclude that the function f is concave.

(13) Can one find d_{11}, d_{22} such that the symmetric matrix over $\mathbb{R}$

$$\begin{pmatrix} d_{11} & 1 \\ 1 & d_{22} \end{pmatrix}$$

admits the eigenvalues $\lambda_1 = 2$, $\lambda_2 = 0$. Show that one has to solve

$$d_{11} + d_{22} = \lambda_1 + \lambda_2 = 2, \quad d_{11}d_{22} - 1 = \lambda_1\lambda_2 = 0.$$

(14) Find the eigenvalues and normalized eigenvectors of the unitary matrix

$$U(\theta, \phi) = \begin{pmatrix} \cos(\theta) & e^{i\phi}\sin(\theta) \\ \sin(\theta) & -e^{i\phi}\cos(\theta) \end{pmatrix}.$$

Note that determinant of the matrix is given by $-e^{i\phi}$. This matrix plays a role for *Majorana neutrinos*.

(15) Find the eigenvalues and normalized eigenvectors of the hermitian 4×4 matrix

$$\begin{pmatrix} 0 & e^{-i\pi\phi} & 0 & 0 \\ e^{i\pi\phi} & 0 & e^{-i\pi\phi} & 0 \\ 0 & e^{i\pi\phi} & 0 & e^{-i\pi\phi} \\ 0 & 0 & e^{i\pi\phi} & 0 \end{pmatrix}.$$

(16) Let A, B, C be $n \times n$ matrices over $\mathbb{R}$ with $B^T = -B$, $C^T = -C$. Study the eigenvalue problem of the $2n \times 2n$ matrix

$$\begin{pmatrix} A & B \\ C & A^T \end{pmatrix}.$$

(17) Let A be a 3×3 hermitian matrix, i.e. the eigenvalues of A are real. Assume that $\operatorname{tr}(A) = 0$ and $\operatorname{tr}(A^3) = 0$. Then the three eigenvalues λ_1, λ_2, λ_3 satisfy

$$\lambda_1 + \lambda_2 + \lambda_3 = 0, \quad \lambda_1^3 + \lambda_2^3 + \lambda_3^3 = 0.$$

Find all solutions of this system of equations. Note that the trivial solution is

$$\lambda_1 = \lambda_2 = \lambda_3 = 0.$$

The system of equation is invariant under all permutations of λ_1, λ_2, λ_3.

1.8   Cayley-Hamilton Theorem

The *Cayley-Hamilton theorem* states that the $n \times n$ matrix A satisfies its own characteristic equation, i.e.

$$(A - \lambda_1 I_n) \cdots (A - \lambda_n I_n) = 0_{n \times n}$$

where $0_{n \times n}$ is the $n \times n$ zero matrix. Notice that the factors commute.

In this section we follow Axler [2].

Definition 1.30. An $n \times n$ matrix A is *upper triangular* with respect to a basis $\{\mathbf{v}_1, \ldots, \mathbf{v}_m\} \subset \mathbb{C}^n$, where $m \leq n$, if

$$A\mathbf{v}_j \in \operatorname{span}\{\mathbf{v}_1, \ldots, \mathbf{v}_j\}, \qquad j = 1, \ldots, m.$$

Theorem 1.6. *For every $n \times n$ matrix A there exists a basis V for $\mathbb{C}^n$ such that A is upper triangular with respect to V.*

Proof.   Let $\mathbf{v}_1$ be an eigenvector of A. The proof that every matrix A is upper triangular is by induction. The case $n = 1$ is obvious. Consider the subspace

$$U := \{(A - x_j I_n)\mathbf{x} : \mathbf{x} \in \mathbb{C}^n\}.$$

For all $\mathbf{u} \in U$

$$A\mathbf{u} = (A - x_j I_n)\mathbf{u} + x_j \mathbf{u} \in U$$

since $(A - x_j I_n)\mathbf{u} \in U$ by definition. Since x_j is an eigenvalue of A we have $\det(A - x_j I_n) = 0$ so that $\dim(U) < n$. The induction hypothesis is that any square matrix is upper triangular with respect to a basis for a (sub)space with dimension less than n. Consequently A has a triangular representation on U. Let $\{\mathbf{v}_1, \ldots, \mathbf{v}_{\dim U}\}$ be a basis for U and $\{\mathbf{v}_1, \ldots, \mathbf{v}_n\}$ be a basis for $\mathbb{C}^n$. We have for $k \in \{\dim(U) + 1, \ldots, n\}$

$$A\mathbf{v}_k = (A - x_j I_n)\mathbf{v}_k + x_j \mathbf{v}_k \in \mathrm{span}\{\mathbf{v}_1, \ldots, \mathbf{v}_{\dim U}, \mathbf{v}_k\}$$

where

$$\mathrm{span}\{\mathbf{v}_1, \ldots, \mathbf{v}_{\dim U}, \mathbf{v}_k\} \subseteq \mathrm{span}\{\mathbf{v}_1, \ldots, \mathbf{v}_k\}$$

and $(A - x_j I_n)\mathbf{v}_k \in U$ by definition. It follows that A is upper triangular with respect to $V = \{\mathbf{v}_1, \ldots, \mathbf{v}_n\}$. $\qquad\square$

If A has a triangular representation with respect to some basis we can find a triangular representation with respect to an orthonormal basis by applying the Gram-Schmidt orthonormalization process (see section 1.3).

Theorem 1.7. *Every $n \times n$ matrix A satisfies its own characteristic equation*

$$(A - \lambda_1 I_n) \cdots (A - \lambda_n I_n) = 0_{n \times n}$$

where $0_{n \times n}$ is the $n \times n$ zero matrix, and λ_1, $\ldots$, λ_n are the eigenvalues of A.

Proof. Let A be triangular with respect to the basis $\{\mathbf{v}_1, \ldots, \mathbf{v}_n\} \subset \mathbb{C}^n$ for $\mathbb{C}^n$. Thus $\mathbf{v}_1$ is an eigenvector of A corresponding to an eigenvalue, say λ_{j_1}. Consequently $(A - \lambda_{j_1} I_n)\mathbf{v}_1 = \mathbf{0}$. Now suppose

$$(A - \lambda_{j_1} I_n)(A - \lambda_{j_2} I_n) \cdots (A - \lambda_{j_k} I_n)\mathbf{v}_k = \left(\prod_{p=1}^{k}(A - \lambda_{j_k} I_n)\right)\mathbf{v}_k = \mathbf{0}$$

for $k = 1, 2, \ldots, r$. Now $A\mathbf{v}_{r+1} \in \mathrm{span}\{\mathbf{v}_1, \mathbf{v}_2, \ldots, \mathbf{v}_{r+1}\}$ so that $A\mathbf{v}_{r+1} = \mathbf{u} + \alpha\mathbf{v}_{r+1}$ for some $\mathbf{u} \in \mathrm{span}\{\mathbf{v}_1, \ldots, \mathbf{v}_r\}, \alpha \in \mathbb{C}$. The supposition above (induction hypothesis) implies

$$\left(\prod_{p=1}^{k}(A - \lambda_{j_k} I_n)\right)\mathbf{u} = \mathbf{0}$$

so that

$$\left(\prod_{p=1}^{k}(A - \lambda_{j_k} I_n)\right) A\mathbf{v}_{r+1} = \alpha \left(\prod_{p=1}^{k}(A - \lambda_{j_k} I_n)\right)\mathbf{v}_{r+1}$$

which simplifies to

$$(A - \alpha I_n)\left(\prod_{p=1}^{k}(A - \lambda_{j_k} I_n)\right)\mathbf{v}_{r+1} = \mathbf{0}$$

since A commutes with $(A - cI_n)$ $(c \in \mathbb{C})$. Thus either

$$\left(\prod_{p=1}^{k} (A - \lambda_{j_k} I_n) \right) \mathbf{v}_{r+1} = \mathbf{0}$$

or α is an eigenvalue, say $\lambda_{j_{r+1}}$, of A. If the first case holds a re-ordering of the basis $\{\mathbf{v}_1, \ldots, \mathbf{v}_n\}$ postpones the arbitrary choice of $\lambda_{j_{r+1}}$. In either case, we have shown by induction that

$$\left(\prod_{p=1}^{n} (A - \lambda_p I_n) \right) \mathbf{v}_k = \mathbf{0}, \qquad k = 1, 2, \ldots, n.$$

Since $\{\mathbf{v}_1, \ldots, \mathbf{v}_r\}$ is a basis we must have

$$\prod_{p=1}^{n} (A - \lambda_p I_n) = (A - \lambda_1 I_n) \cdots (A - \lambda_n I_n) = 0_{n \times n}.$$

$\square$

1.9 Projection Matrices

First we introduce the definition of a projection matrix and give some of its properties. Projection matrices (projection operators) play a central role in finite group theory in the decomposition of Hilbert spaces into invariant subspaces (Steeb [57, 60]).

Definition 1.31. An $n \times n$ matrix Π is called a *projection matrix* if

$$\Pi = \Pi^* \quad \text{and} \quad \Pi^2 = \Pi.$$

The element $\Pi \mathbf{u}$ ($\mathbf{u} \in \mathbb{C}^n$) is called the *projection* of the element $\mathbf{u}$.

Example 1.26. Let $n = 2$ and

$$\Pi_1 = \begin{pmatrix} 1 & 0 \\ 0 & 0 \end{pmatrix}, \qquad \Pi_2 = \begin{pmatrix} 0 & 0 \\ 0 & 1 \end{pmatrix}.$$

Then $\Pi_1^* = \Pi_1$, $\Pi_1^2 = \Pi_1$, $\Pi_2^* = \Pi_2$ and $\Pi_2^2 = \Pi_2$. Furthermore $\Pi_1 \Pi_2 = 0_2$ and

$$\Pi_1 \begin{pmatrix} u_1 \\ u_2 \end{pmatrix} = \begin{pmatrix} u_1 \\ 0 \end{pmatrix}, \qquad \Pi_2 \begin{pmatrix} u_1 \\ u_2 \end{pmatrix} = \begin{pmatrix} 0 \\ u_2 \end{pmatrix}.$$

♣

Theorem 1.8. *Let Π_1 and Π_2 be two $n \times n$ projection matrices. Assume that $\Pi_1 \Pi_2 = 0_n$. Then $\langle \Pi_1 \mathbf{u}, \Pi_2 \mathbf{u} \rangle = 0$.*

Proof.   We find

$$\langle \Pi_1 \mathbf{u}, \Pi_2 \mathbf{u} \rangle = (\Pi_1 \mathbf{u})^*(\Pi_1 \mathbf{u}) = (\mathbf{u}^* \Pi_1^*)(\Pi_2 \mathbf{u}) = \mathbf{u}^*(\Pi_1 \Pi_2)\mathbf{u} = 0. \qquad \square$$

Theorem 1.9. *Let I_n be the $n \times n$ unit matrix and Π be a projection matrix. Then $I_n - \Pi$ is a projection matrix.*

Proof.   Since $(I_n - \Pi)^* = I_n^* - \Pi^* = I_n - \Pi$ and

$$(I_n - \Pi)^2 = (I_n - \Pi)(I_n - \Pi) = I_n - \Pi - \Pi + \Pi = I_n - \Pi$$

we find that $I_n - \Pi$ is a projection matrix. $\qquad \square$

Theorem 1.10. *The eigenvalues λ_j of a projection matrix Π are given by $\lambda_j \in \{0, 1\}$.*

Proof.   From the eigenvalue equation $\Pi \mathbf{u} = \lambda \mathbf{u}$ we find

$$\Pi(\Pi \mathbf{u}) = (\Pi \Pi)\mathbf{u} = \lambda \Pi \mathbf{u}.$$

Using the fact that $\Pi^2 = \Pi$ we obtain $\Pi \mathbf{u} = \lambda^2 \mathbf{u}$. Thus $\lambda = \lambda^2$ since $\mathbf{u} \neq \mathbf{0}$ and hence $\lambda \in \{0, 1\}$. $\qquad \square$

Example 1.27. Consider the 4×4 matrix

$$A = \begin{pmatrix} 0\,0\,0\,1 \\ 0\,0\,1\,0 \\ 0\,1\,0\,0 \\ 1\,0\,0\,0 \end{pmatrix} \equiv \begin{pmatrix} 0 & 1 \\ 1 & 0 \end{pmatrix} \otimes \begin{pmatrix} 0 & 1 \\ 1 & 0 \end{pmatrix}$$

with the eigenvalues $+1$ $(2\times)$ and -1 $(2\times)$ and the corresponding normalized eigenvectors

$$\mathbf{v}_{1,+1} = \frac{1}{2} \begin{pmatrix} 1 \\ 1 \\ 1 \\ 1 \end{pmatrix}, \quad \mathbf{v}_{2,+1} = \frac{1}{2} \begin{pmatrix} 1 \\ -1 \\ -1 \\ 1 \end{pmatrix},$$

$$\mathbf{v}_{1,-1} = \frac{1}{2} \begin{pmatrix} 1 \\ 1 \\ -1 \\ -1 \end{pmatrix}, \quad \mathbf{v}_{2,-1} = \frac{1}{2} \begin{pmatrix} 1 \\ -1 \\ 1 \\ -1 \end{pmatrix}.$$

They form an orthonormal basis in $\mathbb{C}^4$. Then

$$\mathbf{v}_{1,+1}\mathbf{v}_{1,+1}^* + \mathbf{v}_{2,+1}\mathbf{v}_{2,+1}^*, \quad \mathbf{v}_{1,-1}\mathbf{v}_{1,-1}^* + \mathbf{v}_{2,-1}\mathbf{v}_{2,-1}^*.$$

One obtains the projection matrices

$$\mathbf{v}_{1,+1}\mathbf{v}_{1,+1}^* + \mathbf{v}_{2,+1}\mathbf{v}_{2,+1}^* = \frac{1}{2}\begin{pmatrix} 1 & 0 & 0 & 1 \\ 0 & 1 & 1 & 0 \\ 0 & 1 & 1 & 0 \\ 1 & 0 & 0 & 1 \end{pmatrix}$$

$$\mathbf{v}_{1,-1}\mathbf{v}_{1,-1}^* + \mathbf{v}_{2,-1}\mathbf{v}_{2,-1}^* = \frac{1}{2}\begin{pmatrix} 1 & 0 & 0 & -1 \\ 0 & 1 & -1 & 0 \\ 0 & -1 & 1 & 0 \\ -1 & 0 & 0 & 1 \end{pmatrix}.$$

♣

Theorem 1.11. (Projection Theorem.) *Let U be a nonempty, convex, closed subset of the vector space $\mathbb{C}^n$. Given any element $\mathbf{w} \in \mathbb{C}^n$, there exists a unique element $\Pi\mathbf{w}$ such that*

$$\Pi\mathbf{w} \in U \quad and \quad \|\mathbf{w} - \Pi\mathbf{w}\| = \inf_{\mathbf{v} \in U} \|\mathbf{w} - \mathbf{v}\|.$$

This element $\Pi\mathbf{w} \in U$ satisfies

$$\langle \Pi\mathbf{w} - \mathbf{w}, \mathbf{v} - \Pi\mathbf{w} \rangle \geq 0 \quad for\ every\ \mathbf{v} \in U$$

and, conversely, if any element $\mathbf{u}$ satisfies

$$\mathbf{u} \in U \quad and \quad \langle \mathbf{u}, \mathbf{v} - \mathbf{u} \rangle \geq 0 \quad for\ every\ \mathbf{v} \in U$$

then $\mathbf{u} = \Pi\mathbf{w}$. Furthermore $\|\Pi\mathbf{u} - \Pi\mathbf{v}\| \leq \|\mathbf{u} - \mathbf{v}\|$.

For the proof refer to Ciarlet [14].

Let $\mathbf{u}$ be a nonzero normalized column vector. Then $\mathbf{u}\mathbf{u}^*$ is a projection matrix, since $(\mathbf{u}\mathbf{u}^*)^* = \mathbf{u}\mathbf{u}^*$ and

$$(\mathbf{u}\mathbf{u}^*)(\mathbf{u}\mathbf{u}^*) = \mathbf{u}(\mathbf{u}^*\mathbf{u})\mathbf{u}^* = \mathbf{u}\mathbf{u}^*.$$

If $\mathbf{u}$ is the zero column vector then $\mathbf{u}\mathbf{u}^*$ is the square zero matrix which is also a projection matrix.

Let A be an $n \times n$ hermitian matrix with $A^2 = I_n$. Then the matrix A must be unitary with $A = A^* = A^{-1}$. Then it can be easily shown that

$$\Pi_1 = \frac{1}{2}(I_n + A), \qquad \Pi_2 = \frac{1}{2}(I_n - A)$$

are projection matrices with $\Pi_1\Pi_2 = 0_n$.

Exercises. (1) Show that the matrices

$$\Pi_1 = \frac{1}{2}\begin{pmatrix} 1 & 1 \\ 1 & 1 \end{pmatrix}, \qquad \Pi_2 = \frac{1}{2}\begin{pmatrix} 1 & -1 \\ -1 & 1 \end{pmatrix}$$

are projection matrices and that $\Pi_1\Pi_2 = 0_2$.

(2) Is the sum of two $n \times n$ projection matrices an $n \times n$ projection matrix?

(3) Let A be an $n \times n$ matrix with $A^2 = A$. Show that $\det(A)$ is either equal to zero or equal to 1.

(4) Let

$$\Pi = \frac{1}{4}\begin{pmatrix} 1 & 1 & 1 & 1 \\ 1 & 1 & 1 & 1 \\ 1 & 1 & 1 & 1 \\ 1 & 1 & 1 & 1 \end{pmatrix}, \qquad \mathbf{u} = \begin{pmatrix} 1 \\ 0 \\ 0 \\ 0 \end{pmatrix}, \qquad \mathbf{v} = \begin{pmatrix} 0 \\ 0 \\ 0 \\ 1 \end{pmatrix}.$$

Show that $\|\Pi\mathbf{u} - \Pi\mathbf{v}\| \le \|\mathbf{u} - \mathbf{v}\|$.

(5) Show that $\Pi = I_n - \mathbf{v}\mathbf{v}^*$ is a projection matrix, where $\mathbf{v}$ is a normalized vector in $\mathbb{C}^n$.

(6) Consider the matrices

$$A = \begin{pmatrix} 2 & 1 \\ 1 & 2 \end{pmatrix}, \qquad I_2 = \begin{pmatrix} 1 & 0 \\ 0 & 1 \end{pmatrix}, \qquad C = \begin{pmatrix} 0 & 1 \\ 1 & 0 \end{pmatrix}.$$

Show that $[A, I_2] = 0_2$, $[A, C] = 0_2$, $I_2C = C$, $CI_2 = C$, $CC = I_2$. A group theoretical reduction (Steeb [60]) leads to the projection matrices

$$\Pi_1 = \frac{1}{2}\begin{pmatrix} 1 & 1 \\ 1 & 1 \end{pmatrix}, \qquad \Pi_2 = \frac{1}{2}\begin{pmatrix} 1 & -1 \\ -1 & 1 \end{pmatrix}.$$

Apply the projection operators to the standard basis to find a new basis. Show that the matrix A takes the form

$$\tilde{A} = \begin{pmatrix} 3 & 0 \\ 0 & 1 \end{pmatrix}$$

within the new basis. Notice that the new basis must be normalized before the matrix $\tilde{A}$ can be calculated.

(7) Consider an $n \times n$ diagonalizable matrix M with non-degenerate eigenvalues λ_j ($j = 1, \ldots, n$). Let $f : \mathbb{R} \to \mathbb{R}$ be an analytic function. Then

$$f(M) = \sum_{j=1}^{n} f(\lambda_j) \Pi_j.$$

Show that the projection operators Π_j are given by

$$\Pi_j = \prod_{k=1, k \neq j}^{n} (M - \lambda_k I_n)(\lambda_j I_n - \lambda_k I_n).$$

(8) Let $n \geq 2$ and A be an $n \times n$ matrix with $A^n = I_n$. Let

$$T = \frac{1}{n}(I_n + A + A^2 + \cdots + A^{n-1}).$$

Show that $T^2 = T$.

1.10 Unitary, Fourier and Hadamard Matrices

An $n \times n$ matrix U over $\mathbb{C}$ is called a unitary matrix if $U^* = U^{-1}$. It follows that $UU^* = I_n$. If U_1 and U_2 are unitary matrices, then $U_1 U_2$ is a unitary matrix. The $n \times n$ unitary matrices form a group under matrix multiplication called $U(n)$. The column vectors in a unitary matrix form an orthonormal basis in $\mathbb{C}^n$. If H is an $n \times n$ hermitian matrix, then $\exp(iH)$ is an $n \times n$ unitary matrix. The eigenvalues of a unitary matrix are of the form $\exp(i\phi)$ ($\phi \in \mathbb{R}$).

Example 1.28. Let $d \geq 2$. Consider the Hilbert space $\mathbb{C}^d$ and let

$$|v_0\rangle, \ |v_1\rangle, \ \ldots, \ |v_{d-1}\rangle, \qquad |w_0\rangle, \ |w_1\rangle, \ \ldots, \ |w_{d-1}\rangle$$

be two orthonormal bases in $\mathbb{C}^d$. Then

$$U = \sum_{j=0}^{d-1} |w_j\rangle\langle v_j|$$

is a unitary matrix as can be seen as follows. We have

$$U^* = \sum_{k=0}^{d-1} |v_k\rangle\langle w_k|.$$

Then with $\langle v_j|v_k\rangle = \delta_{jk}$ we obtain

$$UU^* = \sum_{j=0}^{d-1}\sum_{k=0}^{d-1} |w_j\rangle\langle v_j|v_k\rangle\langle w_k| = \sum_{j=0}^{d-1}\sum_{k=0}^{d-1} |w_j\rangle\delta_{jk}\langle w_k|$$

$$= \sum_{j=0}^{d-1} |w_j\rangle\langle w_j| = I_d.$$

♣

Important subsets of unitary matrices are the Fourier matrices and the permutation matrices.

Fourier and Hadamard matrices play an important role in spectral analysis (Davis [17], Elliott and Rao [20], Regalia and Mitra [49]). We give a short introduction to these types of matrices. In sections 2.8 and 3.18 we discuss the connection with the Kronecker product.

Let n be a fixed integer ≥ 1. We define

$$w := \exp(2\pi i/n) \equiv \cos(2\pi/n) + i\sin(2\pi/n)$$

where $i = \sqrt{-1}$. w might be taken as any primitive n-th root of unity. It can easily be proved that

$$w^n = 1, \quad w\bar{w} = 1, \quad \bar{w} = w^{-1}, \quad \bar{w}^k = w^{-k} = w^{n-k}$$

and

$$1 + w + w^2 + \cdots + w^{n-1} = 0$$

where $\bar{w}$ is the complex conjugate of w. For $n = 4$ we have $w = i$, $w^2 = -1$, $w^3 = -i$, $w^4 = 1$.

Definition 1.32. By the *Fourier matrix* of order n, we mean the matrix $F(= F_n)$ where

$$F^* := \frac{1}{\sqrt{n}}(w^{(i-1)(j-1)}) \equiv \frac{1}{\sqrt{n}}\begin{pmatrix} 1 & 1 & 1 & \cdots & 1 \\ 1 & w & w^2 & \cdots & w^{n-1} \\ 1 & w^2 & w^4 & \cdots & w^{2(n-1)} \\ \vdots & \vdots & \vdots & & \vdots \\ 1 & w^{n-1} & w^{2(n-1)} & \cdots & w^{(n-1)(n-1)} \end{pmatrix}$$

where F^* is the conjugate transpose of F.

The sequence w^k, $k = 0, 1, 2 \ldots$, is periodic with period n. Consequently there are only n distinct elements in F. Therefore F^* can be written as

$$F^* = \frac{1}{\sqrt{n}}\begin{pmatrix} 1 & 1 & 1 & \cdots & 1 \\ 1 & w & w^2 & \cdots & w^{n-1} \\ 1 & w^2 & w^4 & \cdots & w^{n-2} \\ \vdots & \vdots & \vdots & & \vdots \\ 1 & w^{n-1} & w^{n-2} & \cdots & w \end{pmatrix}.$$

The following theorem can easily be proved

Theorem 1.12. F is unitary, i.e. $FF^* = F^*F = I_n \iff F^{-1} = F^*$.

Proof. This is a result of the *geometric series identity*

$$\sum_{r=0}^{n-1} w^{r(j-k)} \equiv \frac{1 - w^{n(j-k)}}{1 - w^{j-k}} = \begin{cases} n \text{ if } j = k \\ 0 \text{ if } j \neq k \end{cases}.$$

$\square$

A second application of the geometrical identity yields the $n \times n$ permutation matrix

$$F^{*2} \equiv F^* F^* \equiv \begin{pmatrix} 1 & 0 & \cdots & 0 & 0 \\ 0 & 0 & \cdots & 0 & 1 \\ 0 & 0 & \cdots & 1 & 0 \\ \vdots & \vdots & & \vdots & \vdots \\ 0 & 1 & \cdots & 0 & 0 \end{pmatrix} = F^2.$$

Corollary 1.1. *We have* $F^{*4} = I_n$, $F^{*3} = F^{*4}(F^*)^{-1} = I_n F = F$.

Corollary 1.2. *The eigenvalues of F are ± 1, $\pm i$, with appropriate multiplicities.*

The characteristic polynomials $f(\lambda)$ of $F^*(= F_n^*)$ are as follows

$$n \equiv 0 \quad \text{modulo } 4, \ f(\lambda) = (\lambda - 1)^2 (\lambda - i)(\lambda + 1)(\lambda^4 - 1)^{(n/4)-1}$$
$$n \equiv 1 \quad \text{modulo } 4, \ f(\lambda) = (\lambda - 1)(\lambda^4 - 1)^{(1/4)(n-1)}$$
$$n \equiv 2 \quad \text{modulo } 4, \ f(\lambda) = (\lambda^2 - 1)(\lambda^4 - 1)^{(n/4)(n-2)}$$
$$n \equiv 3 \quad \text{modulo } 4, \ f(\lambda) = (\lambda - i)(\lambda^2 - 1)(\lambda^4 - 1)^{(1/4)(n-3)}.$$

Definition 1.33. Let

$$Z = (z_1, z_2, \ldots, z_n)^T, \qquad \hat{Z} = (\hat{z}_1, \hat{z}_2, \ldots, \hat{z}_n)^T$$

where $z_j \in \mathbb{C}$. The linear transformation

$$\hat{Z} = FZ$$

where F is the Fourier matrix is called the discrete *Fourier transform*.

Its inverse transformation exists since F^{-1} exists and is given by

$$Z = F^{-1}\hat{Z} \equiv F^*\hat{Z}.$$

Let

$$p(z) = a_0 + a_1 z + \cdots + a_{n-1} z^{n-1}$$

be a polynomial of degree $\leq n - 1$. It will be determined uniquely by specifying its values $p(z_n)$ at n distinct points z_k, $k = 1, 2, \ldots, n$ in the complex plane $\mathbb{C}$. Select these points z_k as the n roots of unity $1, w, w^2,$ $\ldots, w^{n-1}$. Then

$$\sqrt{n}F^* \begin{pmatrix} a_0 \\ a_1 \\ \vdots \\ a_{n-1} \end{pmatrix} = \begin{pmatrix} p(1) \\ p(w) \\ \vdots \\ p(w^{n-1}) \end{pmatrix} \Rightarrow \begin{pmatrix} a_0 \\ a_1 \\ \vdots \\ a_{n-1} \end{pmatrix} = \frac{1}{\sqrt{n}}F \begin{pmatrix} p(1) \\ p(w) \\ \vdots \\ p(w^{n-1}) \end{pmatrix}.$$

These formulas for interpolation at the roots of unity can be given another form.

Definition 1.34. By a *Vandermonde matrix* $V(z_0, z_1, \ldots, z_{n-1})$ is meant a matrix of the form

$$V(z_0, z_1, \ldots, z_{n-1}) := \begin{pmatrix} 1 & 1 & \cdots & 1 \\ z_0 & z_1 & \cdots & z_{n-1} \\ z_0^2 & z_1^2 & \cdots & z_{n-1}^2 \\ \vdots & \vdots & & \vdots \\ z_0^{n-1} & z_1^{n-1} & \cdots & z_{n-1}^{n-1} \end{pmatrix}.$$

It follows that

$$V(1, w, w^2, \ldots, w^{n-1}) = n^{1/2}F^*, \quad V(1, \bar{w}, \bar{w}^2, \ldots, \bar{w}^{n-1}) = n^{1/2}\bar{F}^* = n^{1/2}F.$$

Furthermore

$$\begin{aligned} p(z) &= (1, z, \ldots, z^{n-1})(a_0, a_1, \ldots, a_{n-1})^T \\ &= (1, z, \ldots, z^{n-1})n^{-1/2}F(p(1), p(w), \ldots, p(w^{n-1}))^T \\ &= n^{-1/2}(1, z, \ldots, z^{n-1})V(1, \bar{w}, \ldots, \bar{w}^{n-1})(p(1), p(w), \ldots, p(w^{n-1}))^T. \end{aligned}$$

Let F'_{2^n} denote the Fourier matrices of order 2^n whose rows have been permuted according to the bit reversing permutation.

Definition 1.35. A sequence in natural order can be arranged in *bit-reversed order* as follows: For an integer expressed in binary notation, reverse the binary form and transform to decimal notation, which is then called bit-reversed notation.

Example 1.29. The number 6 can be written as $6 = 1 \cdot 2^2 + 1 \cdot 2^1 + 0 \cdot 2^0$. Therefore in binary $6 \rightarrow 110$. Reversing the binary digits yields 011. Since

$$3 = 0 \cdot 2^2 + 1 \cdot 2^1 + 1 \cdot 2^0$$

we have $6 \rightarrow 3$. ♣

Since the sequence 0, 1 is the bit reversed order of 0, 1 and 0, 2, 1, 3 is the bit reversed order of 0, 1, 2, 3 we find that the matrices F_2' and F_4' are given by

$$F_2' = \frac{1}{\sqrt{2}} \begin{pmatrix} 1 & 1 \\ 1 & -1 \end{pmatrix} = F_2, \quad F_4' = \frac{1}{\sqrt{4}} \begin{pmatrix} 1 & 1 & 1 & 1 \\ 1 & -1 & 1 & -1 \\ 1 & i & -1 & -i \\ 1 & -i & -1 & i \end{pmatrix}.$$

Definition 1.36. By a *Hadamard matrix* of order n, H ($\equiv H_n$), is meant a matrix whose elements are either $+1$ or -1 and for which

$$HH^T = H^T H = nI_n$$

where I_n is the $n \times n$ unit matrix. Thus, $n^{-1/2}H$ is an orthogonal matrix.

The 1×1, 2×2 and 4×4 Hadamard matrices are given by

$$H_1 = (1), \quad H_2 = \sqrt{2}F_2 = \begin{pmatrix} 1 & 1 \\ 1 & -1 \end{pmatrix},$$

$$H_{4,1} = \begin{pmatrix} 1 & 1 & 1 & 1 \\ -1 & -1 & 1 & 1 \\ -1 & 1 & 1 & -1 \\ 1 & -1 & 1 & -1 \end{pmatrix}, \quad H_{4,2} = \begin{pmatrix} 1 & 1 & 1 & -1 \\ 1 & 1 & -1 & 1 \\ 1 & -1 & 1 & 1 \\ -1 & 1 & 1 & 1 \end{pmatrix}.$$

Note that the columns (or rows) considered as vectors are orthogonal to each other. Sometimes the term Hadamard matrix is limited to the matrices of order 2^n. These matrices have the property

$$H_{2^n} = H_{2^n}^T$$

so that $H_{2^n}^2 = 2^n I$. A recursion relation to find H_{2^n} using the Kronecker product will be given later.

Definition 1.37. The *Walsh-Hadamard transform* is defined as

$$\hat{Z} = HZ$$

where H is an Hadamard matrix, where $Z = (z_1, z_2, \ldots, z_n)^T$. Since H^{-1} exists we have

$$Z = H^{-1}\hat{Z}.$$

For example the inverse of H_2 is given by

$$H_2^{-1} = \frac{1}{2} \begin{pmatrix} 1 & 1 \\ 1 & -1 \end{pmatrix}.$$

Exercises. (1) Is the 2×2 matrix

$$V = \frac{1}{2} \begin{pmatrix} 1-i & 1+i \\ 1+i & 1-i \end{pmatrix}$$

unitary?

(2) Find a 3×3 unitary matrix U such that

$$U \frac{1}{\sqrt{2}} \begin{pmatrix} 1 \\ 0 \\ 1 \end{pmatrix} = \begin{pmatrix} 0 \\ 1 \\ 0 \end{pmatrix}.$$

(3) Show that $F = F^T$, $F^* = (F^*)^T = \bar{F}$, $F = \bar{F}^*$. This means F and F^* are symmetric.

(4) Can one find 2×2 unitary matrices U, V such that

$$\begin{pmatrix} 1 & 1 \\ 0 & 1 \end{pmatrix} U \begin{pmatrix} 1 & 0 \\ 1 & 1 \end{pmatrix} = V ?$$

(5) Show that the 2×2 matrix

$$U(\rho, \phi_1, \phi_2) = \begin{pmatrix} \sqrt{1-\rho^2} e^{i\phi_2} & \rho e^{i\phi_1} \\ -\rho e^{-i\phi_1} & \sqrt{1-\rho^2} e^{-i\phi_2} \end{pmatrix}$$

is unitary, where $0 \leq \rho \leq 1$ and $\phi_1, \phi_2 \in \mathbb{R}$.

(6) Show that the sequence $0, 8, 4, 12, 2, 10, 6, 14, 1, 9, 5, 13, 3, 11, 7, 15$ is the bit reversed order of

$$0, 1, 2, 3, 4, 5, 6, 7, 8, 9, 10, 11, 12, 13, 14, 15.$$

(7) Find the eigenvalues of

$$F_4^* = \frac{1}{2} \begin{pmatrix} 1 & 1 & 1 & 1 \\ 1 & \omega & \omega^2 & \omega^3 \\ 1 & \omega^2 & \omega & \omega^2 \\ 1 & \omega^3 & \omega^2 & \omega \end{pmatrix}.$$

Derive the eigenvalues of F_4.

(8) The discrete Fourier transform in one dimension can also be written as

$$\hat{x}(k) = \frac{1}{N} \sum_{n=0}^{N-1} x(n) \exp(-ik2\pi n/N)$$

where $N \in \mathbb{N}$ and $k = 0, 1, \ldots, N - 1$. Show that

$$x(n) = \sum_{k=0}^{N-1} \hat{x}(k) \exp(ik2\pi n/N).$$

Let $x(n) = \cos(2\pi n/N)$, where $N = 8$ and $n = 0, 1, \ldots, N - 1$. Find $\hat{x}(k)$.

(9) Find all 8×8 Hadamard matrices and their eigenvalues.

(10) Let $|0\rangle$, $|1\rangle$, $|2\rangle$, $|3\rangle$ be an orthonormal basis in the Hilbert space $\mathbb{C}^4$.
(i) Is the 4×4 matrix

$$V = |0\rangle\langle 1| + |1\rangle\langle 2| + |2\rangle\langle 3| + |3\rangle\langle 0|$$

unitary?
(ii) Is the 4×4 matrix

$$W = |0\rangle\langle 1| + e^{-i\pi/4}|1\rangle\langle 2| + e^{i\pi/2}|2\rangle\langle 3| + e^{-i\pi/4}|3\rangle\langle 0|$$

unitary?
(iii) Is the 4×4 matrix

$$T = |0\rangle\langle 0| + e^{-i\pi/4}|1\rangle\langle 1| + e^{i\pi/2}|2\rangle\langle 2| + e^{-i\pi/4}|3\rangle\langle 3|$$

unitary?

1.11 Transformation of Matrices

Let V be a vector space of finite dimension n and let $\mathcal{A} : V \rightarrow V$ be a linear transformation, represented by a (square) matrix $A = (a_{ij})$ relative to a basis (e_i). Relative to another basis (f_i), the same transformation is represented by the matrix

$$B = Q^{-1}AQ$$

where Q is the invertible matrix whose jth column vector consists of the components of the vector f_j in the basis (e_i). Since the same linear transformation $\mathcal{A}$ can in this way be represented by different matrices, depending on the basis that is chosen, the problem arises of finding a basis relative to which the matrix representing the transformation is as simple as possible.

Equivalently, given a matrix A, that is to say, those which are of the form $Q^{-1}AQ$, with Q invertible, those which have a form that is 'as simple as possible'.

Definition 1.38. If there exists an invertible matrix Q such that the matrix $Q^{-1}AQ$ is diagonal, then the matrix A is said to be *diagonalizable*.

In this case, the diagonal elements of the matrix $Q^{-1}AQ$ are the eigenvalues $\lambda_1, \lambda_2, \ldots, \lambda_n$ of the matrix A. The jth column vector of the matrix Q consists of the components (relative to the same basis as that used for the matrix A) of a normalized eigenvector corresponding to λ_j. In other words, a matrix is diagonalizable if and only if there exists a basis of eigenvectors.

Example 1.30. The 2×2 matrix

$$A = \begin{pmatrix} 0 & 1 \\ 1 & 0 \end{pmatrix}$$

is diagonalizable with the Hadamard matrix

$$Q = \frac{1}{\sqrt{2}} \begin{pmatrix} 1 & 1 \\ 1 & -1 \end{pmatrix}.$$

We find $Q^{-1}AQ = \mathrm{diag}(1, -1)$. The 2×2 matrix

$$B = \begin{pmatrix} 0 & 1 \\ 0 & 0 \end{pmatrix}$$

cannot be diagonalized.   ♣

For nondiagonalizable matrices *Jordan's theorem* gives the simplest form among all similar matrices.

Definition 1.39. A matrix $A = (a_{ij})$ of order n is *upper triangular* if $a_{ij} = 0$ for $i > j$ and *lower triangular* if $a_{ij} = 0$ for $i < j$. If there is no need to distinguish between the two, the matrix is simply called *triangular*.

Theorem 1.13. *(1) Given a square matrix A, there exists a unitary matrix U such that the matrix $U^{-1}AU$ is triangular.*
(2) Given a normal matrix A, there exists a unitary matrix U such that the matrix $U^{-1}AU$ is diagonal.
(3) Given a symmetric matrix A, there exists an orthogonal matrix O such that the matrix $O^{-1}AO$ is diagonal.

For the proof refer to Ciarlet [14].

The matrices U satisfying the conditions of the statement are not unique (consider, for example, $A = I$). The diagonal elements of the triangular matrix $U^{-1}AU$ of (1), or of the diagonal matrix $U^{-1}AU$ of (2), or of the diagonal matrix of (3), are the eigenvalues of the matrix A. Consequently, they are real numbers if the matrix A is Hermitian or symmetric and complex numbers of modulus 1 if the matrix is unitary or orthogonal. It follows from (2) that every Hermitian or unitary matrix is diagonalizable by a unitary matrix. The preceding argument shows that if, O is an orthogonal matrix, there exists a unitary matrix U such that $D = U^*OU$ is diagonal (the diagonal elements of D having modulus equal to 1), but the matrix U is not, in general, real, that is to say, orthogonal.

Definition 1.40. The *singular values* of a square matrix A are the positive square roots of the eigenvalues of the Hermitian matrix A^*A (or A^TA, if the matrix A is real).

Example 1.31. Consider the 2×3 matrix

$$A = \begin{pmatrix} 2 & 1 & 0 \\ 0 & 2 & 2 \end{pmatrix}.$$

Then

$$A^*A = \begin{pmatrix} 4 & 2 & 0 \\ 2 & 5 & 4 \\ 0 & 4 & 4 \end{pmatrix}, \quad AA^* = \begin{pmatrix} 5 & 2 \\ 2 & 8 \end{pmatrix}.$$

Obviously A^*A is a positive semidefinite matrix. The eigenvalues of A^*A are 0,4,9. Thus the singular values are 0,2,3. Hence the matrix AA^* has the eigenvalues 4 and 9. ♣

They are always nonnegative, since from the relation $A^*Av = \lambda v$, $v \neq 0$, it follows that $(Av)^*Av = \lambda v^*v$.

The singular values are all strictly positive if and only if the matrix A is invertible. In fact, we have

$$Au = 0 \Rightarrow A^*Au = 0 \Rightarrow u^*A^*Au = (Au)^*Au = 0 \Rightarrow Au = 0.$$

Definition 1.41. Two matrices A and B of type (m,n) are said to be *equivalent* if there exists an invertible matrix Q of order m and an invertible matrix R of order n such that $B = QAR$.

This is a more general notion than that of the similarity of matrices. In fact, it can be shown that every square matrix is equivalent to a diagonal matrix.

Theorem 1.14. *If A is a real, square matrix, there exist two orthogonal matrices U and V such that*

$$U^T AV = diag(\mu_i)$$

and, if A is a complex, square matrix, there exist two unitary matrices U and V such that

$$U^* AV = diag(\mu_i).$$

In either case, the numbers $\mu_i \geq 0$ are the singular values of the matrix A.

The proof of this theorem follows from the proofs in section 1.15.

If A is an $n \times n$ matrix and U is an $n \times n$ unitary matrix, then $(m \in \mathbb{N})$

$$U A^m U^* = (U A U^*)^m$$

since $UU^* = I_n$.

Exercises. (1) Find the eigenvalues and normalized eigenvectors of the symmetric matrix over $\mathbb{R}$

$$A = \begin{pmatrix} 2 & 1 \\ 1 & 2 \end{pmatrix}.$$

Then use the normalized eigenvectors to construct the matrix Q^{-1} such that $Q^{-1}AQ$ is a diagonal matrix.

(2) Consider the skew-symmetric matrix

$$A = \begin{pmatrix} 0 & 0 & 1 \\ 0 & 0 & 0 \\ -1 & 0 & 0 \end{pmatrix}.$$

Find the eigenvalues and the corresponding normalized eigenvectors. Can one find an invertible 3×3 matrix S such that SAS^{-1} is a diagonal matrix?

(3) Show that the matrix

$$\begin{pmatrix} 0 & 1 & 0 & 0 \\ 0 & 0 & 1 & 0 \\ 0 & 0 & 0 & 1 \\ 0 & 0 & 0 & 0 \end{pmatrix}$$

is not diagonalizable. Is the matrix nonnormal?

(4) Let O be an orthogonal matrix. Show that there exists an orthogonal matrix Q such that $Q^{-1}OQ$ is given by

$$(1)\oplus\cdots\oplus(1)\oplus(-1)\oplus(-1)\oplus\begin{pmatrix}\cos(\theta_1) & \sin(\theta_1)\\ -\sin(\theta_1) & \cos(\theta_1)\end{pmatrix}\oplus\cdots\oplus\begin{pmatrix}\cos(\theta_r) & \sin(\theta_r)\\ -\sin(\theta_r) & \cos(\theta_r)\end{pmatrix}$$

where $\oplus$ denotes the direct sum.

(5) Let A be a real matrix of order n. Show that a necessary and sufficient condition for the existence of a unitary matrix U of the same order and of a real matrix B (of the same order) such that $U = A + iB$ (in other word, such that the matrix A is the 'real part' of the matrix U) is that all the singular values of the matrix A should be not greater than 1.

(6) Let $a, b \in \mathbb{R}$. Find the eigenvalues and normalized eigenvectors of the symmetric 6×6 matrix

$$M(a,b) = \begin{pmatrix} a & b & 0 & 0 & 0 & 0\\ b & a & b & 0 & 0 & 0\\ 0 & b & a & b & 0 & 0\\ 0 & 0 & b & a & b & 0\\ 0 & 0 & 0 & b & a & b\\ 0 & 0 & 0 & 0 & b & a \end{pmatrix}$$

by calculating the eigenvalues and normalized eigenvectors of the 6×6 matrix

$$C = \begin{pmatrix} 0 & 1 & 0 & 0 & 0 & 0\\ 1 & 0 & 1 & 0 & 0 & 0\\ 0 & 1 & 0 & 1 & 0 & 0\\ 0 & 0 & 1 & 0 & 1 & 0\\ 0 & 0 & 0 & 1 & 0 & 1\\ 0 & 0 & 0 & 0 & 1 & 0 \end{pmatrix}.$$

1.12 Cayley Transform

Let M_n be the algebra of all $n \times n$ matrices over the complex field and $H_n \subset M_n$ the subalgebra of Hermitian matrices. As usual, a conjugate transpose of a complex matrix $A \in M_n$ will be denoted by A^*. Now, suppose that $A \in H_n$, i.e. $A^* = A$, and let I_n be the $n \times n$ identity matrix. Then $(A + iI_n)^{-1}$ exists and

$$U_A = (A - iI_n)(A + iI_n)^{-1}$$

is called a Cayley transform of A. It is easy to see that U_A is a unitary matrix and the inverse transform is given by

$$A = i(I_n + U_A)(I_n - U_A)^{-1}.$$

Furthermore, $+1$ cannot be an eigenvalue of U_A. We give some basic examples.

(1) If $A = I_n$, then $U_A = -iI_n$.
(2) If A is the $n \times n$ zero matrix, i.e., $A = 0_n$, then $U_A = -I_n$.
(3) If A is a diagonal matrix, then U_A is also a diagonal matrix.
(4) If A has degenerate eigenvalues, then U_A has degenerate eigenvalues as well.
(5) If $A \in H_n$ is unitary, i.e., $A^2 = I_n$, then $U_A = -iA$. The Pauli spin matrices σ_1, σ_2 and σ_3 satisfy these conditions.

The Cayley transform is actually a generalization of a mapping of the complex plane to itself, given by

$$U(z) = \frac{z - i}{z + i}, \qquad z \in \mathbb{C} \setminus \{-i\}.$$

In particular, U maps the upper half plane of $\mathbb{C}$ conformally onto the unit disc of $\mathbb{C}$ and the real line $\mathbb{R}$ injectively into the unit circle. Moreover, no finite point on the real line can be mapped to $+1$ on the unit circle.

Some useful properties of the Cayley transform are

(1) If $V \in M_n$ is invertible, then $U_{VAV^{-1}} = VU_AV^{-1}$ for $A \in H_n$.
(2) If $\mathbf{x} \in \mathbb{C}^n$ is an eigenvector for an eigenvalue $\lambda \in \mathbb{R}$ of a matrix $A \in H_n$, then $\mathbf{x}$ is an eigenvector of a Cayley transform U_A and

$$U(\lambda) = (\lambda - i)/(\lambda + i)$$

is its eigenvalue.
(3) If $A \in H_m$ and $B \in H_n$, then $U_{B \otimes A} = PU_{A \otimes B}P^t$, where $P \in M_{mn}$ is the permutation matrix satisfying $P(A \otimes B)P^t = B \otimes A$. Here, P^t denotes the transpose of a matrix P and $\otimes$ denotes the Kronecker product.
(4) If $A, B \in H_n$ such that $[A, B] = AB - BA = 0_n$ then $[U_A, U_B] = 0_n$.

The Cayley transform is named after Arthur Cayley.

Let $A \in H_m$ and $B \in H_n$ be two Hermitian matrices. Then the Cayley transform provides the unitary matrices U_A and U_B, respectively. Now, $A \otimes B$ is again a Hermitian matrix and the Cayley transform gives us another unitary matrix $U_{A \otimes B}$.

Consider two Hermitian matrices A and B such that the Cayley transform $U_{A \otimes B}$ cannot be presented as a Kronecker product of two complex matrices. Let

$$A = B = \begin{pmatrix} 1 & 0 \\ 0 & 0 \end{pmatrix}$$

be diagonal 2×2 Hermitian matrices. Then $U_{A \otimes B} = \text{diag}(-i, -1, -1, -1)$ is a diagonal 4×4 matrix which cannot be presented as a Kronecker product of two 2×2 complex matrices.

Now, let us write one simple example showing that $U_{A \otimes B} = U_A \otimes U_B$ does not hold in general. In this example the Cayley transform $U_{A \otimes B}$ can be presented as a Kronecker product of two complex matrices.

Let $A = 0_m$ be the $m \times m$ zero matrix and $B = 0_n$ be the $n \times n$ zero matrix. Then $U_A = -I_m$, $U_B = -I_m$, $U_{A \otimes B} = -I_{mn} = U_{I_m} \otimes U_{I_n}$, and $U_A \otimes U_B = I_{mn}$. Obviously, $U_{A \otimes B} \neq U_A \otimes U_B$.

1.13 Permutation Matrices

In this section we introduce permutation matrices and discuss their properties. The connection with the Kronecker product is described in section 2.4. By a *permutation* σ of the set

$$N := \{ 1, 2, \ldots, n \}$$

is meant a one-to-one mapping of N onto itself. Including the identity permutation there are $n!$ distinct permutations of N. We indicate a permutation by $\sigma(1) = i_1$, $\sigma(2) = i_2$, $\ldots$, $\sigma(n) = i_n$ which is written as

$$\sigma : \begin{pmatrix} 1 & 2 & \cdots & n \\ i_1 & i_2 & \cdots & i_n \end{pmatrix}.$$

The inverse permutation is designated by σ^{-1}. Thus $\sigma^{-1}(i_k) = k$. Let $\mathbf{e}_{j,n}^T$ denote the unit (row) vector of n components which has a 1 in the j-th position and 0's elsewhere

$$\mathbf{e}_{j,n}^T := (0, \ldots, 0, 1, 0, \ldots, 0).$$

Definition 1.42. By a *permutation matrix* of order n is meant a matrix of the form

$$P = P_\sigma = \begin{pmatrix} \mathbf{e}_{i_1,n}^T \\ \mathbf{e}_{i_2,n}^T \\ \vdots \\ \mathbf{e}_{i_n,n}^T \end{pmatrix}.$$

The i-th row of P has a 1 in the $\sigma(i)$-th column and 0's elsewhere. The j-th column of P has a 1 in the $\sigma^{-1}(j)$-th row and 0's elsewhere. Thus each row and each column of P has precisely one 1 in it. We have

$$P_\sigma \begin{pmatrix} x_1 \\ x_2 \\ \vdots \\ x_n \end{pmatrix} = \begin{pmatrix} x_{\sigma(1)} \\ x_{\sigma(2)} \\ \vdots \\ x_{\sigma(n)} \end{pmatrix}.$$

Example 1.32. We have

$$\sigma : \begin{pmatrix} 1\ 2\ 3\ 4 \\ 4\ 1\ 3\ 2 \end{pmatrix} \Rightarrow P_\sigma = \begin{pmatrix} 0\ 0\ 0\ 1 \\ 1\ 0\ 0\ 0 \\ 0\ 0\ 1\ 0 \\ 0\ 1\ 0\ 0 \end{pmatrix}.$$

♣

The set of all 3×3 permutation matrices are given by the six matrices

$$\begin{pmatrix} 1\ 0\ 0 \\ 0\ 1\ 0 \\ 0\ 0\ 1 \end{pmatrix}, \begin{pmatrix} 1\ 0\ 0 \\ 0\ 0\ 1 \\ 0\ 1\ 0 \end{pmatrix}, \begin{pmatrix} 0\ 1\ 0 \\ 1\ 0\ 0 \\ 0\ 0\ 1 \end{pmatrix}, \begin{pmatrix} 0\ 1\ 0 \\ 0\ 0\ 1 \\ 1\ 0\ 0 \end{pmatrix}, \begin{pmatrix} 0\ 0\ 1 \\ 1\ 0\ 0 \\ 0\ 1\ 0 \end{pmatrix}, \begin{pmatrix} 0\ 0\ 1 \\ 0\ 1\ 0 \\ 1\ 0\ 0 \end{pmatrix}.$$

It can easily be proved that

$$P_\sigma P_\tau = P_{\sigma\tau}$$

where the product of the permutations σ, τ is applied from left to right. Furthermore, $(P_\sigma)^* = P_{\sigma^{-1}}$. Hence

$$(P_\sigma)^* P_\sigma = P_{\sigma^{-1}} P_\sigma = P_I = I_n$$

where I_n is the $n \times n$ unit matrix. It follows that

$$(P_\sigma)^* = P_{\sigma^{-1}} = (P_\sigma)^{-1}.$$

Consequently, the permutation matrices form a group under matrix multiplication. We find that the permutation matrices are unitary, forming a finite subgroup of the unitary group.

The determinant of a permutation matrix is either $+1$ or -1. The trace of an $n \times n$ permutation matrix is in the set $\{0, 1, \ldots, n-1, n\}$. For any permutation matrix an eigenvectors is given by

$$(1\ 1\ \cdots\ 1)^T$$

with the corresponding eigenvalue $+1$.

Exercises. (1) Show that the rank of an $n \times n$ permutation matrix is n. Show that the determinant of a permutation matrix is either $+1$ or -1. Show that at least one eigenvalue of a permutation matrix is equal to $+1$.

(2) Consider the set of all $n \times n$ permutation matrices. How many of the elements are their own inverses, i.e. $P = P^{-1}$? Note that $P^{-1} = P^T$.

(3) Find the 4×4 permutation matrix P such that $(1\ 2\ 3\ 4) = (3\ 1\ 4\ 2)\,P$.

(4) Show that the number of $n \times n$ permutation matrices is given by $n!$.

(5) Find all 4×4 permutation matrices. Show that they form a group under matrix multiplication. Find all subgroups.

(6) A 3×3 permutation matrix P has $\mathrm{tr}(P) = 1$ and $\det(P) = -1$. What can be said about the eigenvalues of P?

(7) Consider the 4×4 permutation matrix

$$P = \begin{pmatrix} 0\,0\,0\,1 \\ 0\,0\,1\,0 \\ 0\,1\,0\,0 \\ 1\,0\,0\,0 \end{pmatrix} \equiv \begin{pmatrix} 0 & 1 \\ 1 & 0 \end{pmatrix} \otimes \begin{pmatrix} 0 & 1 \\ 1 & 0 \end{pmatrix}.$$

Find all the eigenvalues and normalized eigenvectors. From the normalized eigenvectors construct an invertible matrix Q such that $Q^{-1}PQ$ is a diagonal matrix.

(8) Let P_1 and P_2 be two $n \times n$ permutation matrices. Is $[P_1, P_2] = 0_n$?

(9) Is it possible to find $\mathbf{v} \in \mathbb{R}^n$ such that $\mathbf{v}\mathbf{v}^T$ is an $n \times n$ permutation matrix?

(10) Let I_n be the $n \times n$ identity matrix and 0_n be the $n \times n$ zero matrix. Is the $2n \times 2n$ matrix

$$\begin{pmatrix} 0_n & I_n \\ I_n & 0_n \end{pmatrix}$$

a permutation matrix? Find the eigenvalues.

(11) Consider the 4×4 permutation matrix

$$P = \begin{pmatrix} 0 & 1 & 0 & 0 \\ 0 & 0 & 0 & 1 \\ 1 & 0 & 0 & 0 \\ 0 & 0 & 1 & 0 \end{pmatrix}.$$

Show that $P^2 \neq I_4$, $P^3 \neq I_4$, $P^4 = I_4$. Use this result to find the eigenvalues of P.

(12) Consider the permutation matrix

$$P = \begin{pmatrix} 0 & 1 & 0 \\ 0 & 0 & 1 \\ 1 & 0 & 0 \end{pmatrix}.$$

Find all 3×3 matrices A such that $[A, P] = 0_3$ or $PAP^{-1} = A$.

1.14   Spectral Theorem

Normal matrices have a *spectral decomposition*. Let A be upper triangular with respect to an orthonormal basis $\{\mathbf{v}_1, \ldots, \mathbf{v}_n\} \subset \mathbb{C}^n$. Then

$$A\mathbf{v}_j = \sum_{k=1}^{j} a_{j,k} \mathbf{v}_k$$

for some $a_{1,1}, a_{2,1}, a_{2,2}, \ldots, a_{n,n} \in \mathbb{C}$. Since $\{\mathbf{v}_1, \ldots, \mathbf{v}_n\}$ is orthonormal we have $a_{j,k} = \mathbf{v}_k^* A \mathbf{v}_j$. It follows that

$$\overline{a}_{k,j} = \mathbf{v}_j^* A^* \mathbf{v}_k$$

or equivalently

$$A^* \mathbf{v}_k = \sum_{j=k}^{n} \overline{a}_{k,j} \mathbf{v}_j.$$

We have

$$(A\mathbf{v}_j)^*(A\mathbf{v}_j) = \sum_{k=1}^{j} |a_{j,k}|^2, \quad (A^*\mathbf{v}_j)^*(A^*\mathbf{v}_j) = \sum_{k=j}^{n} |a_{k,j}|^2.$$

However

$$(A^*\mathbf{v}_1)^*(A^*\mathbf{v}_1) = \mathbf{v}_1 AA^*\mathbf{v}_1 = \mathbf{v}_1 A^*A\mathbf{v}_1 = (A\mathbf{v}_j)^*(A\mathbf{v}_j)$$

$$= \sum_{k=1}^{j} |a_{j,k}|^2 = \sum_{k=j}^{n} |a_{k,j}|^2$$

i.e.

$$\sum_{k=1}^{j-1} |a_{j,k}|^2 = \sum_{k=j+1}^{n} |a_{k,j}|^2.$$

For $j = 1$ we find $a_{2,1} = a_{3,1} = \cdots = 0$, for $j = 2$ we find $a_{3,2} = a_{4,2} = \cdots = 0$, etc. Thus A is diagonal with respect to $\{\mathbf{v}_1, \ldots, \mathbf{v}_n\}$. As a consequence $\{\mathbf{v}_1, \ldots, \mathbf{v}_n\}$ form an orthonormal set of eigenvectors of A (and span $\mathbb{C}^n$). Thus we can write (*spectral theorem*)

$$A = \sum_{j=1}^{n} \lambda_j \mathbf{v}_j \mathbf{v}_j^*$$

where $A\mathbf{v}_j = \lambda_j \mathbf{v}_j$. Since

$$\left(\sum_{k=1}^{n} \mathbf{v}_k \mathbf{e}_{k,n}^* \right) \mathbf{e}_{j,n} = \mathbf{v}_j$$

it follows that

$$\left[\left(\sum_{k=1}^{n} \mathbf{v}_k \mathbf{e}_{k,n}^* \right) \mathbf{e}_{j,n} \right]^* = \mathbf{e}_{j,n}^* \left(\sum_{k=1}^{n} \mathbf{v}_k \mathbf{e}_{l,n}^* \right)^* = \mathbf{v}_k^*$$

and consequently we write

$$A = \left(\sum_{k=1}^{n} \mathbf{v}_k \mathbf{e}_{k,n}^* \right) \left(\sum_{j=1}^{n} \lambda_j \mathbf{e}_{j,n} \mathbf{e}_{j,n}^* \right) \left(\sum_{k=1}^{n} \mathbf{v}_k \mathbf{e}_{k,n}^* \right)^* = VDV^*$$

where

$$V = \sum_{k=1}^{n} \mathbf{v}_k \mathbf{e}_{k,n}^*$$

is a unitary matrix. The columns are the orthonormal eigenvectors of A and

$$D = \sum_{k=1}^{n} \lambda_k \mathbf{e}_{k,n} \mathbf{e}_{k,n}^*$$

is a diagonal matrix of corresponding eigenvalues, i.e the eigenvalue λ_k in the k-th entry on the diagonal of D corresponds the eigenvector $\mathbf{v}_k$ which is the k-th column of V. This decomposition is known as a *spectral decomposition* or *diagonalization* of A.

Example 1.33. Consider the 2×2 hermitian matrix (Pauli spin matrix)

$$A = \sigma_2 = \begin{pmatrix} 0 & -i \\ i & 0 \end{pmatrix}.$$

The eigenvalues are $\lambda_1 = 1$, $\lambda_2 = -1$ with the corresponding normalized eigenvectors

$$\mathbf{u}_1 = \frac{1}{\sqrt{2}} \begin{pmatrix} 1 \\ i \end{pmatrix}, \qquad \mathbf{u}_2 = \frac{1}{\sqrt{2}} \begin{pmatrix} 1 \\ -i \end{pmatrix}.$$

Then

$$A \equiv \lambda_1 \mathbf{u}_1 \mathbf{u}_1^* + \lambda_2 \mathbf{u}_2 \mathbf{u}_2^* \equiv \frac{1}{2} \begin{pmatrix} 1 \\ i \end{pmatrix} (1, -i) - \frac{1}{2} \begin{pmatrix} 1 \\ -i \end{pmatrix} (1, i)$$

$$\equiv \frac{1}{2} \begin{pmatrix} 1 & -i \\ i & 1 \end{pmatrix} - \frac{1}{2} \begin{pmatrix} 1 & i \\ -i & 1 \end{pmatrix} \equiv \begin{pmatrix} 0 & -i \\ i & 0 \end{pmatrix}.$$

♣

Example 1.34. Consider a normal 3×3 matrix A with the eigenvalues $\lambda_1 = 0, \lambda_2 = \sqrt{2}, \lambda_3 = -\sqrt{2}$ and the corresponding normalized eigenvectors

$$\mathbf{v}_1 = \frac{1}{\sqrt{2}} \begin{pmatrix} 1 \\ 0 \\ -1 \end{pmatrix}, \qquad \mathbf{v}_2 = \frac{1}{2} \begin{pmatrix} 1 \\ \sqrt{2} \\ 1 \end{pmatrix}, \qquad \mathbf{v}_3 = \frac{1}{2} \begin{pmatrix} 1 \\ -\sqrt{2} \\ 1 \end{pmatrix}.$$

Then the matrix A can be reconstructed as

$$A = \lambda_1 \mathbf{v}_1 \mathbf{v}_1^* + \lambda_2 \mathbf{v}_2 \mathbf{v}_2^* + \lambda_3 \mathbf{v}_3 \mathbf{v}_3^* = \sqrt{2} \mathbf{v}_2 \mathbf{v}_2^* - \sqrt{2} \mathbf{v}_3 \mathbf{v}_3^* = \begin{pmatrix} 0 & 1 & 0 \\ 1 & 0 & 1 \\ 0 & 1 & 0 \end{pmatrix}.$$

♣

If the normal matrix A has multiple eigenvalues with corresponding nonorthogonal eigenvectors, we proceed as follows. Let λ be an eigenvalue of multiplicity m. Then the eigenvalues with their corresponding eigenvectors can be ordered as

$$\lambda, \lambda, \ldots, \lambda, \lambda_{m+1}, \ldots, \lambda_n, \qquad \mathbf{v}_1, \mathbf{v}_2, \ldots, \mathbf{v}_m, \mathbf{v}_{m+1}, \ldots, \mathbf{v}_n.$$

The vectors $\mathbf{v}_{m+1}, \ldots, \mathbf{v}_n$ are orthogonal to each other and to the rest. What is left is to find a new set of orthogonal vectors $\mathbf{v}_1', \mathbf{v}_2', \ldots, \mathbf{v}_m'$ each being orthogonal to $\mathbf{v}_{m+1}, \ldots, \mathbf{v}_n$ together with each being an eigenvector of A. The procedure we use is the Gram-Schmidt algorithm.

Let $\mathbf{v}_1' = \mathbf{v}_1$, $\mathbf{v}_2' = \mathbf{v}_2 + \alpha\mathbf{v}_1$. Then $\mathbf{v}_2'$ is an eigenvector of A, for it is a combination of eigenvectors corresponding to the same eigenvalue λ. Also $\mathbf{v}_2'$ is orthogonal to $\mathbf{v}_{m+1}, \ldots, \mathbf{v}_n$ since the latter are orthogonal to $\mathbf{v}_1$ and $\mathbf{v}_2$. What remains is to make $\mathbf{v}_2'$ orthogonal to $\mathbf{v}_1'$ i.e. to $\mathbf{v}_1$. We obtain

$$\alpha = -\frac{\langle \mathbf{v}_1, \mathbf{v}_2 \rangle}{\langle \mathbf{v}_1, \mathbf{v}_1 \rangle}.$$

Next we set

$$\mathbf{v}_3' = \mathbf{v}_3 + \alpha\mathbf{v}_1 + \beta\mathbf{v}_2$$

where α and β have to be determined. Using the same reasoning, we obtain the linear equation for α and β

$$\begin{pmatrix} \langle \mathbf{v}_1, \mathbf{v}_1 \rangle & \langle \mathbf{v}_1, \mathbf{v}_2 \rangle \\ \langle \mathbf{v}_2, \mathbf{v}_1 \rangle & \langle \mathbf{v}_2, \mathbf{v}_2 \rangle \end{pmatrix} \begin{pmatrix} \alpha \\ \beta \end{pmatrix} = - \begin{pmatrix} \langle \mathbf{v}_1, \mathbf{v}_3 \rangle \\ \langle \mathbf{v}_2, \mathbf{v}_3 \rangle \end{pmatrix}.$$

The approach can be repeated until we obtain $\mathbf{v}_m'$. The Gramian matrix of the above equations is nonsingular, since the eigenvectors of a Hermitian matrix are linearly independent.

Example 1.35. Consider the symmetric matrix over $\mathbb{R}$

$$A = \begin{pmatrix} 5 & -2 & -4 \\ -2 & 2 & 2 \\ -4 & 2 & 5 \end{pmatrix}.$$

The eigenvalues are $\lambda_1 = 1$, $\lambda_2 = 1$, $\lambda_3 = 10$. This means the eigenvalue $\lambda = 1$ is twofold. The eigenvectors are

$$\mathbf{v}_1 = \begin{pmatrix} -1 \\ -2 \\ 0 \end{pmatrix}, \qquad \mathbf{v}_2 = \begin{pmatrix} -1 \\ 0 \\ -1 \end{pmatrix}, \qquad \mathbf{v}_3 = \begin{pmatrix} 2 \\ -1 \\ -2 \end{pmatrix}.$$

We find that $\langle \mathbf{v}_1, \mathbf{v}_3 \rangle = 0$, $\langle \mathbf{v}_2, \mathbf{v}_3 \rangle = 0$, $\langle \mathbf{v}_1, \mathbf{v}_2 \rangle = 1$. However, the two vectors $\mathbf{v}_1$ and $\mathbf{v}_2$ are linearly independent. Now we use the *Gram-Schmidt algorithm* to find orthogonal eigenvectors (see section 1.3). We choose

$$\mathbf{v}_1' = \mathbf{v}_1, \qquad \mathbf{v}_2' = \mathbf{v}_2 + \alpha\mathbf{v}_1$$

such that $\alpha := -\langle \mathbf{v}_1, \mathbf{v}_2 \rangle / \langle \mathbf{v}_1, \mathbf{v}_1 \rangle = -1/5$. Then

$$\mathbf{v}_2' = \frac{1}{5} \begin{pmatrix} -4 \\ 2 \\ -5 \end{pmatrix}.$$

The normalized eigenvectors are

$$\mathbf{v}_1 = \frac{1}{\sqrt{5}} \begin{pmatrix} -1 \\ -2 \\ 0 \end{pmatrix}, \qquad \mathbf{v}_2' = \frac{1}{3\sqrt{5}} \begin{pmatrix} -4 \\ 2 \\ -5 \end{pmatrix}, \qquad \mathbf{v}_3 = \frac{1}{3} \begin{pmatrix} 2 \\ -1 \\ -2 \end{pmatrix}.$$

Consequently A can be written as (spectral representation)

$$A = \lambda_1 \mathbf{v}_1 \mathbf{v}_1^* + \lambda_2 \mathbf{v}_2' \mathbf{v}_2'^* + \lambda_3 \mathbf{v}_3 \mathbf{v}_3^*.$$

From the normalized eigenvectors we obtain the orthogonal matrix

$$O = \begin{pmatrix} \frac{-1}{\sqrt{5}} & \frac{-4}{3\sqrt{5}} & \frac{2}{3} \\ \frac{-2}{\sqrt{5}} & \frac{2}{3\sqrt{5}} & -\frac{1}{3} \\ 0 & \frac{-5}{3\sqrt{5}} & -\frac{2}{3} \end{pmatrix}.$$

Thus we obtain the diagonal matrix $O^T A O = \mathrm{diag}(1, 1, 10)$, where $O^T = O^{-1}$ and the eigenvalues of A are 1, 1, 10.   ♣

Exercises. (1) Consider the symmetric 4×4 matrix

$$A = \begin{pmatrix} 1\,1\,1\,1 \\ 1\,1\,1\,1 \\ 1\,1\,1\,1 \\ 1\,1\,1\,1 \end{pmatrix} \equiv \begin{pmatrix} 1 & 1 \\ 1 & 1 \end{pmatrix} \otimes \begin{pmatrix} 1 & 1 \\ 1 & 1 \end{pmatrix}$$

with $\mathrm{tr}(A) = 4$. Find A^2. Use these properties to find all eigenvalues of A. Find the normalized eigenvectors. Reconstruct the matrix from the eigenvalues and normalized eigenvectors.

(2) Let A be a 4×4 symmetric matrix over $\mathbb{R}$ with eigenvalues $\lambda_1 = 0$, $\lambda_2 = 1$, $\lambda_3 = 2$, $\lambda_4 = 3$ and the corresponding normalized eigenvectors

$$\mathbf{v}_1 = \frac{1}{\sqrt{2}}(1, 0, 0, 1)^T, \qquad \mathbf{v}_2 = \frac{1}{\sqrt{2}}(1, 0, 0, -1)^T,$$

$$\mathbf{v}_3 = \frac{1}{\sqrt{2}}(0, 1, 1, 0)^T, \qquad \mathbf{v}_4 = \frac{1}{\sqrt{2}}(0, 1, -1, 0)^T.$$

Find the matrix A.

(3) Let A and B be the skew-symmetric 2×2 matrices

$$A = \begin{pmatrix} 0 & 1 \\ -1 & 0 \end{pmatrix}, \qquad B = \begin{pmatrix} 0 & i \\ i & 0 \end{pmatrix}.$$

Find the eigenvalues λ_1 and λ_2 of A and the corresponding normalized eigenvectors $\mathbf{u}_1$ and $\mathbf{u}_2$. Show that A is given by $A = \lambda_1 \mathbf{u}_1 \mathbf{u}_1^* + \lambda_2 \mathbf{u}_2 \mathbf{u}_2^*$. Find the eigenvalues μ_1 and μ_2 of B and the corresponding normalized eigenvectors $\mathbf{v}_1$ and $\mathbf{v}_2$. Show that B is given by $B = \mu_1 \mathbf{v}_1 \mathbf{v}_1^* + \mu_2 \mathbf{v}_2 \mathbf{v}_2^*$.

(4) Let A be a real symmetric 3×3 matrix. Can A be reconstructed from $\text{tr}(A)$, $\text{tr}(A^2)$, $\text{tr}(A^3)$?

(5) Explain why the matrix

$$A = \begin{pmatrix} 0 & 1 & 1 \\ 0 & 0 & 1 \\ 0 & 0 & 0 \end{pmatrix}$$

cannot be reconstructed from the eigenvalues and eigenvectors. Can the matrix

$$B = \begin{pmatrix} 0 & 1 & 1 \\ 0 & 0 & 1 \\ 1 & 0 & 0 \end{pmatrix}$$

be reconstructed from the eigenvalues and eigenvectors? Is the matrix normal?

1.15 Singular Value Decomposition

We consider the *singular value decomposition* (SVD). Let A be $m \times n$ over $\mathbb{C}$. The matrices A^*A and AA^* are Hermitian and therefore normal. The nonzero eigenvalues of A^*A and AA^* are identical, for if

$$(A^*A)\mathbf{x}_\lambda = \lambda \mathbf{x}_\lambda, \qquad \lambda \neq 0, \quad \mathbf{x}_\lambda \neq \mathbf{0}$$

for some eigenvector $\mathbf{x}_\lambda$, then

$$A(A^*A)\mathbf{x}_\lambda = \lambda A \mathbf{x}_\lambda \Rightarrow (AA^*)(A\mathbf{x}_\lambda) = \lambda(A\mathbf{x}_\lambda)$$

so, since $A\mathbf{x}_\lambda \neq \mathbf{0}$, λ is an eigenvalue of AA^*. Similarly the nonzero eigenvalues of AA^* are eigenvalues of A^*A.

Since A^*A is normal the spectral theorem provides

$$A^*A = \sum_{j=1}^{n} \lambda_j \mathbf{v}_j \mathbf{v}_j^* = VDV^*$$

where $\{\mathbf{v}_1, \ldots, \mathbf{v}_n\}$ are the orthonormal eigenvectors of A^*A and

$$\lambda_1 \geq \lambda_2 \geq \cdots \geq \lambda_n$$

are the corresponding eigenvalues. Here we choose a convenient ordering of the eigenvalues. From

$$(A\mathbf{v}_k)^*(A\mathbf{v}_k) \geq 0 \;\Rightarrow\; \mathbf{v}_k^*(A^*A)\mathbf{v}_k = \lambda_k \mathbf{v}_k^* \mathbf{v}_k = \lambda_k \geq 0$$

we find that all of the eigenvalues are real and nonnegative. Define the *singular values* $\sigma_1 \geq \cdots \geq \sigma_r > 0$ by

$$\sigma_k := \sqrt{\lambda_k}, \quad \lambda_k \neq 0, \quad k = 1, \ldots, r.$$

Now we define the $m \times n$ matrix Σ by

$$(\Sigma)_{j,k} = \begin{cases} \sigma_j & j = k, j < r \\ 0 & \text{otherwise} \end{cases}$$

so that $A^*A = (\Sigma V^*)^*(\Sigma V)$. Let U be an arbitrary $m \times m$ unitary matrix, then $A^*A = (U\Sigma V^*)^*(U\Sigma V)$.

Let $\mathbf{v}_j$ be an eigenvector corresponding to the singular value σ_j (i.e. $A\mathbf{v}_j \neq \mathbf{0}$) and similarly for σ_k and $\mathbf{v}_k$. Then

$$(A\mathbf{v}_j)^*(A\mathbf{v}_k) = \mathbf{v}_j^*(A^*A)\mathbf{v}_k = \sigma_k^2 \mathbf{v}_j^* \mathbf{v}_k = \sigma_k^2 \delta_{j,k}.$$

Thus $\{A\mathbf{v}_1, \ldots, A\mathbf{v}_r\}$ are orthogonal. Let $\{\mathbf{u}_1, \ldots, \mathbf{u}_m\}$ be an orthonormal basis in $\mathbb{C}^m$ where

$$\mathbf{u}_j = \frac{1}{\sigma_j} A\mathbf{v}_j, \quad j = 1, \ldots, r.$$

By construction, for $j, k = 1, \ldots, r$

$$\mathbf{u}_j^* \mathbf{u}_k = \frac{1}{\sigma_j \sigma_k}(A\mathbf{v}_j)^*(A\mathbf{v}_k) = \frac{1}{\sigma_j \sigma_k}\mathbf{v}_j^*(A^*A)\mathbf{v}_k = \frac{\sigma_k^2}{\sigma_j \sigma_k}\mathbf{v}_j^* \mathbf{v}_k$$

$$= \frac{\sigma_k^2}{\sigma_j \sigma_k}\delta_{j,k} = \frac{\sigma_j^2}{\sigma_j \sigma_j}\delta_{j,k} = \delta_{j,k}$$

and for $j = 1, \ldots, r$ and $k = 1, \ldots, m - r$

$$\mathbf{u}_{r+k}^* \mathbf{u}_j = 0 = \frac{1}{\sigma_j}\mathbf{u}_{r+k}^* A\mathbf{v}_j \;\Rightarrow\; \mathbf{u}_{r+k}^* A\mathbf{v}_j = 0.$$

We also have $A\mathbf{v}_{r+k} = \mathbf{0}$. Since $\{\mathbf{v}_1, \ldots, \mathbf{v}_n\}$ is an orthonormal basis we must have

$$A^*\mathbf{u}_{r+k} = \sum_{j=1}^{n}[\mathbf{v}_k^*(A^*\mathbf{u}_{r+k})]\mathbf{v}_k = \sum_{j=1}^{n}[\mathbf{u}_{r+k}^* A\mathbf{v}_k]^*\mathbf{v}_k = \mathbf{0}$$

so that $\mathbf{u}_{r+k}^* A = \mathbf{0}$. Choosing

$$U = \sum_{j=1}^{m} \mathbf{u}_j \mathbf{e}_{j,m}^*$$

(as an exercise, verify that U is unitary) we find

$$U^* A = \sum_{j=1}^{r} \mathbf{e}_{j,m} \frac{1}{\sigma_j} \mathbf{v}_j^* A^* A + \sum_{j=r+1}^{m} \mathbf{e}_{j,m} \mathbf{u}_j^* A$$

$$= \sum_{j=1}^{r} \mathbf{e}_{j,m} \frac{1}{\sigma_j} (A^* A \mathbf{v}_j)^* = \sum_{j=1}^{r} \mathbf{e}_{j,m} \sigma_j \mathbf{v}_j^*$$

$$= \sum_{j=1}^{r} \sigma_j \mathbf{e}_{j,m} \mathbf{e}_{j,n}^* \mathbf{e}_{j,n} \mathbf{v}_j^* = \left(\sum_{j=1}^{r} \sigma_j \mathbf{e}_{j,m} \mathbf{e}_{j,n}^* \right) \left(\sum_{k=1}^{n} \mathbf{e}_{k,n} \mathbf{v}_k^* \right)$$

$$= \left(\sum_{j=1}^{r} \sigma_j \mathbf{e}_{j,m} \mathbf{e}_{j,n}^* \right) \left(\sum_{k=1}^{n} \mathbf{v}_k \mathbf{e}_{k,n}^* \right)^* = \Sigma V^*.$$

Thus $A = U\Sigma V^*$. This is a *singular value decomposition*.

Next we consider the *polar decomposition*. Let $m = n$ and $A = U\Sigma V^*$ be a singular value decomposition of A. Here the order of the singular values in Σ does not matter. Then $H = V\Sigma V^*$ is Hermitian and $\tilde{U} = UV^*$ is unitary. Consequently

$$A = (UV^*)(V\Sigma V^*) = \tilde{U} H$$

is a polar decomposition of A (a product of a unitary and a Hermitian matrix).

Exercises. (1) Find the singular value decomposition of the 3×2 matrix

$$A = \begin{pmatrix} 1 & 0 \\ 1 & 1 \\ 0 & 1 \end{pmatrix} \quad \Rightarrow \quad A^T = \begin{pmatrix} 1 & 1 & 0 \\ 0 & 1 & 1 \end{pmatrix} \quad \Rightarrow \quad A^T A = \begin{pmatrix} 2 & 1 \\ 1 & 2 \end{pmatrix}.$$

(2) Find the singular value decomposition of the 3×2 matrix

$$A = \begin{pmatrix} 1 & 1 \\ 2 & 2 \\ 3 & 3 \end{pmatrix} \quad \Rightarrow \quad A^T = \begin{pmatrix} 1 & 2 & 3 \\ 1 & 2 & 3 \end{pmatrix} \quad \Rightarrow \quad A^T A = \begin{pmatrix} 14 & 14 \\ 14 & 14 \end{pmatrix}.$$

(3) Find the polar decomposition of

$$A = \begin{pmatrix} 1 & 1 \\ 1 & 1 \end{pmatrix}.$$

1.16   Pseudo Inverse

A *1-inverse* of the $m \times n$ matrix A is an $n \times m$ matrix A^- such that $AA^- A = A$. If $m = n$ and A^{-1} exists we find

$$AA^- A = A \;\Rightarrow\; A^{-1}(AA^- A)A^{-1} = A^{-1}AA^{-1} \;\Rightarrow\; A^- = A^{-1}.$$

Consider

$$A = \begin{pmatrix} 1 & 0 \\ 0 & 0 \end{pmatrix}.$$

Then both A and the 2×2 identity matrix I_2 are 1-inverses of A. Consequently, the 1-inverse is in general not unique.

The *Moore-Penrose pseudo inverse* of the $m \times n$ matrix A is the 1-inverse A^- of A which additionally satisfies

$$A^- AA^- = A^- \qquad (AA^-)^* = AA^- \qquad (A^- A)^* = A^- A.$$

With these properties the matrices $A^- A$ and AA^- are projection matrices.

Let $A = U\Sigma V^*$ be a singular value decomposition of A. Then

$$A^- = V\Sigma^- U^*$$

is a Moore-Penrose pseudo inverse of A, where

$$(\Sigma^-)_{jk} = \begin{cases} \frac{1}{(\Sigma)_{kj}} & (\Sigma)_{kj} \neq 0 \\ 0 & (\Sigma)_{kj} = 0 \end{cases}.$$

We verify this by first calculating ($j, k \in \{1, 2, \ldots, n\}$)

$$(\Sigma^- \Sigma)_{jk} = \sum_{l=1}^{m} (\Sigma^-)_{jl}(\Sigma)_{lk} = \sum_{\substack{l=1 \\ (\Sigma)_{lj} \neq 0}}^{m} \frac{(\Sigma)_{lk}}{(\Sigma)_{lj}}.$$

Since $(\Sigma)_{lj} = 0$ when $l \neq j$ we find

$$(\Sigma^- \Sigma)_{jk} = \begin{cases} \frac{(\Sigma)_{jk}}{(\Sigma)_{jj}} & (\Sigma)_{jj} \neq 0 \\ 0 & (\Sigma)_{jj} = 0 \end{cases} = \begin{cases} \delta_{jk} & (\Sigma)_{jj} \neq 0 \\ 0 & (\Sigma)_{jj} = 0 \end{cases}.$$

Similarly ($j, k \in \{1, 2, \ldots, m\}$)

$$(\Sigma\Sigma^-)_{jk} = \begin{cases} \delta_{jk} & (\Sigma)_{jj} \neq 0 \\ 0 & (\Sigma)_{jj} = 0 \end{cases}.$$

The matrix $\Sigma^- \Sigma$ is a diagonal $n \times n$ matrix while $\Sigma\Sigma^-$ is a diagonal $m \times m$ matrix. All the entries of these matrices are 0 or 1 so that

$$(\Sigma\Sigma^-)^* = \Sigma\Sigma^-, \qquad (\Sigma^- \Sigma)^* = \Sigma^- \Sigma.$$

Now

$$(\Sigma\Sigma^-\Sigma)_{jk} = \sum_{l=1}^{n} (\Sigma)_{jl}(\Sigma^-\Sigma)_{lk} = (\Sigma)_{jj}(\Sigma^-\Sigma)_{jk}$$

$$= \begin{cases} \delta_{jk}(\Sigma)_{jj} & (\Sigma)_{jj} \neq 0 \\ 0 & (\Sigma)_{jj} = 0 \end{cases} = (\Sigma)_{jk}$$

and

$$(\Sigma^-\Sigma\Sigma^-)_{jk} = \sum_{l=1}^{m} (\Sigma^-)_{jl}(\Sigma\Sigma^-)_{lk} = (\Sigma^-)_{jj}(\Sigma\Sigma^-)_{jk}$$

$$= \begin{cases} \delta_{jk}(\Sigma^-)_{jj} & (\Sigma)_{jj} \neq 0 \\ 0 & (\Sigma)_{jj} = 0 \end{cases} = (\Sigma^-)_{jk}$$

i.e. $\Sigma\Sigma^-\Sigma = \Sigma$, $\Sigma^-\Sigma\Sigma^- = \Sigma^-$ so that Σ^- is the Moore-Penrose pseudo inverse of Σ. The remaining properties are easy to show

$$AA^-A = (U\Sigma V^*)(V\Sigma^-U^*)(U\Sigma V^*) = U\Sigma\Sigma^-\Sigma V^* = U\Sigma V^* = A,$$
$$A^-AA^- = (V\Sigma^-U^*)(U\Sigma V^*)(V\Sigma^-U^*) = V\Sigma^-\Sigma\Sigma^-U^* = V\Sigma^-U^* = A^-,$$
$$(AA^-)^* = (U\Sigma V^*V\Sigma^-U^*)^* = I_m^* = I_m = AA^-,$$
$$(A^-A)^* = (V\Sigma^-U^*U\Sigma V^*)^* = I_n^* = I_n = AA^-.$$

Thus $A^- = V\Sigma^-U^*$ is a Moore-Penrose pseudo inverse of A.

Example 1.36. The nonnormal matrix

$$\begin{pmatrix} 0 & 0 \\ 1 & 0 \end{pmatrix}$$

has the singular value decomposition

$$\begin{pmatrix} 0 & 0 \\ 1 & 0 \end{pmatrix} = \begin{pmatrix} 0 & 1 \\ 1 & 0 \end{pmatrix}\begin{pmatrix} 1 & 0 \\ 0 & 0 \end{pmatrix}\begin{pmatrix} 1 & 0 \\ 0 & 1 \end{pmatrix}$$

so that

$$\begin{pmatrix} 0 & 0 \\ 1 & 0 \end{pmatrix}^- = \begin{pmatrix} 1 & 0 \\ 0 & 1 \end{pmatrix}\begin{pmatrix} 1 & 0 \\ 0 & 0 \end{pmatrix}\begin{pmatrix} 0 & 1 \\ 1 & 0 \end{pmatrix} = \begin{pmatrix} 0 & 1 \\ 0 & 0 \end{pmatrix}.$$

Analogously

$$\begin{pmatrix} 0 & 1 \\ 0 & 0 \end{pmatrix}^- = \begin{pmatrix} 0 & 0 \\ 1 & 0 \end{pmatrix}.$$

♣

Exercise. (1) Find the pseudo-inverse of the 3×3 nonnormal matrix

$$\begin{pmatrix} 0 & 1 & 1 \\ 0 & 0 & 1 \\ 0 & 0 & 0 \end{pmatrix}.$$

Is the pseudo-inverse nonnormal?

(2) Find the pseudo-inverse of the 2×2 normal matrices

$$\begin{pmatrix} 1 & 1 \\ 1 & 1 \end{pmatrix} \quad \text{and} \quad \begin{pmatrix} 1 & -1 \\ -1 & 1 \end{pmatrix}.$$

Is the pseudo-inverse nonnormal?

1.17   Vec Operator

Let A be an $m \times n$ matrix. A matrix operation is that of stacking the columns of a matrix one under the other to form a single column. This operation is called vec (Neudecker [45], Brewer [12], Graham [25], Searle [50]). Thus $\mathrm{vec}(A)$ is a column vector of order $m \times n$.

Example 1.37. Let

$$A = \begin{pmatrix} 1 & 2 & 3 \\ 4 & 5 & 6 \end{pmatrix}.$$

Then we obtain the column vector

$$\mathrm{vec}(A) = \begin{pmatrix} 1 \\ 4 \\ 2 \\ 5 \\ 3 \\ 6 \end{pmatrix}.$$

♣

Let A, B be $m \times n$ matrices. We can prove that

$$\mathrm{vec}(A + B) = \mathrm{vec}(A) + \mathrm{vec}(B).$$

It is also easy to see that

$$\mathrm{vec}(\alpha A) = \alpha \mathrm{vec}(A), \qquad \alpha \in \mathbb{C}.$$

This means the vec operation is linear. The operation vech(A) is defined in the same way that vec(A) is, except that for each column of A only that part of it which is on or below the diagonal of A is put into vech(A) (vector-half of A). In this way, for A symmetric, vech(A) contains only the distinct elements of A.

Example 1.38. Consider the square matrix

$$A = \begin{pmatrix} 1 & 7 & 6 \\ 7 & 3 & 8 \\ 6 & 8 & 2 \end{pmatrix} = A^T.$$

Then we obtain the column vector

$$\text{vech}(A) = \begin{pmatrix} 1 \\ 7 \\ 6 \\ 3 \\ 8 \\ 2 \end{pmatrix}.$$

♣

The following theorems give useful properties of the vec operator. Proofs of the first depend on the elementary vector $\mathbf{e}_{j,n}$, the j-th column of the $n \times n$ unit matrix, i.e.

$$\mathbf{e}_{1,n} = \begin{pmatrix} 1 \\ 0 \\ \vdots \\ 0 \end{pmatrix}, \quad \mathbf{e}_{2,n} = \begin{pmatrix} 0 \\ 1 \\ \vdots \\ 0 \end{pmatrix}, \dots, \mathbf{e}_{n,n} = \begin{pmatrix} 0 \\ 0 \\ \vdots \\ 1 \end{pmatrix}$$

and

$$\mathbf{e}_{1,n}^T = (1, 0, \dots, 0), \quad \mathbf{e}_{2,n}^T = (0, 1, \dots, 0), \quad \dots, \quad \mathbf{e}_{n,n}^T = (0, \dots, 0, 1).$$

Theorem 1.15. *Let A, B be $n \times n$ matrices. Then*

$$\text{tr}(AB) \equiv (\text{vec}(A^T))^T \text{vec}(B).$$

Proof. We have

$$\text{tr}(AB) = \sum_{i=1}^{n} \mathbf{e}_{i,n}^T AB\mathbf{e}_{i,n} = (\mathbf{e}_{1,n}^T A \cdots \mathbf{e}_{r,n}^T A) \begin{pmatrix} B\mathbf{e}_{1,n} \\ B\mathbf{e}_{2,n} \\ \vdots \\ B\mathbf{e}_{r,n} \end{pmatrix} = (\text{vec}(A^T))^T \text{vec}(B).$$

□

Theorem 1.16. *Let A be an $m \times m$ matrix. Then there is a permutation matrix P such that $\mathrm{vec}(A) = P\mathrm{vec}(A^T)$.*

The proof is left to the reader as an exercise.

Let

$$A = \begin{pmatrix} 1 & 2 \\ 3 & 4 \end{pmatrix} \quad \Rightarrow \quad A^T = \begin{pmatrix} 1 & 3 \\ 2 & 4 \end{pmatrix}.$$

Then $\mathrm{vec}(A) = P\mathrm{vec}(A^T)$, where

$$P = \begin{pmatrix} 1 & 0 & 0 & 0 \\ 0 & 0 & 1 & 0 \\ 0 & 1 & 0 & 0 \\ 0 & 0 & 0 & 1 \end{pmatrix} \equiv (1) \oplus \begin{pmatrix} 0 & 1 \\ 1 & 0 \end{pmatrix} \oplus (1).$$

The full power of the vec operator will be seen when we consider the Kronecker product and the vec operator.

1.18   Vector and Matrix Norms

Definition 1.43. Let V be an n dimensional vector space over the field $\mathbb{F}$ of scalars. A *norm* on V is a function $\| \cdot \| : V \to \mathbb{R}$ which satisfies the following properties

$$\|\mathbf{v}\| = 0 \Leftrightarrow \mathbf{v} = 0, \text{ and } \|\mathbf{v}\| \geq 0 \text{ for every } \mathbf{v} \in V$$

$$\|\alpha\mathbf{v}\| = |\alpha|\|\mathbf{v}\| \text{ for every } \alpha \in \mathbb{F} \text{ and } \mathbf{v} \in V$$

$$\|\mathbf{u} + \mathbf{v}\| \leq \|\mathbf{u}\| + \|\mathbf{v}\| \text{ for every } \mathbf{u}, \mathbf{v} \in V.$$

The last property is known as the *triangle inequality*. A norm on V will also be called a *vector norm.*. We call a vector space which is provided with a norm a *normed vector space.*

The norm induces a metric (distance) $\|\mathbf{u} - \mathbf{v}\|$ on the vector space V, where $\mathbf{u}, \mathbf{v} \in V$. We have

$$\|\mathbf{u} - \mathbf{v}\| \geq |\|\mathbf{u}\| - \|\mathbf{v}\||.$$

Let V be a finite dimensional space. The following three norms are the ones most commonly used in practice

$$\|\mathbf{v}\|_1 := \sum_{j=1}^{n} |v_j|$$

$$\|\mathbf{v}\|_2 := \left(\sum_{j=1}^{n} |v_j|^2\right)^{1/2} = \langle \mathbf{v}, \mathbf{v}\rangle^{1/2}$$

$$\|\mathbf{v}\|_\infty := \max_{1 \le j \le n} |v_j|.$$

The norm $\|\cdot\|_2$ is called the *Euclidean norm*. It is easy to verify directly that the two functions $\|\cdot\|_1$ and $\|\cdot\|_\infty$ are indeed norms. As for the function $\|\cdot\|_2$, it is a particular case of the following more general result.

Theorem 1.17. *Let V be a finite dimensional vector space and $\mathbf{v} \in V$. For every real number $p \ge 1$, the function $\|\cdot\|_p$ defined by*

$$\|\mathbf{v}\|_p := \left(\sum_{j=1}^{n} |v_j|^p\right)^{1/p}$$

is a norm.

For the proof refer to Ciarlet [14].

The proof uses the following inequalities: For $p > 1$ and $1/p + 1/q = 1$, the inequality

$$\sum_{i=1}^{n} |u_j v_j| \le \left(\sum_{j=1}^{n} |u_j|^p\right)^{1/p} \left(\sum_{j=1}^{n} |v_i|^q\right)^{1/q}$$

is called *Hölder's inequality*. Hölder's inequality for $p = 2$,

$$\sum_{j=1}^{n} |u_j v_j| \le \left(\sum_{j=1}^{n} |u_j|^2\right)^{1/2} \left(\sum_{j=1}^{n} |v_i|^2\right)^{1/2}$$

is called *the Cauchy-Schwarz inequality*. The triangle inequality for the norm $\|\cdot\|_p$,

$$\left(\sum_{j=1}^{n} |u_j + v_j|^p\right)^{1/p} \le \left(\sum_{j=1}^{n} |u_j|^p\right)^{1/p} + \left(\sum_{j=1}^{n} |v_j|^p\right)^{1/p}$$

is called *Minkowski's inequality.*

The norms defined above are *equivalent*, this property being a particular case of the equivalence of norms in a finite dimensional space.

Definition 1.44. Two norms $\|\cdot\|$ and $\|\cdot\|'$, defined over the same vector space V, are equivalent if there exist two constants C and C' such that

$$\|\mathbf{v}\|' \leq C\|\mathbf{v}\| \quad \text{and} \quad \|\mathbf{v}\| \leq C'\|\mathbf{v}\|' \quad \text{for every} \quad \mathbf{v} \in V.$$

Let $\mathcal{A}_n$ be the ring of matrices of order n, with elements in the field $\mathbb{F}$.

Definition 1.45. A *matrix norm* is a function $\|\cdot\| : \mathcal{A}_n \to \mathbb{R}$ which satisfies the following properties

$$\|A\| = 0 \Leftrightarrow A = 0 \text{ and } \|A\| \geq 0 \text{ for every } A \in \mathcal{A}_n$$
$$\|\alpha A\| = |\alpha|\|A\| \text{ for every } \alpha \in \mathbb{F}, \quad A \in \mathcal{A}_n$$
$$\|A + B\| \leq \|A\| + \|B\| \text{ for every } A, B \in \mathcal{A}_n$$
$$\|AB\| \leq \|A\| \cdot \|B\| \text{ for every } A, B \in \mathcal{A}_n.$$

The ring $\mathcal{A}_n$ is itself a vector space of dimension n^2. Thus the first three properties above are nothing other than those of a vector norm, considering a matrix as a vector with n^2 components. The last property is evidently special to square matrices.

The result which follows gives a particularly simple means of constructing matrix norms.

Definition 1.46. Given a vector norm $\|\cdot\|$ on $\mathbb{C}^n$, the function $\|\cdot\| : \mathcal{A}_n(\mathbb{C}) \to \mathbb{R}$ defined by

$$\|A\| := \sup_{\substack{\mathbf{v} \in \mathbb{C}^n \\ \|\mathbf{v}\| = 1}} \|A\mathbf{v}\|$$

is a matrix norm, called the *subordinate matrix norm* (subordinate to the given vector norm). Sometimes it is also called the *induced matrix norm.*

This is just one particular case of the usual definition of the norm of a linear transformation.

Example 1.39. Consider the nonnormal 2×2 matrix

$$A = \begin{pmatrix} 1 & 1 \\ 2 & 2 \end{pmatrix}.$$

Then we find

$$\|A\| = \sup_{\substack{\mathbf{v} \in \mathbb{C}^2 \\ \|\mathbf{v}\|=1}} \|A\mathbf{v}\| = \sqrt{10}.$$

This result can be found by using the method of the *Lagrange multiplier*. The constraint is $\|\mathbf{v}\| = 1$. Furthermore we note that the eigenvalues of the matrix

$$A^T A = \begin{pmatrix} 1 & 2 \\ 1 & 2 \end{pmatrix} \begin{pmatrix} 1 & 1 \\ 2 & 2 \end{pmatrix} = \begin{pmatrix} 5 & 5 \\ 5 & 5 \end{pmatrix}$$

are given by $\lambda_1 = 10$ and $\lambda_2 = 0$. Thus the norm of A is the square root of the largest eigenvalue of $A^T A$, i.e. $\sqrt{10}$. ♣

Example 1.40. Let U_1 and U_2 be unitary $n \times n$ matrices with

$$\|U_1 - U_2\| \le \epsilon.$$

Let $\mathbf{v}$ be a normalized vector in $\mathbb{C}^n$. We have

$$\|U_1\mathbf{v} - U_2\mathbf{v}\| = \|(U_1 - U_1)\mathbf{v}\| \le \max_{\|\mathbf{y}\|=1} \|(U_1 - U_2)\mathbf{y}\| = \|U_1 - U_2\| \le \epsilon.$$
♣

It follows from the definition of a subordinate norm that

$$\|A\mathbf{v}\| \le \|A\| \|\mathbf{v}\| \text{ for every } \mathbf{v} \in \mathbb{C}^n.$$

A subordinate norm always satisfies $\|I_n\| = 1$, where I_n is the $n \times n$ unit matrix. Let us now calculate each of the subordinate norms of the vector norms $\|\cdot\|_1, \|\cdot\|_2, \|\cdot\|_\infty$.

Theorem 1.18. *Let $A = (a_{ij})$ be a square matrix. Then*

$$\|A\|_1 := \sup_{\|\mathbf{v}\|=1} \|A\mathbf{v}\|_1 = \max_{1 \le j \le n} \sum_{i=1}^{n} |a_{ij}|$$

$$\|A\|_2 := \sup_{\|\mathbf{v}\|=1} \|A\mathbf{v}\|_2 = \sqrt{\varrho(A^*A)} = \|A^*\|_2$$

$$\|A\|_\infty := \sup_{\|\mathbf{v}\|=1} \|A\mathbf{v}\|_\infty = \max_{1 \le i \le n} \sum_{j=1}^{n} |a_{ij}|$$

where $\varrho(A^*A)$ is the spectral radius of A^*A. The norm $\|\cdot\|_2$ is invariant under unitary transformations

$$UU^* = I \Rightarrow \|A\|_2 = \|AU\|_2 = \|UA\|_2 = \|U^*AU\|_2.$$

Furthermore, if the matrix A is normal, i.e. $AA^* = A^*A$

$$\|A\|_2 = \varrho(A).$$

The invariance of the norm $\| \cdot \|_2$ under unitary transformations is nothing more than the interpretation of the equalities

$$\varrho(A^*A) = \varrho(U^*A^*AU) = \varrho(A^*U^*UA) = \varrho(U^*A^*UU^*AU).$$

If the matrix A is normal, there exists a unitary matrix U such that

$$U^*AU = \text{diag}(\lambda_i(A)) := D.$$

Accordingly, $A^*A = (UDU^*)^*UDU^* = UD^*DU^*$ which proves that

$$\varrho(A^*A) = \varrho(D^*D) = \max_{1 \le i \le n} |\lambda_i(A)|^2 = (\varrho(A))^2.$$

The norm $\|A\|_2$ is nothing other than the largest singular value of the matrix A. If a matrix A is Hermitian, or symmetric (and hence normal), we have $\|A\|_2 = \varrho(A)$. If a matrix A is unitary, or orthogonal (and hence normal), we have

$$\|A\|_2 = \sqrt{\varrho(A^*A)} = \sqrt{\varrho(I)} = 1.$$

There exist matrix norms which are not subordinate to any vector norm. An example of a matrix norm which is not subordinate is given in the following theorem.

Theorem 1.19. *The function* $\| \cdot \|_E : \mathcal{A}_n \to \mathbb{R}$ *defined by*

$$\|A\|_E := \left(\sum_{i=1}^{n} \sum_{j=1}^{n} |a_{ij}|^2 \right)^{1/2} = (\text{tr}(A^*A))^{1/2}$$

for every matrix $A = (a_{ij})$ *of order* n *is a matrix norm which is not subordinate (for* $n \ge 2$). *Furthermore, the function is invariant under unitary transformations,*

$$UU^* = I \Rightarrow \|A\|_E = \|AU\|_E = \|UA\|_E = \|U^*AU\|_E$$

and satisfies

$$\|A\|_2 \le \|A\|_E \le \sqrt{n}\|A\|_2 \text{ for every } A \in \mathcal{A}_n.$$

Proof.   The fact that $\|A\|_E$ is invariant under unitary transformation of A follows from the cyclic invariance of the trace. The eigenvalues of A^*A are real and nonnegative. Let

$$\lambda_n \ge \lambda_{n-1} \ge \cdots \ge \lambda_1 \ge 0$$

be the eigenvalues of A^*A. Then $\varrho(A^*A) = \lambda_n$. Since

$$\|A\|_2 = \sqrt{\varrho(A^*A)} = \sqrt{\lambda_n}$$

and

$$\|A\|_E = \sqrt{\lambda_n + \lambda_{n-1} + \cdots + \lambda_1} \geq \sqrt{\lambda_n}$$

we have $\|A\|_2 \leq \|A\|_E$. Also

$$\|A\|_E = \sqrt{\lambda_n + \lambda_{n-1} + \cdots + \lambda_1} \leq \sqrt{n\lambda_n} = \sqrt{n}\sqrt{\varrho(A^*A)}$$

so that $\|A\|_E \leq \sqrt{n}\|A\|_2$. □

Example 1.41. Let I_2 be the 2×2 unit matrix. Then

$$\|I_2\|_E = \sqrt{2} = \sqrt{2}\|I_2\|_2.$$ ♣

Let M be an $m \times n$ matrix over $\mathbb{C}$. We consider the *rank-k approximation* problem, i.e. to find an $m \times n$ matrix A with rank$(A) = k \leq$ rank(M) such that

$$\|M - A\|$$

is a minimum and $\|\cdot\|$ denotes some norm. Different norms lead to different matrices A. For the remainder of this section we consider the Frobenius norm $\|\cdot\|_F$.

Certain transformations leave the Frobenius norm invariant, for example unitary transformations. Let U_m and U_n be $m \times m$ and $n \times n$ unitary matrices respectively. Using the property that tr$(AB) =$ tr(BA) we find

$$\|U_m M\|_F = \sqrt{\text{tr}((U_m M)^*(U_m M))} = \sqrt{\text{tr}(M^* U_m^* U_m M)}$$
$$= \sqrt{\text{tr}(M^* M)} = \|M\|_F$$

and

$$\|M U_n\|_F = \sqrt{\text{tr}((M U_n)^*(M U_n))} = \sqrt{\text{tr}(U_n^* M^* M U_n)}$$
$$= \sqrt{\text{tr}(U_n U_n^* M^* M)} = \sqrt{\text{tr}(M^* M)} = \|M\|_F.$$

Using the singular value decomposition $M = U\Sigma V^*$ of M we obtain

$$\|M - A\|_F = \|U^*(M - A)V\|_F = \|\Sigma - A'\|_F$$

where $A' := U^*AV$. Since

$$\|\Sigma - A'\|_F = \sqrt{\sum_{i=1}^{m}\sum_{j=1}^{n} |\Sigma_{ij} - (A')_{ij}|^2} = \sqrt{\sum_{i=1}^{m}\sum_{j=1}^{n} |\delta_{ij}\sigma_i - (A')_{ij}|^2}$$

we find that minimizing $\|M - A\|$ implies that A' must be "diagonal"

$$\|\Sigma - A'\|_F = \sqrt{\sum_{i=1}^{\min\{m,n\}} |\sigma_i - (A')_{ii}|^2}.$$

Since A must have rank k, A' must have rank k (since unitary operators are rank preserving). Since A' is "diagonal" only k entries can be nonzero. Thus to minimize $\|M - A\|_F$ we set

$$A' = \sum_{j=1}^{k} \sigma_j \mathbf{e}_{j,m} \mathbf{e}_{j,n}^T$$

where $\sigma_1 \geq \cdots \geq \sigma_k$ are the k largest singular values of M by convention. Finally $A = U A' V^*$.

Exercises. (1) Let A, B be $n \times n$ matrices over $\mathbb{C}$. Show that

$$\langle A, B \rangle := \mathrm{tr}(AB^*)$$

defines a scalar product. Then $\|A\| = \sqrt{(A, A)}$ defines a norm. Find $\|A\|$ for

$$A = \begin{pmatrix} 0 & i \\ -i & 0 \end{pmatrix}.$$

(2) Given a diagonalizable matrix A. Does a matrix norm $\| \cdot \|$ exist for which $\varrho(A) = \|A\|$?

(3) Let $\mathbf{v} = \exp(i\alpha)\mathbf{u}$, where $\alpha \in \mathbb{R}$. Show that $\|\mathbf{u}\| = \|\mathbf{v}\|$.

(4) (i) What can be said about the norm of a nilpotent matrix?
(ii) What can be said about the norm of an idempotent matrix?

(5) Let A be a Hermitian matrix. Find a necessary and sufficient condition for the function $\mathbf{v} \to (\mathbf{v}^* A \mathbf{v})^{1/2}$ to be a norm.

(6) Let $\| \cdot \|$ be a subordinate matrix norm and A an $n \times n$ matrix satisfying $\|A\| < 1$. Show that the matrix $I_n + A$ is invertible and

$$\|(I_n + A)^{-1}\| \leq \frac{1}{1 - \|A\|}.$$

(7) Prove that the function

$$\mathbf{v} \in \mathbb{C}^n \to \|\mathbf{v}\|_p = \left(\sum_{i=1}^{n} |v_i|^p \right)^{1/p}$$

is not a norm when $0 < p < 1$ (unless $n = 1$).

(8) Find the smallest constants C for which
$$\|\mathbf{v}\| \leq C\|\mathbf{v}\|' \text{ for every } \mathbf{v} \in \mathbb{F}^n$$
when the distinct norms $\|\cdot\|$ and $\|\cdot\|'$ are chosen from the set
$$\{\|\cdot\|_1, \quad \|\cdot\|_2, \quad \|\cdot\|_\infty\}.$$

(9) Consider the four points in the Euclidean plane
$$\mathbf{v}_0 = \begin{pmatrix} 0 \\ 0 \end{pmatrix}, \quad \mathbf{v}_1 = \begin{pmatrix} 1 \\ 0 \end{pmatrix}, \quad \mathbf{v}_2 = \begin{pmatrix} 1 \\ 2 \end{pmatrix}, \quad \mathbf{v}_3 = \begin{pmatrix} 0 \\ 1 \end{pmatrix}.$$
Find $\mathbf{v} \in \mathbb{R}^2$ such that $\sum_{j=0}^{3} \|\mathbf{v}_j - \mathbf{v}\|^2$ is a minimum.

(10) Consider the two-dimensional Euclidean space $\mathbb{E}^2$ and $\mathbf{v}_1, \mathbf{v}_2, \mathbf{v}_3 \in \mathbb{E}^2$. Maximize
$$\|\mathbf{v}_1 - \mathbf{v}_2\|^2 + \|\mathbf{v}_2 - \mathbf{v}_3\|^2 + \|\mathbf{v}_3 - \mathbf{v}_1\|^2$$
subject to $\|\mathbf{v}_1\| = \|\mathbf{v}_2\| = \|\mathbf{v}_2\| = 1$.

1.19 Sequences of Vectors and Matrices

Definition 1.47. In a vector space V, equipped with a norm $\|\cdot\|$, a sequence (x_k) of elements of V is said to converge to an element $x \in V$, which is the limit of the sequence (x_k), if
$$\lim_{k\to\infty} \|x_k - x\| = 0$$
and one writes
$$x = \lim_{k\to\infty} x_k.$$

If the space is finite dimensional, the equivalence of the norms shows that the convergence of a sequence is independent of the norm chosen. The particular choice of the norm $\|\cdot\|_\infty$ shows that the convergence of a sequence of vectors is equivalent to the convergence of n sequences (n being equal to the dimension of the space) of scalars consisting of the components of the vectors.

Example 1.42. Let $V = \mathbb{C}^2$ and
$$\mathbf{u}_k := \begin{pmatrix} \exp(-k) \\ 1/(1+k) \end{pmatrix}, \qquad k = 0, 1, 2, \ldots.$$
Then $\mathbf{u} = \lim_{k\to\infty} \mathbf{u}_k = \begin{pmatrix} 0 & 0 \end{pmatrix}^T$. ♣

By considering the set $\mathcal{A}_{m,n}(K)$ of matrices of type (m, n) as a vector space of dimension mn, one sees in the same way that the convergence of a sequence of matrices of type (m, n) is independent of the norm chosen, and that it is equivalent to the convergence of mn sequences of scalars consisting of the elements of these matrices. The following result gives necessary and sufficient conditions for the convergence of the particular sequence consisting of the successive powers of a given (square) matrix to the null matrix. From these conditions can be derived the fundamental criterion for the convergence of iterative methods for the solution of linear systems of equations.

Theorem 1.20. *Let B be an $n \times n$ matrix. The following conditions are equivalent:*

(1) $$\lim_{k \to \infty} B^k = 0_n$$

(2) $$\lim_{k \to \infty} B^k \mathbf{v} = \mathbf{0} \text{ for every vector } \mathbf{v}$$

(3) $$\varrho(B) < 1$$

(4) $$\|B\| < 1 \text{ for at least one subordinate matrix norm } \| \cdot \|.$$

For the proof of the theorem refer to Ciarlet [14].

Example 1.43. Consider the matrix

$$B = \begin{pmatrix} 1/4 & 1/3 \\ 1/3 & 1/4 \end{pmatrix}.$$

Then $\lim_{k \to \infty} B^k = 0_2$. ♣

The following theorem (Ciarlet [14]) is useful for the study of iterative methods, as regards the rate of convergence.

Theorem 1.21. *Let A be a square matrix and let $\| \cdot \|$ be any subordinate matrix norm. Then*

$$\lim_{k \to \infty} \|A^k\|^{1/k} = \varrho(A).$$

Example 1.44. Let $A = I_n$, where I_n is the $n \times n$ unit matrix. Then $I_n^k = I_n$ and therefore $\|I_n^k\|^{1/k} = 1$. Moreover $\varrho(A) = 1$. ♣

In theoretical physics and in particular in quantum mechanics a very important role is played by the exponential function of a square matrix. Let A be an $n \times n$ matrix. We set

$$A_k := I_n + \frac{A}{1!} + \frac{A^2}{2!} + \cdots + \frac{A^k}{k!}, \qquad k \geq 1.$$

The sequence (A_k) converges. Its limit is denoted by $\exp(A)$. We have

$$\exp(A) := \sum_{k=0}^{\infty} \frac{A^k}{k!}.$$

Let $z \in \mathbb{C}$. Then we have

$$\exp(zA) = \sum_{k=0}^{\infty} \frac{(zA)^k}{k!}.$$

We can also calculate $\exp(zA)$ using

$$\exp(zA) = \lim_{n \to \infty} \left(I_n + \frac{zA}{n} \right)^n.$$

Example 1.45. Let A be an $n \times n$ matrix with $A^2 = I_n$. Then

$$\exp(zA) = I_n \cosh(z) + A \sinh(z). \qquad \clubsuit$$

Theorem 1.22. *Let A be an $n \times n$ matrix. Then*

$$\det(\exp(A)) \equiv \exp(\mathrm{tr}(A)).$$

For the proof refer to Steeb [56]. The theorem shows that the matrix $\exp(A)$ is always invertible. If A is the zero matrix, then we have $\exp(A) = I_n$, where I_n is the identity matrix.

We can also define

$$\sin(A) := \sum_{k=0}^{\infty} \frac{(-1)^k A^{2k+1}}{(2k+1)!}, \qquad \cos(A) := \sum_{k=0}^{\infty} \frac{(-1)^k A^{2k}}{(2k)!}.$$

Example 1.46. Let $x, y \in \mathbb{R}$ and

$$A = \begin{pmatrix} x & y \\ 0 & x \end{pmatrix}.$$

Then we find

$$\sin(A) = \begin{pmatrix} \sin(x) & y\cos(x) \\ 0 & \sin(x) \end{pmatrix}, \qquad \cos(A) = \begin{pmatrix} \cos(x) & -y\sin(x) \\ 0 & \cos(x) \end{pmatrix}.$$

$\clubsuit$

Exercises. (1) Let

$$A = \begin{pmatrix} 0 & 1 & 1 \\ 0 & 0 & 1 \\ 0 & 0 & 0 \end{pmatrix}.$$

Calculate $\exp(\epsilon A)$, where $\epsilon \in \mathbb{R}$.

(2) Let

$$A = \begin{pmatrix} 1 & 1 \\ 1 & 1 \end{pmatrix}.$$

Find $\cos(A)$, $\sin(A)$ and $\sin^2(A) + \cos^2(A)$.

(3) Let A be an $n \times n$ matrix such that $A^3 = I_n$, where I_n is the $n \times n$ unit matrix. Let $\epsilon \in \mathbb{R}$. Find $\exp(\epsilon A)$.

(4) Let A be a square matrix such that the sequence $(A^k)_{k \geq 1}$ converges to an invertible matrix. Find A.

(5) Let B be a square matrix satisfying $\|B\| < 1$. Prove that the sequence $(C_k)_{k \geq 1}$, where

$$C_k = I + B + B^2 + \cdots + B^k$$

converges and that $\lim_{k \to \infty} C_k = (I - B)^{-1}$.

(6) Prove that $AB = BA \Rightarrow \exp(A + B) = \exp(A)\exp(B)$.

(7) Let A and B be $n \times n$ matrices. Assume that $\exp(A)\exp(B) = \exp(A + B)$. Show that in general $[A, B] \neq 0_n$.

(8) Let (A_k) be a sequence of $n \times n$ matrices. Show that the following conditions are equivalent:
(i) the sequence (A_k) converges
(ii) for every vector $\mathbf{v} \in \mathbb{R}^n$, the sequence of vectors $(A_k \mathbf{v})$ converges in $\mathbb{R}^n$.

(9) Extend the *Taylor expansion* for $\ln(1 + x)$

$$\ln(1 + x) = x - \frac{x^2}{2} + \frac{x^3}{3} - \frac{x^4}{4} + \cdots \qquad -1 < x \leq 1$$

to $n \times n$ matrices.

1.20 Commutators and Anti-Commutators

Let A and B be $n \times n$ matrices. Then we define the *commutator* of A and B as

$$[A, B] := AB - BA.$$

For all $n \times n$ matrices A, B, C we have the *Jacobi identity*

$$[A, [B, C]] + [C, [A, B]] + [B, [C, A]] = 0_n$$

where 0_n is the $n \times n$ zero matrix. Since $\text{tr}(AB) = \text{tr}(BA)$ we have

$$\text{tr}([A, B]) = 0.$$

If $[A, B] = 0_n$ we say that the matrices A and B *commute*. For example, if A and B are diagonal matrices then the commutator is the zero matrix 0_n.

Let A and B be $n \times n$ matrices. Then we define the *anticommutator* of A and B as $[A, B]_+ := AB + BA$. We have $\text{tr}([A, B]_+) = 2\text{tr}(AB)$. The anticommutator plays a role for Fermi operators.

Example 1.47. Consider the *Pauli spin matrices*

$$\sigma_1 = \begin{pmatrix} 0 & 1 \\ 1 & 0 \end{pmatrix}, \qquad \sigma_2 = \begin{pmatrix} 0 & -i \\ i & 0 \end{pmatrix}, \qquad \sigma_3 = \begin{pmatrix} 1 & 0 \\ 0 & -1 \end{pmatrix}.$$

Then

$$[\sigma_1, \sigma_3] = \begin{pmatrix} 0 & -2 \\ 2 & 0 \end{pmatrix} = 2i\sigma_2, \qquad [\sigma_1, \sigma_3]_+ = \begin{pmatrix} 0 & 0 \\ 0 & 0 \end{pmatrix}.$$

♣

Example 1.48. Consider the vectors in $\mathbb{R}^3$

$$\mathbf{u} = \begin{pmatrix} u_1 \\ u_2 \\ u_3 \end{pmatrix}, \qquad \mathbf{v} = \begin{pmatrix} v_1 \\ v_2 \\ v_3 \end{pmatrix}$$

and the skew-symmetric matrices

$$M(\mathbf{u}) = \begin{pmatrix} 0 & u_3 & -u_2 \\ -u_3 & 0 & u_1 \\ u_2 & -u_1 & 0 \end{pmatrix}, \qquad M(\mathbf{v}) = \begin{pmatrix} 0 & v_3 & -v_2 \\ -v_3 & 0 & v_1 \\ v_2 & -v_1 & 0 \end{pmatrix}.$$

Then

$$[M(\mathbf{u}), M(\mathbf{v})] = -M(\mathbf{u} \times \mathbf{v})$$

where $\times$ denotes the *vector product*.

♣

Exercises. (1) Consider the 3×3 matrices

$$A = \begin{pmatrix} a_{11} & a_{12} & 0 \\ a_{21} & a_{22} & 0 \\ 0 & 0 & a_{33} \end{pmatrix}, \quad B = \begin{pmatrix} 0 & 0 & b_{13} \\ 0 & 0 & b_{23} \\ b_{31} & b_{32} & 0 \end{pmatrix}.$$

Find the commutators $[A, B]$, $[A, [A, B]]$, $[B, [A.B]]$. Discuss.

(2) Consider the Hilbert space $\mathbb{C}^3$ and the 3×3 symmetric matrices A and B over $\mathbb{R}$ and the commutator

$$A = \begin{pmatrix} 0 & 1 & 0 \\ 1 & 0 & 1 \\ 0 & 1 & 0 \end{pmatrix}, \quad B = \begin{pmatrix} 1 & 0 & 1 \\ 0 & 1 & 0 \\ 1 & 0 & 1 \end{pmatrix}, \quad [A, B] = \begin{pmatrix} 0 & 1 & 0 \\ -1 & 0 & -1 \\ 0 & 1 & 0 \end{pmatrix}.$$

Show that the three matrices admit the eigenvalue 0 with the normalized eigenvector

$$\frac{1}{\sqrt{2}} \begin{pmatrix} 1 \\ 0 \\ -1 \end{pmatrix}.$$

Show that the two other eigenvalues of A are $-\sqrt{2}$, $\sqrt{2}$, of B are 1 and 2 and of $[A, B]$ are i, $-i$.

(3) Let A, B, C be $n \times n$ matrices over $\mathbb{C}$. Show that

$$\operatorname{tr}(A[B, C]) = \operatorname{tr}([A, B]C).$$

(4) Consider the three 3×3 antisymmetric real matrices

$$T_1 = \begin{pmatrix} 0 & 0 & 0 \\ 0 & 0 & -1 \\ 0 & 1 & 0 \end{pmatrix}, \quad T_2 = \begin{pmatrix} 0 & 0 & 1 \\ 0 & 0 & 0 \\ -1 & 0 & 0 \end{pmatrix}, \quad T_3 = \begin{pmatrix} 0 & -1 & 0 \\ 1 & 0 & 0 \\ 0 & 0 & 0 \end{pmatrix}.$$

Show that

$$[T_j, T_k] = \sum_{\ell=1}^{3} \epsilon_{jk\ell} T_\ell$$

where $\epsilon_{jk\ell}$ is totally antisymmetric with $\epsilon_{123} = +1$.

(5) Consider the four 4×4 matrices

$$X_1 = \begin{pmatrix} 0 & 0 & 0 & i \\ 0 & 0 & 0 & 0 \\ 0 & 0 & 0 & 0 \\ 0 & 0 & 0 & 0 \end{pmatrix}, \quad X_2 = \begin{pmatrix} 0 & 0 & 1 & 0 \\ 0 & 0 & 0 & -1 \\ 0 & 0 & 0 & 0 \\ 0 & 0 & 0 & 0 \end{pmatrix},$$

$$X_3 = \begin{pmatrix} 0 & 0 & i & 0 \\ 0 & 0 & 0 & i \\ 0 & 0 & 0 & 0 \\ 0 & 0 & 0 & 0 \end{pmatrix}, \quad X_4 = \begin{pmatrix} 0 & 0 & 0 & 0 \\ 0 & 0 & i & 0 \\ 0 & 0 & 0 & 0 \\ 0 & 0 & 0 & 0 \end{pmatrix}.$$

Let J be the *counter identity matrix*

$$J = \begin{pmatrix} 0 & 0 & 0 & 1 \\ 0 & 0 & 1 & 0 \\ 0 & 1 & 0 & 0 \\ 1 & 0 & 0 & 0 \end{pmatrix}.$$

Show that $X_j J + J X_j^* = 0_4$.

1.21 Groups and Lie Groups

In the representation of groups as $n \times n$ matrices the Kronecker product plays a central role. We give a short introduction to group theory and then discuss the connection with the Kronecker product. For further reading in group theory we refer to the books of Miller [44], Baumslag and Chandler [6] and Steeb [56]. In sections 2.13 and 2.14 we give a more detailed introduction to representation theory and the connection with the Kronecker product.

Definition 1.48. A *group* G is a set of objects $\{g, h, k, \ldots\}$ (not necessarily countable) together with a binary operation which associates with any ordered pair of elements $g, h \in G$ a third element $gh \in G$. The binary operation (called *group multiplication*) is subject to the following requirements:
(1) There exists an element e in G called the *identity element* such that $ge = eg = g$ for all $g \in G$.
(2) For every $g \in G$ there exists in G an *inverse element* g^{-1} such that $gg^{-1} = g^{-1}g = e$.
(3) *Associative law.* The identity $(gh)k = g(hk)$ is satisfied for all $g, h, k \in G$.

Thus, any set together with a binary operation which satisfies conditions (1)-(3) is called a group.

If $gh = hg$ we say that the elements g and h *commute*. If all elements of G commute then G is a *commutative* or *abelian* group. If G has a finite number of elements it has *finite order* $n(G)$, where $n(G)$ is the number of elements. Otherwise, G has *infinite order*.

A *subgroup* H of G is a subset which is itself a group under the group multiplication defined in G. The subgroups G and $\{e\}$ are called *improper* subgroups of G. All other subgroups are *proper*.

If a group G consists of a finite number of elements, then G is called a *finite group*; otherwise, G is called an *infinite group*.

Example 1.49. The set of integers $\mathbb{Z}$ with addition as group composition is an infinite additive group with $e = 0$.   ♣

Example 1.50. The set $\{1, -1, i, -i\}$ with multiplication as group composition is a finite abelian group with $e = 1$.   ♣

Definition 1.49. Let G be a finite group. The number of elements of G is called the *dimension* or *order* of G.

Example 1.51. The order of the group of the $n \times n$ permutation matrices under matrix multiplication is $n!$.   ♣

Theorem 1.23. *The order of a subgroup of a finite group divides the order of the group.*

This theorem is called *Lagrange's theorem*. For the proof we refer to the literature (Miller [44]).

A way to partition G is by means of *conjugacy classes*.

Definition 1.50. A group element h is said to be conjugate to the group element k, $h \sim k$, if there exists a $g \in G$ such that

$$k = ghg^{-1}.$$

It is easy to show that conjugacy is an *equivalence relation*, i.e. (1) $h \sim h$ (reflexive), (2) $h \sim k$ implies $k \sim h$ (symmetric), and (3) $h \sim k, k \sim j$ implies $h \sim j$ (transitive). Thus, the elements of G can be divided into *conjugacy classes* of mutually conjugate elements. The class containing e consists of just one element since

$$geg^{-1} = e$$

for all $g \in G$. Different conjugacy classes do not necessarily contain the same number of elements.

Let G be an abelian group. Then each conjugacy class consists of one group element each, since

$$ghg^{-1} = h, \quad \text{for all} \quad g \in G.$$

Let us now give a number of examples to illustrate the definitions given above.

Example 1.52. A *field* $\mathbb{F}$ is an (infinite) abelian group with respect to addition. The set of nonzero elements of a field forms a group with respect to multiplication, which is called a multiplicative group of the field. ♣

Example 1.53. A *linear vector space* over a field $\mathbb{F}$ (such as the real numbers $\mathbb{R}$) is an abelian group with respect to the usual addition of vectors. The group composition of two elements (vectors) $\mathbf{a}$ and $\mathbf{b}$ is their vector sum $\mathbf{a} + \mathbf{b}$. The identity element is the zero vector and the inverse of an element is its negative. ♣

Example 1.54. Let N be an integer with $N \geq 1$. The set

$$\left\{ e^{2\pi i n/N} : n = 0, 1, \ldots, N-1 \right\}$$

is an abelian (finite) group under multiplication since

$$\exp(2\pi i n/N)\exp(2\pi i m/N) = \exp(2\pi i(n+m)/N)$$

where $n, m = 0, 1, \ldots, N-1$. Note that $\exp(2\pi i n) = 1$ for $n \in \mathbb{N}$. We consider some special cases of N: For $N = 2$ we find the set $\{1, -1\}$ and for $N = 4$ we find $\{1, i, -1, -i\}$. These are elements on the unit circle in the complex plane. For $N \to \infty$ the number of points on the unit circle increases. As $N \to \infty$ we find the unitary group

$$U(1) := \left\{ e^{i\alpha} : \alpha \in \mathbb{R} \right\}.$$ ♣

Example 1.55. Let N be a positive integer. The set of all matrices

$$Z_{2\pi k/N} = \begin{pmatrix} \cos(2k\pi/N) & -\sin(2k\pi/N) \\ \sin(2k\pi/N) & \cos(2k\pi/N) \end{pmatrix}$$

where $k = 0, 1, \ldots, N-1$, forms an abelian group under matrix multiplication. The elements of the group can be generated from the transformation

$$Z_{2k\pi/N} = \left(Z_{2\pi/N} \right)^k, \qquad k = 0, 1, \ldots, N-1.$$

For example, if $N = 2$ the group consists of the elements

$$\left\{ (Z_\pi)^0, (Z_\pi)^1 \right\} \equiv \{ -I_2, +I_2 \}$$

where I_2 is the 2×2 unit matrix. This is an example of a cyclic group. ♣

Example 1.56. The two matrices

$$\left\{ \begin{pmatrix} 1 & 0 \\ 0 & 1 \end{pmatrix}, \begin{pmatrix} 0 & 1 \\ 1 & 0 \end{pmatrix} \right\}$$

form a finite abelian group of order two with matrix multiplication as group composition. The closure can easily be verified

$$\begin{pmatrix} 1 & 0 \\ 0 & 1 \end{pmatrix}\begin{pmatrix} 1 & 0 \\ 0 & 1 \end{pmatrix} = \begin{pmatrix} 1 & 0 \\ 0 & 1 \end{pmatrix}, \quad \begin{pmatrix} 1 & 0 \\ 0 & 1 \end{pmatrix}\begin{pmatrix} 0 & 1 \\ 1 & 0 \end{pmatrix} = \begin{pmatrix} 0 & 1 \\ 1 & 0 \end{pmatrix}, \quad \begin{pmatrix} 0 & 1 \\ 1 & 0 \end{pmatrix}\begin{pmatrix} 0 & 1 \\ 1 & 0 \end{pmatrix} = \begin{pmatrix} 1 & 0 \\ 0 & 1 \end{pmatrix}.$$

The identity element is the 2×2 unit matrix.   ♣

Example 1.57. Let $M = \{1, 2, \ldots, n\}$. Let $Bi(M, M)$ be the set of bijective mappings $\sigma : M \to M$ so that

$$\sigma : \{1, 2, \ldots, n\} \to \{p_1, p_2, \ldots, p_n\}$$

forms a group S_n under the composition of functions. Let S_n be the set of all the permutations

$$\sigma = \begin{pmatrix} 1 & 2 & \cdots & n \\ p_1 & p_2 & \cdots & p_n \end{pmatrix}.$$

We say 1 is mapped into p_1, 2 into p_2, ..., n into p_n. The numbers $p_1, p_2, \ldots, p_n$ are a reordering of $1, 2, \ldots, n$ and no two of the p_j's $j = 1, 2 \ldots, n$ are the same. The inverse permutation is given by

$$\sigma^{-1} = \begin{pmatrix} p_1 & p_2 & \cdots & p_n \\ 1 & 2 & \cdots & n \end{pmatrix}.$$

The product of two permutations σ and τ, with

$$\tau = \begin{pmatrix} q_1 & q_2 & \cdots & q_n \\ 1 & 2 & \cdots & n \end{pmatrix}$$

is given by the permutation

$$\sigma \circ \tau = \begin{pmatrix} q_1 & q_2 & \cdots & q_n \\ p_1 & p_2 & \cdots & p_n \end{pmatrix}.$$

That is, the integer q_i is mapped to i by τ and i is mapped to p_i by σ, so q_i is mapped to p_i by $\sigma \circ \tau$. The identity permutation is

$$e = \begin{pmatrix} 1 & 2 & \cdots & n \\ 1 & 2 & \cdots & n \end{pmatrix}.$$

S_n has order $n!$. The group of all permutations on M is called the *symmetric group* on M which is nonabelian, if $n > 2$.   ♣

Example 1.58. The set of all invertible $n \times n$ matrices form a group with respect to the usual multiplication of matrices. The group is called the *general linear group* over the real numbers $GL(n, \mathbb{R})$, or over the complex numbers $GL(n, \mathbb{C})$. This group together with its subgroups are the so-called *classical groups* which are Lie groups. ♣

Example 1.59. Let $\mathbb{C}$ be the complex plane. Let $z \in \mathbb{C}$. The set of Möbius transformations in $\mathbb{C}$ form a group called the *Möbius group* denoted by M where $m : \mathbb{C} \to \mathbb{C}$,

$$M := \{ m(a, b, c, d) : a, b, c, d \in \mathbb{C}, \ ad - bc \neq 0 \}$$

and

$$m : z \mapsto z' = \frac{az + b}{cz + d}.$$

The condition $ad - bc \neq 0$ must hold for the transformation to be invertible. Here, $z = x + iy$, where $x, y \in \mathbb{R}$. This forms a group under the composition of functions: Let

$$m(z) = \frac{az + b}{cz + d}, \qquad \widetilde{m}(z) = \frac{ez + f}{gz + h}$$

where $ad - bc \neq 0$ and $eh - fg \neq 0$ ($e, f, g, h \in \mathbb{C}$). Consider the composition

$$m(\widetilde{m}(z)) = \frac{a(ez + f)/(gz + h) + b}{c(ez + f)/(gz + h) + d} = \frac{aez + af + bgz + hb}{cez + cf + dgz + hd}$$

$$= \frac{(ae + bg)z + (af + hb)}{(ce + dg)z + (cf + hd)}.$$

Thus $m(\widetilde{m}(z))$ has the form of a Möbius transformation, since

$$(ae + bg)(cf + hd) - (af + hb)(ce + dg)$$
$$= aecf + aehd + bgcf + bghd - afce - afdg - hbce - hbdg$$
$$= ad(eh - fg) + bc(gf - eh)$$
$$= (ad - bc)(eh - fg) \neq 0.$$

Thus we conclude that m is closed under composition. Associativity holds since we consider the multiplication of complex numbers. The identity element is given by $m(1, 0, 0, 1) = z$. To find the inverse of $m(z)$ we assume that

$$m(\widetilde{m}(z)) = \frac{(ae + bg)z + (af + hb)}{(ce + dg)z + (cf + hd)} = z$$

so that

$$ae + bg = 1, \qquad af + hb = 0, \qquad ce + dg = 0, \qquad cf + hd = 1$$

and we find

$$e = \frac{d}{ad - bc}, \qquad f = -\frac{b}{ad - bc}, \qquad g = -\frac{c}{ad - bc}, \qquad h = \frac{a}{ad - bc}.$$

The inverse is thus given by

$$(z')^{-1} = \frac{dz - b}{-cz + a}.$$

♣

Example 1.60. Let $\mathbb{Z}$ be the abelian group of integers. Let E be the set of even integers. Obviously, E is an abelian group under addition and is a subgroup of $\mathbb{Z}$. Let C_2 be the cyclic group of order 2. Then $\mathbb{Z}/E \cong C_2$. ♣

Example 1.61. The compact Lie group $SO(2, \mathbb{R})$ is described by the rotation matrices

$$R(\alpha) = \begin{pmatrix} \cos(\alpha) & -\sin(\alpha) \\ \sin(\alpha) & \cos(\alpha) \end{pmatrix}.$$

Then

$$R^{-1}(\alpha) = \begin{pmatrix} \cos(\alpha) & \sin(\alpha) \\ -\sin(\alpha) & \cos(\alpha) \end{pmatrix}, \qquad dR(\alpha) = \begin{pmatrix} -\sin(\alpha)d\alpha & -\cos(\alpha)d\alpha \\ \cos(\alpha)d\alpha & -\sin(\alpha)d\alpha \end{pmatrix}.$$

It follows that

$$R^{-1}(\alpha)dR(\alpha) = \begin{pmatrix} 0 & -d\alpha \\ d\alpha & 0 \end{pmatrix} \equiv \begin{pmatrix} 0 & -1 \\ 1 & 0 \end{pmatrix} d\alpha.$$

♣

We denote the mapping between two groups by ρ and present the following definitions

Definition 1.51. A mapping of a group G into another group G' is called a *homomorphism* if it preserves all combinatorial operations associated with the group G so that

$$\rho(a \cdot b) = \rho(a) * \rho(b)$$

$a, b \in G$ and $\rho(a)$, $\rho(b) \in G'$. Here $\cdot$ and $*$ are the group compositions in G and G', respectively.

Example 1.62. There is a homomorphism ρ from $GL(2, \mathbb{C})$ into the Möbius group M given by

$$\rho : \begin{pmatrix} a & b \\ c & d \end{pmatrix} \mapsto m(z) = \frac{az + b}{cz + d}.$$

We now check that ρ is indeed a homomorphism: Consider the 2×2 matrix

$$A = \begin{pmatrix} a & b \\ c & d \end{pmatrix}$$

where $a, b, c, d \in \mathbb{C}$ and $ad - bc \neq 0$. The matrices A form a group with matrix multiplication as group composition. We find

$$AB = \begin{pmatrix} a & b \\ c & d \end{pmatrix} \begin{pmatrix} e & f \\ g & h \end{pmatrix} = \begin{pmatrix} ae + bg & af + bh \\ ce + dg & cf + dh \end{pmatrix}$$

where $e, f, g, h \in \mathbb{C}$. Thus

$$\rho(AB) = \frac{(ae + bg)z + (af + bh)}{(ce + dg)z + (cf + dh)}$$

and

$$\rho(A) = \frac{az + b}{cz + d}, \qquad \rho(B) = \frac{ez + f}{gz + h}$$

so that

$$\rho(A) * \rho(B) = \frac{(ae + bg)z + (af + bh)}{(ce + dg)z + (cf + dh)}.$$

We have shown that $\rho(A \cdot B) = \rho(A) * \rho(B)$ and thus that ρ is a homomorphism. ♣

An extension of the Möbius group is as follows. Consider the transformation

$$\mathbf{v} = \frac{A\mathbf{w} + B}{C\mathbf{w} + D}$$

where $\mathbf{v} = (v_1, \ldots, v_n)^T$, $\mathbf{w} = (w_1, \ldots, w_n)^T$ (T transpose). A is an $n \times n$ matrix, B an $n \times 1$ matrix, C a $1 \times n$ matrix and D a 1×1 matrix. The $(n + 1) \times (n + 1)$ matrix

$$\begin{pmatrix} A & B \\ C & D \end{pmatrix}$$

is invertible.

Example 1.63. An $n \times n$ permutation matrix is a matrix that has in each row and each column precisely one 1. There are $n!$ permutation matrices. The $n \times n$ permutation matrices form a group under matrix multiplication. Consider the symmetric group S_n given above. It is easy to see that the two groups are isomorphic. *Cayley's theorem* tells us that every finite

group is isomorphic to a subgroup (or the group itself) of these permutation
matrices. The six 3×3 permutation matrices are given by

$$A = \begin{pmatrix} 1 & 0 & 0 \\ 0 & 1 & 0 \\ 0 & 0 & 1 \end{pmatrix}, \qquad B = \begin{pmatrix} 1 & 0 & 0 \\ 0 & 0 & 1 \\ 0 & 1 & 0 \end{pmatrix}, \qquad C = \begin{pmatrix} 0 & 1 & 0 \\ 1 & 0 & 0 \\ 0 & 0 & 1 \end{pmatrix},$$

$$D = \begin{pmatrix} 0 & 1 & 0 \\ 0 & 0 & 1 \\ 1 & 0 & 0 \end{pmatrix}, \qquad E = \begin{pmatrix} 0 & 0 & 1 \\ 1 & 0 & 0 \\ 0 & 1 & 0 \end{pmatrix}, \qquad F = \begin{pmatrix} 0 & 0 & 1 \\ 0 & 1 & 0 \\ 1 & 0 & 0 \end{pmatrix}.$$

We have

$$AA = A \quad AB = B \quad AC = C \quad AD = D \quad AE = E \quad AF = F$$
$$BA = B \quad BB = A \quad BC = D \quad BD = C \quad BE = F \quad BF = E$$
$$CA = C \quad CB = E \quad CC = A \quad CD = F \quad CE = B \quad CF = D$$
$$DA = D \quad DB = F \quad DC = B \quad DD = E \quad DE = A \quad DF = C$$
$$EA = E \quad EB = C \quad EC = F \quad ED = A \quad EE = D \quad EF = B$$
$$FA = F \quad FB = D \quad FC = E \quad FD = B \quad FE = C \quad FF = A.$$

For the inverse we find

$$A^{-1} = A, \quad B^{-1} = B, \quad C^{-1} = C, \quad D^{-1} = E, \quad E^{-1} = D, \quad F^{-1} = F.$$

The order of a finite group is the number of elements of the group. Thus our
group has order 6. *Lagrange's theorem* tells us that the order of a subgroup
of a finite group divides the order of the group. Thus the subgroups must
have order 3, 2, 1. From the group table we find the subgroups

$$\{A, \ D, \ E\}$$

$$\{A, \ B\}, \qquad \{A, \ C\}, \qquad \{A, \ F\}$$

$$\{A\}.$$

Cayley's theorem tells us that every finite group is isomorphic to a subgroup
(or the group itself) of these permutation matrices. The *order of an element*
$g \in G$ is the order of the cyclic subgroup generated by $\{g\}$, i.e. the smallest
positive integer m such that

$$g^m = e$$

where e is the identity element of the group. The integer m divides the
order of G. Consider, for example, the element D of our group. Then

$$D^2 = E, \qquad D^3 = A, \qquad A \quad \text{identity element.}$$

Thus $m = 3$.   ♣

Example 1.64. Let $c \in \mathbb{R}$ and $c \neq 0$. The 2×2 matrices

$$\begin{pmatrix} c & c \\ c & c \end{pmatrix}$$

form a commutative group under matrix multiplication. Multiplication of two such matrices yields

$$\begin{pmatrix} c_1 & c_1 \\ c_1 & c_1 \end{pmatrix} \begin{pmatrix} c_2 & c_2 \\ c_2 & c_2 \end{pmatrix} = \begin{pmatrix} 2c_1c_2 & 2c_1c_2 \\ 2c_1c_2 & 2c_1c_2 \end{pmatrix}.$$

The neutral element is the matrix

$$\begin{pmatrix} 1/2 & 1/2 \\ 1/2 & 1/2 \end{pmatrix}$$

and the inverse element is the matrix

$$\begin{pmatrix} 1/(4c) & 1/(4c) \\ 1/(4c) & 1/(4c) \end{pmatrix}.$$

♣

Exercises. (1) Let S be the set of even integers. Show that S is a group under addition of integers. Consider $n \times n$ matrices with even integers as entries. Show that they form a group under addition.

(2) Let S be the set of real numbers of the form $a + b\sqrt{2}$, where $a, b \in \mathbb{Q}$ and are not simultaneously zero. Show that S is a group under the usual multiplication of real numbers.

(3) Find all subgroups of the group of the 4×4 permutation matrices. Apply Lagrange's theorem.

(4) Consider the matrices

$$B(\alpha) = \begin{pmatrix} \cosh(\alpha) & \sinh(\alpha) \\ \sinh(\alpha) & \cosh(\alpha) \end{pmatrix}, \qquad \alpha \in \mathbb{R}.$$

Show that these matrices form a group under matrix multiplication.

(5) Consider *rotation matrix*

$$R(\alpha) = \begin{pmatrix} \cos(\alpha) & \sin(\alpha) \\ -\sin(\alpha) & \cos(\alpha) \end{pmatrix}, \qquad \alpha \in \mathbb{R}.$$

(i) Show that these matrices form a group under matrix multiplication.

(ii) Find

$$X = \frac{d}{d\alpha} R(\alpha) \Big|_{\alpha=0} .$$

Show that $\exp(\alpha X) = R(\alpha)$.

(6) Show that all 2×2 matrices

$$\begin{pmatrix} 1 & a \\ 0 & 1 \end{pmatrix}, \qquad a \in \mathbb{R}$$

form a group under matrix multiplication.

(7) (i) Let $x \in \mathbb{R}$. Show that the 4×4 matrix

$$A(x) = \begin{pmatrix} \cos(x) & 0 & -\sin(x) & 0 \\ 0 & \cos(x) & 0 & -\sin(x) \\ \sin(x) & 0 & \cos(x) & 0 \\ 0 & \sin(x) & 0 & \cos(x) \end{pmatrix}$$

is invertible. Find the inverse. Do these matrices form a group under matrix multiplication?

(ii) Let $x \in \mathbb{R}$. Show that the 4×4 matrix

$$B(x) = \begin{pmatrix} \cosh(x) & 0 & \sinh(x) & 0 \\ 0 & \cosh(x) & 0 & \sinh(x) \\ \sinh(x) & 0 & \cosh(x) & 0 \\ 0 & \sinh(x) & 0 & \cosh(x) \end{pmatrix}$$

is invertible. Find the inverse. Do these matrices form a group under matrix multiplication?

(8) (i) Find the group generated by the matrix

$$A = \begin{pmatrix} 0 & 1 \\ e^{i\pi/2} & 0 \end{pmatrix}$$

under matrix multiplication. Find the group generated by $A \otimes A$ under matrix multiplication.

(ii) Find the group generated by the 6×6 permutation matrices

$$P_1 = \begin{pmatrix} 0&0&1&0&0&0 \\ 1&0&0&0&0&0 \\ 0&1&0&0&0&0 \\ 0&0&0&0&0&1 \\ 0&0&0&1&0&0 \\ 0&0&0&0&1&0 \end{pmatrix}, \quad P_2 = \begin{pmatrix} 0&0&0&1&0&0 \\ 0&0&0&0&0&1 \\ 0&0&0&0&1&0 \\ 1&0&0&0&0&0 \\ 0&0&1&0&0&0 \\ 0&1&0&0&0&0 \end{pmatrix} .$$

(iii) Find the group generated by 3×3 the permutation matrices

$$P_1 = \begin{pmatrix} 0 & 1 & 0 \\ 0 & 0 & 1 \\ 1 & 0 & 0 \end{pmatrix}, \quad P_2 = \begin{pmatrix} 0 & 0 & 1 \\ 0 & 1 & 0 \\ 1 & 0 & 0 \end{pmatrix}$$

under matrix multiplication.

(iv) Find the group generated by the 4×4 permutation matrices

$$\begin{pmatrix} 0 & 1 & 0 & 0 \\ 1 & 0 & 0 & 0 \\ 0 & 0 & 1 & 0 \\ 0 & 0 & 0 & 1 \end{pmatrix}, \quad \begin{pmatrix} 1 & 0 & 0 & 0 \\ 0 & 0 & 1 & 0 \\ 0 & 1 & 0 & 0 \\ 0 & 0 & 0 & 1 \end{pmatrix}, \quad \begin{pmatrix} 1 & 0 & 0 & 0 \\ 0 & 1 & 0 & 0 \\ 0 & 0 & 0 & 1 \\ 0 & 0 & 1 & 0 \end{pmatrix}$$

under matrix multiplication.

(v) Find the group generated by the 3×3 matrices

$$\begin{pmatrix} 0 & 0 & -1 \\ 1 & 0 & 0 \\ 0 & 1 & 0 \end{pmatrix}, \quad \begin{pmatrix} 1 & 0 & 0 \\ 0 & 0 & -1 \\ 0 & -1 & 0 \end{pmatrix}$$

under matrix multiplication.

(9) Do the eight 2×2 matrices

$$\begin{pmatrix} 1 & 0 \\ 0 & 1 \end{pmatrix}, \quad \begin{pmatrix} 1 & 0 \\ 0 & -1 \end{pmatrix}, \quad \begin{pmatrix} -1 & 0 \\ 0 & 1 \end{pmatrix}, \quad \begin{pmatrix} -1 & 0 \\ 0 & -1 \end{pmatrix},$$

$$\begin{pmatrix} 0 & 1 \\ 1 & 0 \end{pmatrix}, \quad \begin{pmatrix} 0 & -1 \\ 1 & 0 \end{pmatrix}, \quad \begin{pmatrix} 0 & 1 \\ -1 & 0 \end{pmatrix}, \quad \begin{pmatrix} 0 & -1 \\ -1 & 0 \end{pmatrix}$$

form a group under matrix multiplication?

(10) The numbers $+1$, -1, i, $-i$ form a commutative group under multiplication. The two 2×2 matrices

$$\begin{pmatrix} 1 & 0 \\ 0 & 1 \end{pmatrix}, \quad \begin{pmatrix} 0 & 1 \\ 1 & 0 \end{pmatrix}$$

also form a group under matrix multiplication. Do the eight 2×2 matrices

$$1 \cdot \begin{pmatrix} 1 & 0 \\ 0 & 1 \end{pmatrix}, \quad -1 \cdot \begin{pmatrix} 1 & 0 \\ 0 & 1 \end{pmatrix}, \quad i \cdot \begin{pmatrix} 1 & 0 \\ 0 & 1 \end{pmatrix}, \quad -i \cdot \begin{pmatrix} 1 & 0 \\ 0 & 1 \end{pmatrix}$$

$$1 \cdot \begin{pmatrix} 0 & 1 \\ 1 & 0 \end{pmatrix}, \quad -1 \cdot \begin{pmatrix} 0 & 1 \\ 1 & 0 \end{pmatrix}, \quad i \cdot \begin{pmatrix} 0 & 1 \\ 1 & 0 \end{pmatrix}, \quad -i \cdot \begin{pmatrix} 0 & 1 \\ 1 & 0 \end{pmatrix}$$

also form a group under matrix multiplication?

(11) The $n \times n$ permutation matrices P form a group under matrix multiplication. The numbers $+1$, -1, i, $-i$ form a group under multiplication. Do the $4 \cdot (n!)$ $n \times n$ matrices $+1 \cdot P$, $-1 \cdot P$, $i \cdot P$, $-i \cdot P$ form a group under matrix multiplication?

(12) The *Weyl reflection matrices* W_1, W_2, W_3 are given by

$$W_1 = \begin{pmatrix} 0 & -1 & 0 \\ -1 & 0 & 0 \\ 0 & 0 & -1 \end{pmatrix}, \quad W_2 = \begin{pmatrix} -1 & 0 & 0 \\ 0 & 0 & -1 \\ 0 & -1 & 0 \end{pmatrix}, \quad W_3 = \begin{pmatrix} 0 & 0 & -1 \\ 0 & -1 & 0 \\ -1 & 0 & 0 \end{pmatrix}.$$

Show that $W_1^2 = W_2^2 = W_3^2 = I_3$ and $W_3 = W_1 W_2 W_1$. Find the group generated by W_1, W_2, W_3 under matrix multiplication. Is the group generated isomorphic to the symmetric group S_3?

(13) Find all 3×3 permutation matrices P such that

$$P \begin{pmatrix} 0 & 1 & 0 \\ 1 & 1 & 1 \\ 0 & 1 & 0 \end{pmatrix} P^{-1} = \begin{pmatrix} 0 & 1 & 0 \\ 1 & 1 & 1 \\ 0 & 1 & 0 \end{pmatrix}.$$

Do they form a group under matrix multiplication?

(14) Consider the *general linear group*

$$GL(n, \mathbb{R}) := \{\, g \in \mathbb{R}^{n \times n} : \det(g) \neq 0 \,\}.$$

Show that the *Gâteaux derivative* of the function $f = \det : \mathbb{R}^{n \times n} \to \mathbb{R}$ is given by

$$df(g)v := \frac{d}{d\epsilon} f(g + \epsilon v)\Big|_{\epsilon=0} = \det(g)\operatorname{tr}(g^{-1}v), \quad v \in \mathbb{R}^{n \times n}.$$

(15) Consider the Lie group $U(n)$. Then

$$U(n)_X := \{\, g \in U(n) : gXg^{-1} = X \,\}$$

is the *centralizer* of X in $U(n)$. Let $n = 2$ and

$$X = \begin{pmatrix} 0 & 1 \\ 1 & 0 \end{pmatrix}.$$

Find the centralizer. Note that

$$\begin{pmatrix} \cos(\alpha) & \sin(\alpha) \\ -\sin(\alpha) & \cos(\alpha) \end{pmatrix} \begin{pmatrix} 0 & 1 \\ 1 & 0 \end{pmatrix} \begin{pmatrix} \cos(\alpha) & -\sin(\alpha) \\ \sin(\alpha) & \cos(\alpha) \end{pmatrix}$$

$$= \begin{pmatrix} 2\cos(\alpha)\sin(\alpha) & \cos^2(\alpha) - \sin^2(\alpha) \\ \cos^2(\alpha) - \sin^2(\alpha) & -2\sin(\alpha)\cos(\alpha) \end{pmatrix}.$$

(16) The numbers $+1, -1, i, -i$ form a group under multiplication. Consider the *Bell basis* in the Hilbert space $\mathbb{C}^4$

$$|0\rangle = \frac{1}{\sqrt{2}}\begin{pmatrix}1\\0\\0\\1\end{pmatrix}, \quad |1\rangle = \frac{1}{\sqrt{2}}\begin{pmatrix}0\\1\\1\\0\end{pmatrix}, \quad |2\rangle = \frac{1}{\sqrt{2}}\begin{pmatrix}0\\1\\-1\\0\end{pmatrix}, \quad |3\rangle = \frac{1}{\sqrt{2}}\begin{pmatrix}1\\0\\0\\-1\end{pmatrix}.$$

Show that the 4×4 matrix

$$V = +1|0\rangle\langle 0| - 1|1\rangle\langle 1| + i|2\rangle\langle 2| - i|3\rangle\langle 3|$$

is unitary and given by

$$V = \frac{1}{2}\begin{pmatrix} 1-i & 0 & 0 & 1+i \\ 0 & 1+i & 1-i & 0 \\ 0 & 1-i & 1+i & 0 \\ 1+i & 0 & 0 & 1-i \end{pmatrix}.$$

Find the group generated by V.

(17) Let $z_1, z_2 \in \mathbb{C}$ and $z_1 \neq 0$. Consider the group under matrix multiplication of the 2×2 matrices

$$g = \begin{pmatrix} z_1 & z_2 \\ 0 & 1 \end{pmatrix} \quad \Rightarrow \quad g^{-1} = \begin{pmatrix} 1/z_1 & -z_2/z_1 \\ 0 & 1 \end{pmatrix}.$$

Show that the left-invariant *volume element* is given by

$$-\frac{1}{|z_1|}(dz_1 \wedge d\bar{z}_1 \wedge dz_2 \wedge d\bar{z}_2)$$

where $z_j = x_j + iy_j$, $\bar{z}_j = x_j - iy_j$, $dz_j = dx_j + dy_j$, $d\bar{z}_j = dx_j - idy_j$ with $x_j, y_j \in \mathbb{R}$.

1.22   Lie Algebras

Lie algebras (Bäuerle and de Kerf [7], Humphreys [34], Jacobson [35]) play a central role in theoretical physics. They are also linked to Lie groups and Lie transformation groups (Steeb [56]). In this section we give the definition of a Lie algebra and some applications.

Definition 1.52. A vector space L over a field $\mathbb{F}$, with an operation $L \times L \to L$ denoted by

$$(x, y) \to [x, y]$$

and called the commutator of x and y, is called a *Lie algebra* over $\mathbb{F}$ if the following axioms are satisfied:

(L1)   The bracket operation is bilinear.

(L2) $[x, x] = 0$ for all $x \in L$

(L3) $[x, [y, z]] + [y, [z, x]] + [z, [x, y]] = 0$ $(x, y, z \in L)$.

Remark: Axiom (L3) is called the *Jacobi identity*.

Notice that (L1) and (L2), applied to $[x+y, x+y]$, imply anticommutativity:
(L2')

$$[x, y] = -[y, x].$$

Conversely, if char$(\mathbb{F}) \neq 2$ (for $\mathbb{R}$ and $\mathbb{C}$ we have char$(\mathbb{F}) = 0$), then (L2') will imply (L2).

Definition 1.53. Let X and Y be $n \times n$-matrices. Then the *commutator* $[X, Y]$ of X and Y is defined as

$$[X, Y] := XY - YX.$$

The $n \times n$ matrices over $\mathbb{R}$ or $\mathbb{C}$ form a Lie algebra under the commutator. This means we have the following properties (X, Y, V, W $n \times n$ matrices and $c \in \mathbb{C}$)

$$[cX, Y] = c[X, Y], \qquad [X, cY] = c[X, Y]$$

$$[X, Y] = -[Y, X]$$

$$[X + Y, V + W] = [X, V] + [X, W] + [Y, V] + [Y, W]$$

and

$$[X, [Y, V]] + [V, [X, Y]] + [Y, [V, X]] = 0_n.$$

The last equation is called the *Jacobi identity.*.

Example 1.65. The Lie group $SU(n)$ ($n \geq 2$) consists of all $n \times n$ unitary matrices with determinant equal to 1. The corresponding Lie algebra $su(n)$ consists of all $n \times n$ skew-hermitian matrices with trace equal to 0. For $n = 2$ a basis of $su(2)$ would be

$$\begin{pmatrix} i & 0 \\ 0 & -i \end{pmatrix}, \quad \begin{pmatrix} 0 & 1 \\ -1 & 0 \end{pmatrix}, \quad \begin{pmatrix} 0 & i \\ i & 0 \end{pmatrix}.$$

♣

Definition 1.54. Two Lie algebras L, L' are called *isomorphic* if there exists a vector space isomorphism $\phi : L \to L'$ satisfying

$$\phi([x,y]) = [\phi(x), \phi(y)]$$

for all $x, y \in L$.

The Lie algebra of all $n \times n$ matrices over $\mathbb{C}$ is also called $g\ell(n, \mathbb{C})$. A basis is given by the *elementary matrices*

$$(E_{ij}), \qquad i, j = 1, 2, \ldots, n$$

where (E_{ij}) is the matrix having 1 in the (i, j) position and 0 elsewhere. Since $(E_{ij})(E_{kl}) = \delta_{jk}(E_{il})$ it follows that the commutator is given by

$$[(E_{ij}), (E_{kl})] = \delta_{jk}(E_{il}) - \delta_{li}(E_{kj}).$$

Thus the coefficients are all ± 1 or 0.

The classical Lie algebras are sub-Lie algebras of $g\ell(n, \mathbb{C})$. For example, $s\ell(n, \mathbb{R})$ is the Lie algebra with the condition

$$\text{tr}(X) = 0$$

for all $X \in g\ell(n, \mathbb{R})$. Furthermore, $so(n, \mathbb{R})$ is the Lie algebra with the condition that $X^T = -X$ and $\text{tr}(X) = 0$ for all $X \in so(n, \mathbb{R})$. For $n = 2$ a basis element is given by

$$X = \begin{pmatrix} 0 & 1 \\ -1 & 0 \end{pmatrix}.$$

For $n = 3$ we have a basis (skew-symmetric matrices)

$$X_1 = \begin{pmatrix} 0 & 0 & 0 \\ 0 & 0 & -1 \\ 0 & 1 & 0 \end{pmatrix}, \qquad X_2 = \begin{pmatrix} 0 & 0 & 1 \\ 0 & 0 & 0 \\ -1 & 0 & 0 \end{pmatrix}, \qquad X_3 = \begin{pmatrix} 0 & -1 & 0 \\ 1 & 0 & 0 \\ 0 & 0 & 0 \end{pmatrix}.$$

Exercises. (1) Show that the 2×2 matrices with trace zero form a Lie algebra under the commutator. Show that

$$\begin{pmatrix} 0 & 1 \\ 0 & 0 \end{pmatrix}, \qquad \begin{pmatrix} 0 & 0 \\ 1 & 0 \end{pmatrix}, \qquad \begin{pmatrix} 1 & 0 \\ 0 & -1 \end{pmatrix}$$

form a basis. Hint: Show that $\text{tr}([A, B]) = 0$ for any two $n \times n$ matrices.

(2) Find all Lie algebras with dimension 2.

(3) Show that the set of all $n \times n$ diagonal matrices form a Lie algebra under the commutator.

(4) (i) Do all Hermitian matrices form a Lie algebra under the commutator?
(ii) Do all skew-Hermitian matrices form a Lie algebra under the commutator?

(5) An *automorphism* of L is an isomorphism of L onto itself. Let $L = s\ell(n, \mathbb{R})$. Show that if $g \in GL(n, \mathbb{R})$ and if $gLg^{-1} = L$ then the map $x \mapsto gxg^{-1}$ is an automorphism.

(6) The *center* of a Lie algebra L is defined as
$$Z(L) := \{ z \in L : [z, x] = 0 \quad \text{for all } x \in L \}.$$
Find the center for the Lie algebra $s\ell(2, \mathbb{R})$.

(7) Show that if a Lie algebra L is nilpotent, then the Killing form of L is identically zero.

(8) Find all 2×2 matrices X, Y such that
$$[X, Y] = \sigma_3 = \begin{pmatrix} 1 & 0 \\ 0 & -1 \end{pmatrix}.$$
Show that
$$X = \begin{pmatrix} 0 & \cosh(\alpha) \\ \sinh(\alpha) & 0 \end{pmatrix}, \quad Y = \begin{pmatrix} 0 & \sinh(\alpha) \\ \cosh(\alpha) & 0 \end{pmatrix}$$
satisfy this condition.

(9) Let L be a finite dimensional Lie algebra and $X, Y \in L$. Show that
$$e^X Y e^{-X} \in L, \quad e^Y X e^{-Y} \in L.$$

(10) (i) Consider the four dimensional Lie algebra with the basis A_1, A_2, A_3, A_4 and the commutation relations
$$[A_1, A_2] = [A_2, A_4] = A_2, \quad [A_3, A_1] = [A_4, A_3] = A_3,$$
$$[A_2, A_3] = A_1 - A_4, \quad [A_1, A_4] = 0.$$
Find the adjoint representation.
(ii) Consider the three dimensional Lie algebra with the basis e_1, e_2, e_3 and the commutation relations $[e_1, e_3] = e_1$, $[e_2, e_3] = -e_2$, $[e_1, e_2] = 0$. Find the adjoint representation.

1.23 Functions of Matrices

Let A be an $n \times n$ matrix over $\mathbb{C}$. Let $f : \mathbb{R} \to \mathbb{R}$ be an analytic function. Then we can consider the matrix valued function $f(A)$. The simplest example of such a function is a polynomial. Let p be a polynomial

$$p(x) = \sum_{j=0}^{n} c_j x^j$$

then the corresponding matrix function of an $n \times n$ matrix A can be defined by

$$p(A) = \sum_{j=0}^{n} c_j A^j$$

with the convention $A^0 = I_n$. If a function f of a complex variable z has a *MacLaurin series expansion*

$$f(z) = \sum_{j=0}^{\infty} c_j z^j$$

which converges for $|z| < R$ $(R > 0)$, then the matrix series

$$f(A) = \sum_{j=0}^{\infty} c_j A^j$$

converges, provided A is square and each of its eigenvalues has absolute value less than R. More generally if a function f of a complex variable z has a *power series expansion*

$$f(z) = \sum_{j=0}^{\infty} c_j z^j$$

which converges for $z \in Z \subseteq \mathbb{C}$, then the matrix series

$$f(A) = \sum_{j=0}^{\infty} c_j A^j$$

converges, provided A is square and each of its eigenvalues are in Z.

Example 1.66. The most used matrix function is $\exp(A)$ defined by

$$\exp(A) := \sum_{j=0}^{\infty} \frac{A^j}{j!} \equiv \lim_{n \to \infty} \left(I_n + \frac{A}{n} \right)^n$$

which converges for all square matrices A. ♣

Example 1.67. Let $\epsilon \in \mathbb{R}$. Consider

$$A(\epsilon) = \begin{pmatrix} e^\epsilon & \epsilon \\ 0 & e^{-\epsilon} \end{pmatrix} \Rightarrow \frac{dA(\epsilon)}{d\epsilon} = \begin{pmatrix} e^\epsilon & 1 \\ 0 & -e^{-\epsilon} \end{pmatrix} \Rightarrow \frac{dA(\epsilon)}{d\epsilon}\bigg|_{\epsilon=0} = \begin{pmatrix} 1 & 1 \\ 0 & -1 \end{pmatrix} = X.$$

Then

$$e^{\epsilon X} = \begin{pmatrix} e^\epsilon & \sinh(\epsilon) \\ 0 & e^{-\epsilon} \end{pmatrix}.$$

♣

Example 1.68. The matrix functions $\sin(A)$ and $\cos(A)$ are defined by

$$\sin(A) := \sum_{j=0}^{\infty} (-1)^j \frac{A^{2j+1}}{(2j+1)!}, \qquad \cos(A) := \sum_{j=0}^{\infty} (-1)^j \frac{A^{2j}}{(2j)!}$$

converge for all square matrices A.

♣

Example 1.69. The matrix function $\arctan(A)$ defined by

$$\arctan(A) := \sum_{j=0}^{\infty} (-1)^j \frac{A^{2j+1}}{2j+1}$$

converges for square matrices A with eigenvalues λ satisfying $|\lambda| < 1$.

♣

Example 1.70. The matrix function $\ln(A)$ defined by

$$\ln(A) := -\sum_{j=1}^{\infty} \frac{(I_n - A)^j}{j}$$

converges for square matrices A with eigenvalues λ, where $|\lambda| \in (0, 2)$.

♣

Example 1.71. We calculate

$$\exp \begin{pmatrix} 0 & i\pi & 0 \\ 0 & 0 & 0 \\ 0 & -i\pi & 0 \end{pmatrix}$$

using MacLaurin series expansion to obtain the exponential. Since the square of the matrix is the 3×3 zero matrix we find from the MacLaurin series expansion

$$\exp \begin{pmatrix} 0 & i\pi & 0 \\ 0 & 0 & 0 \\ 0 & -i\pi & 0 \end{pmatrix} = \begin{pmatrix} 1 & 0 & 0 \\ 0 & 1 & 0 \\ 0 & 0 & 1 \end{pmatrix} + \begin{pmatrix} 0 & i\pi & 0 \\ 0 & 0 & 0 \\ 0 & -i\pi & 0 \end{pmatrix} = \begin{pmatrix} 1 & i\pi & 0 \\ 0 & 1 & 0 \\ 0 & -i\pi & 1 \end{pmatrix}.$$

♣

We can also use the *Cayley-Hamilton theorem*. Since A satisfies its own characteristic equation

$$A^n - \text{tr}(A)A^{n-1} + \cdots + (-1)^n \det(A)A^0 = 0_n$$

where 0_n is the $n \times n$ zero matrix, we find that A^n can be expressed in terms of $A^{n-1}, \ldots, A$ and I_n

$$A^n = \text{tr}(A)A^{n-1} - \cdots - (-1)^n \det(A)I_n.$$

Then the MacLaurin series contracts to

$$f(A) = \sum_{j=0}^{\infty} c_j A^j = \sum_{j=0}^{n-1} \alpha_j A^j.$$

The values $\alpha_0, \ldots, \alpha_{n-1}$ can be determined from the eigenvalues of A. Let λ be an eigenvalue of A corresponding to the eigenvector $\mathbf{x}$. Then $f(A)\mathbf{x} = f(\lambda)\mathbf{x}$ and

$$f(A)\mathbf{x} = \sum_{j=0}^{\infty} c_j A^j \mathbf{x} = \sum_{j=0}^{n-1} \alpha_j A^j \mathbf{x} = \sum_{j=0}^{n-1} \alpha_j \lambda_j^j \mathbf{x}.$$

Since $\mathbf{x}$ is nonzero we have the equation

$$f(\lambda) = \sum_{j=0}^{n-1} \alpha_j \lambda_j^j.$$

Thus we have a system of linear equations for the α_j. If an eigenvalue λ of A has multiplicity greater than 1, these equations will be insufficient to determine all of the α_j. Since f is analytic the derivative f' is also analytic and

$$f'(A) = \sum_{j=1}^{\infty} j c_j A^{j-1} = \sum_{j=1}^{n-1} j \alpha_j A^{j-1}$$

which provides the equation

$$f'(\lambda) = \sum_{j=1}^{n-1} j \alpha_j \lambda_j^j.$$

If the multiplicity is of λ is k then we have the equations

$$f(\lambda) = \sum_{j=0}^{n-1} \alpha_j \lambda_j^j$$

$$f'(\lambda) = \sum_{j=1}^{n-1} j \alpha_j \lambda_j^{j-1}$$

$$\vdots$$

$$f^{(k-1)}(\lambda) = \sum_{j=k-1}^{n-1} \left(\prod_{m=0}^{k-1} (j-m) \right) \alpha_j \lambda_j^{j-k-1}.$$

Example 1.72. We calculate

$$\exp \begin{pmatrix} 0 & i\pi & 0 \\ 0 & 0 & 0 \\ 0 & -i\pi & 0 \end{pmatrix}$$

using the Cayley-Hamilton theorem. The eigenvalue is 0 with multiplicity 3. Thus we solve the three equations

$$e^\lambda = \alpha_0 + \alpha_1\lambda + \alpha_2\lambda^2, \quad e^\lambda = \alpha_1 + 2\alpha_2\lambda, \quad e^\lambda = 2\alpha_2$$

where $\lambda = 0$. Thus $\alpha_2 = \frac{1}{2}$ and $\alpha_0 = \alpha_1 = 1$. Consequently

$$\exp \begin{pmatrix} 0 & i\pi & 0 \\ 0 & 0 & 0 \\ 0 & -i\pi & 0 \end{pmatrix} = \begin{pmatrix} 1 & 0 & 0 \\ 0 & 1 & 0 \\ 0 & 0 & 1 \end{pmatrix} + \begin{pmatrix} 0 & i\pi & 0 \\ 0 & 0 & 0 \\ 0 & -i\pi & 0 \end{pmatrix} + \frac{1}{2}\begin{pmatrix} 0 & i\pi & 0 \\ 0 & 0 & 0 \\ 0 & -i\pi & 0 \end{pmatrix}^2 = \begin{pmatrix} 1 & i\pi & 0 \\ 0 & 1 & 0 \\ 0 & -i\pi & 1 \end{pmatrix}.$$

♣

Example 1.73. Consider the matrix

$$A = \begin{pmatrix} \frac{\pi}{\sqrt{2}} & 0 & 0 & \frac{\pi}{\sqrt{2}} \\ 0 & \frac{\pi}{\sqrt{2}} & \frac{\pi}{\sqrt{2}} & 0 \\ 0 & \frac{\pi}{\sqrt{2}} & -\frac{\pi}{\sqrt{2}} & 0 \\ \frac{\pi}{\sqrt{2}} & 0 & 0 & -\frac{\pi}{\sqrt{2}} \end{pmatrix}$$

with eigenvalues are π, π, $-\pi$ and $-\pi$. We wish to calculate $\sec(A)$ where $\sec(z)$ which is analytic on the intervals $(\pi/2, 3\pi/2)$ and $(-3\pi/2, -\pi/2)$ amongst others. The eigenvalues lie within these intervals. However, the power series for $\sec(z)$ on these two intervals are different, the methods above do not apply directly. We can restrict ourselves to the action on subspaces. The eigenvalue π corresponds to the eigenspace spanned by

$$\left\{ \frac{1}{\sqrt{4 - 2\sqrt{2}}}(1, 0, 0, \sqrt{2} - 1)^T, \ \frac{1}{\sqrt{4 - 2\sqrt{2}}}(0, 1, \sqrt{2} - 1, 0)^T \right\}.$$

The projection Π_1 onto this eigenspace is given by

$$\Pi_1 := \frac{1}{4 - 2\sqrt{2}} \begin{pmatrix} 1 & 0 & 0 & \sqrt{2} - 1 \\ 0 & 1 & \sqrt{2} - 1 & 0 \\ 0 & \sqrt{2} - 1 & 3 - 2\sqrt{2} & 0 \\ \sqrt{2} - 1 & 0 & 0 & 3 - 2\sqrt{2} \end{pmatrix}.$$

On this two dimensional subspace, A has one eigenvalue with multiplicity 2. We solve the equation $\sec(\pi) = -1 = \alpha_0 + \alpha_1\pi$ and $\sec'(\pi) = \sec(\pi)\tan(\pi) = 0 = \alpha_1$. Thus $\alpha_1 = 0$ and $\alpha_0 = -1$. The solution on this subspace is $(\alpha_0 I_4 + \alpha_1 A)\Pi_1 = -\Pi_1$. We perform a similar calculation for the eigenvalue $-\pi$ and the projection onto the eigenspace Π_2. We solve the equation $\sec(-\pi) = -1 = \beta_0 - \beta_1\pi$ and $\sec'(-\pi) = \sec(-\pi)\tan(-\pi) = 0 = -\beta_1$. Thus $\beta_1 = 0$ and $\beta_0 = -1$. The solution on this subspace is $(\beta_0 I_4 + \beta_1 A)\Pi_2 = -\Pi_2$. Thus the solution on $\mathbb{C}^4$ is given by

$$\sec(A) = (-\Pi_1) + (-\Pi_2) = -I_4.$$

♣

Let A be an $n \times n$ matrix over $\mathbb{C}$. An $n \times n$ matrix B is called a *square root* of A if $B^2 = A$. For example the matrix

$$A = \begin{pmatrix} 0 & 1 \\ 0 & 0 \end{pmatrix}$$

does not admit a square root. Consider the 2×2 identity matrix I_2. Then obviously

$$I_2 = \begin{pmatrix} 1 & 0 \\ 0 & 1 \end{pmatrix}, \quad \sigma_3 = \begin{pmatrix} 1 & 0 \\ 0 & -1 \end{pmatrix}, \quad \begin{pmatrix} -1 & 0 \\ 0 & 1 \end{pmatrix}, \quad \begin{pmatrix} -1 & 0 \\ 0 & -1 \end{pmatrix},$$

$$\sigma_1 = \begin{pmatrix} 0 & 1 \\ 1 & 0 \end{pmatrix}, \quad \sigma_2 = \begin{pmatrix} 0 & -i \\ i & 0 \end{pmatrix}$$

are square roots of I_2. However we also find the square roots

$$\begin{pmatrix} 1 & b_{12} \\ 0 & -1 \end{pmatrix}, \quad \begin{pmatrix} -1 & b_{12} \\ 0 & 1 \end{pmatrix}, \quad \begin{pmatrix} 1 & 0 \\ b_{12} & -1 \end{pmatrix}, \quad \begin{pmatrix} -1 & 0 \\ b_{12} & 1 \end{pmatrix}.$$

With $b_{12} \neq 0$ the matrices are nonnormal.

A bounded linear operator K on the Hilbert space $\mathcal{H}$ is self-adjoint if and only if the bounded linear operators

$$U(\tau) := \sum_{n=0}^{\infty} \frac{(iK\tau)^n}{n!}$$

are unitary for all $\tau \in \mathbb{R}$. So it applies to $n \times n$ hermitian matrices acting in the Hilbert space $\mathbb{C}^n$.

Exercises. (1) Consider the matrices

$$A = \begin{pmatrix} 0 & 1 & 0 & 0 \\ 0 & 0 & 1 & 0 \\ 0 & 0 & 0 & 1 \\ 0 & 0 & 0 & 0 \end{pmatrix}, \quad B = \begin{pmatrix} 0 & 1 & 0 & 0 \\ 0 & 0 & 1 & 0 \\ 0 & 0 & 0 & 1 \\ 1 & 0 & 0 & 0 \end{pmatrix}.$$

Find $\exp(A)$ and $\exp(B)$. Note that $A^4 = 0_4$ and $B^4 = I_4$.

(2) Find the *square roots* of the nonnormal 3×3 matrix
$$A = \begin{pmatrix} 0 & 1 & 0 \\ 0 & 0 & 0 \\ 0 & 0 & 0 \end{pmatrix}$$
i.e. find the matrices B such that $B^2 = A$. Find the square roots of A^T.

(3) Let $c > 1$ and
$$H = \begin{pmatrix} 0 & 1 \\ 1 & 0 \end{pmatrix}.$$
Show that
$$(H - cI_2)^{-1} = \int_0^\infty e^{-c\tau} e^{-\tau H} d\tau.$$

(4) Let $\epsilon \in \mathbb{R}$. Consider the matrix
$$B(\epsilon) = \begin{pmatrix} \cosh(\epsilon) & \epsilon \\ 0 & 1/\cosh(\epsilon) \end{pmatrix}.$$
Is $B(\epsilon) \in SL(2, \mathbb{R})$? Find
$$X = \frac{d}{d\epsilon} B(\epsilon) \Big|_{\epsilon=0}.$$
Calculate $\exp(\epsilon X)$. Is $B(\epsilon) = \exp(\epsilon X)$?

(5) Find the square roots of the Hadamard matrix
$$\begin{pmatrix} 1 & 1 \\ 1 & -1 \end{pmatrix}.$$

(6) Let H and K be hermitian $n \times n$ matrices. Then there exist unitary $n \times n$ matrices U and V such that
$$\exp(iH) \exp(iK) = \exp(iUHU^{-1} + iVKV^{-1}).$$
Note that $\exp(iH)$ and $\exp(iK)$ are unitary matrices. Let $n = 2$ and
$$H = \sigma_1 = \begin{pmatrix} 0 & 1 \\ 1 & 0 \end{pmatrix}, \quad K = \sigma_3 = \begin{pmatrix} 1 & 0 \\ 0 & -1 \end{pmatrix}.$$
Find U and V.

(7) Let σ_1, σ_2, σ_3 be the Pauli spin matrices. Show that the map
$$\mathbb{R}^3 \ni (\alpha_1, \alpha_2, \alpha_3) \mapsto \exp(i(\alpha_1\sigma_1 + \alpha_2\sigma_2 + \alpha_3\sigma)) \in SU(2)$$
is surjective.

(8) Let A be an $n \times n$ invertible matrix over $\mathbb{C}$. Is the matrix
$$B = \frac{1}{2}(A + A^{-1})$$
invertible? Next consider the case that the underlying field is $\mathbb{R}$.

1.24   Nonnormal Matrices

A square matrix M over $\mathbb{C}$ is called normal if

$$MM^* = M^*M.$$

A square matrix M over $\mathbb{C}$ is called nonnormal if

$$MM^* \neq M^*M.$$

Examples of normal matrices are: hermitian, skew-hermitian, unitary, projection matrices. Examples for nonnormal matrices are

$$\begin{pmatrix} 0 & 1 \\ 0 & 0 \end{pmatrix}, \quad \begin{pmatrix} 0 & 1 & 1 \\ 0 & 0 & 1 \\ 0 & 0 & 0 \end{pmatrix}.$$

If M is a nonnormal matrix, then M^* and M^T are nonnormal matrices. If M is nonnormal and invertible, then M^{-1} is nonnormal. If A and B are nonnormal, then $A \oplus B$ is nonnormal.

If M is nonnormal, then $M + M^*$, MM^*, the commutator $[M, M^*]$ and the anti-commutator $[M, M^*]_+$ are normal matrices.

The Kronecker product of two normal matrices is normal again. The direct sum of two normal matrices is normal again. The star product of two normal matrices is normal again.

Any nondiagonalizeable matrix M is nonnormal. Every normal matrix is diagonalizable by the spectral decomposition. There are nonnormal matrices which are diagonalizable.

If A is nonnormal, then $A \bullet A^*$ is normal, where $\bullet$ denotes the entrywise product. Notice that $A \bullet B = B \bullet A$ and $(A \bullet B)^* = A^* \bullet B^*$.

We cannot conclude that the commutator of two nonnormal matrices is nonnormal. For example

$$A = \begin{pmatrix} 0 & 1 \\ 0 & 0 \end{pmatrix}, \quad B = \begin{pmatrix} 0 & 0 \\ 1 & 0 \end{pmatrix} \quad \Rightarrow \quad [A, B] = \begin{pmatrix} 1 & 0 \\ 0 & -1 \end{pmatrix}$$

which is a normal matrix.

Example 1.74. Let $\alpha, \beta \in \mathbb{C}$. Consider the matrix

$$M(\alpha, \beta) = \begin{pmatrix} \alpha & \beta \\ 0 & 1 \end{pmatrix} \Rightarrow M^*(\alpha, \beta) = \begin{pmatrix} \bar{\alpha} & 0 \\ \bar{\beta} & 1 \end{pmatrix}.$$

Then

$$M^*(\alpha, \beta)M(\alpha, \beta) = \begin{pmatrix} \alpha\bar{\alpha} & \bar{\alpha}\beta \\ \alpha\bar{\beta} & \beta\bar{\beta} + 1 \end{pmatrix}, \quad M(\alpha, \beta)M^*(\alpha, \beta) = \begin{pmatrix} \alpha\bar{\alpha} + \beta\bar{\beta} & \beta \\ \bar{\beta} & 1 \end{pmatrix}.$$

Thus the condition that $M(\alpha, \beta)$ is normal is given by $\beta\bar{\beta} = 0$ which implies that $\beta = 0$. ♣

Example 1.75. Let $\theta \in \mathbb{R}$. Consider the matrix

$$A(\theta) = \begin{pmatrix} 0 & \sinh(\theta) \\ \cosh(\theta) & 0 \end{pmatrix} \Rightarrow A^*(\theta) = \begin{pmatrix} 0 & \cosh(\theta) \\ \sinh(\theta) & 0 \end{pmatrix}.$$

Then

$$A(\theta)A^*(\theta) = \begin{pmatrix} \sinh^2(\theta) & 0 \\ 0 & \cosh^2(\theta) \end{pmatrix}, \quad A^*(\theta)A(\theta) = \begin{pmatrix} \cosh^2(\theta) & 0 \\ 0 & \sinh^2(\theta) \end{pmatrix}.$$

Thus the condition for the matrix to be normal is $\cosh^2(\theta) = \sinh^2(\theta)$. There are no solutions to this equation and $A(\theta)$ is nonnormal for all θ. ♣

We cannot conclude that an invertible matrix A is normal. Consider the example

$$A = \begin{pmatrix} 1 & 1 \\ 0 & 1 \end{pmatrix} \Rightarrow A^{-1} = \begin{pmatrix} 1 & -1 \\ 0 & 1 \end{pmatrix}.$$

Both are nonnormal. Is the matrix $A \otimes A^{-1}$ normal?

Not all nonnormal matrices are nondiagonalizable. Consider the 2×2 matrix

$$A = \begin{pmatrix} 1 & 2 \\ 1 & 1 \end{pmatrix}$$

which is nonnormal and the invertible matrix

$$B = \begin{pmatrix} 1 & 1 \\ 1/\sqrt{2} & -1/\sqrt{2} \end{pmatrix} \Rightarrow B^{-1} = \begin{pmatrix} 1/2 & 1/\sqrt{2} \\ 1/2 & -1/\sqrt{2} \end{pmatrix}.$$

Then the matrix $B^{-1}AB$ is diagonal. Is B nonnormal?

Let A, B be normal matrices. Can we conclude that AB is a normal matrix? We have $AA^* = A^*A$ and $BB^* = B^*B$. We have to show that

$$(AB)(AB)^* = (AB)^*(AB).$$

It follows that $ABB^*A^* = B^*A^*AB$. This equation can only be satisfied if $AB = BA$, i.e. A and B commute. For example, let

$$A = \begin{pmatrix} 1 & 0 \\ 0 & 0 \end{pmatrix}, \quad B = \begin{pmatrix} 1 & 1 \\ 1 & 1 \end{pmatrix}.$$

Then A and B do not commute and we find the nonnormal matrices

$$AB = \begin{pmatrix} 1 & 1 \\ 0 & 0 \end{pmatrix}, \qquad BA = \begin{pmatrix} 1 & 0 \\ 1 & 0 \end{pmatrix}.$$

Exercises. (1) Are all nonzero nilpotent matrices nonnormal?

(2) (i) Let $a_{11}, a_{22}, \epsilon \in \mathbb{R}$. What is the condition on a_{11}, a_{22}, ϵ such that the 2×2 matrix

$$A(\epsilon) = \begin{pmatrix} a_{11} & e^{\epsilon} \\ e^{-\epsilon} & a_{22} \end{pmatrix}$$

is normal?
(ii) Let $a > 0$, $b \geq 0$ and $\phi \in [0, \pi]$. What are the conditions a, b, ϕ such that

$$B(a, b, \phi) = \begin{pmatrix} 0 & a \\ e^{i\phi}b & 0 \end{pmatrix}$$

is a normal matrix?
(iii) Let $z \in \mathbb{C}$. Is the 2×2 matrix

$$C(z) = \begin{pmatrix} 1 & 0 \\ 1 - e^z & e^{-z} \end{pmatrix}$$

nonnormal?

(3) (i) Let $x_1, x_2, x_3 \in \mathbb{R}$. What is the condition such that the 3×3 matrix

$$A(x_1, x_2, x_3) = \begin{pmatrix} 0 & x_1 & 0 \\ 0 & 0 & x_2 \\ x_3 & 0 & 0 \end{pmatrix}$$

is normal? Show that $\lambda^3 = x_1 x_2 x_3$ for $\det(A - \lambda I_3) = 0$. Find the eigenvalues and normalized eigenvectors of A.
(ii) Let $x \in \mathbb{R}$. For which values of x is the 3×3 matrix

$$A(x) = \begin{pmatrix} -1 + ix & ix & ix \\ ix & ix & ix \\ ix & ix & 1 + ix \end{pmatrix}$$

nonnormal?

(4) Find all nonnormal 2×2 matrices A such that $AA^* + A^*A = I_2$. An example is

$$A = \begin{pmatrix} 0 & 1 \\ 0 & 0 \end{pmatrix} \quad \Rightarrow \quad A^* = \begin{pmatrix} 0 & 0 \\ 1 & 0 \end{pmatrix}.$$

(5) Let $z \in \mathbb{C}$ and $z \neq 0$. Show that the 2×2 matrix

$$A(z) = \begin{pmatrix} 1 & z \\ 0 & -1 \end{pmatrix}$$

is nonnormal. Show that the eigenvalues are given by $+1$ and -1.

(6) Let $k > 0$. Is the 2×2 matrix

$$A(k, \alpha) = \begin{pmatrix} \cos(\alpha) & -k\sin(\alpha) \\ \sin(\alpha)/k & \cos(\alpha) \end{pmatrix}$$

with $\det(A(k, \alpha)) = 1$ nonnormal?

(7) (i) Show that the 2×2 matrices over $\mathbb{R}$

$$M(a, b, c) = \begin{pmatrix} a & b \\ -c & -a \end{pmatrix}$$

can be diagonalized whenever $bc \neq a^2$.
(ii) Consider the 2×2 matrix over the real numbers

$$A = \begin{pmatrix} a_1 & b \\ -b & a_2 \end{pmatrix}.$$

Show that the eigenvalues of A are degenerate when $a_2 = a_1 \pm 2b$. Show that with this condition the matrix A cannot be diagonalized.

Chapter 2

Kronecker Product

2.1 Definitions and Notations

We introduce the Kronecker product of two matrices and give a number of examples (Henderson, Pukelsheim and Searle [31], van Loan [70], Horn and Johnson [32]).

Definition 2.1. Let A be an $m \times n$ matrix and let B be a $p \times q$ matrix. Then the *Kronecker product* $\otimes$ of A and B is that $(mp) \times (nq)$ matrix defined by

$$A \otimes B := \begin{pmatrix} a_{11}B & a_{12}B & \cdots & a_{1n}B \\ a_{21}B & a_{22}B & \cdots & a_{2n}B \\ \vdots & \vdots & \ddots & \vdots \\ a_{m1}B & a_{m2}B & \cdots & a_{mn}B \end{pmatrix}.$$

Remark 1. Sometimes the Kronecker product is also called *direct product* or *tensor product*.

Remark 2. Instead of the definition given above some authors (Regalia and Mitra [49]) use the definition

$$A \otimes B := \begin{pmatrix} b_{11}A & b_{12}A & \cdots & b_{1q}A \\ b_{21}A & b_{22}A & \cdots & b_{2q}A \\ \vdots & & & \\ b_{p1}A & b_{p2}A & \cdots & b_{pq}A \end{pmatrix}.$$

Throughout this book we use the first definition.

Let us now give some examples.

Example 2.1. Let

$$A = \begin{pmatrix} 2 & 3 \\ 0 & 1 \end{pmatrix}, \qquad B = \begin{pmatrix} 0 & -1 \\ -1 & 1 \end{pmatrix}.$$

Then

$$A \otimes B = \begin{pmatrix} 0 & -2 & 0 & -3 \\ -2 & 2 & -3 & 3 \\ 0 & 0 & 0 & -1 \\ 0 & 0 & -1 & 1 \end{pmatrix}, \qquad B \otimes A = \begin{pmatrix} 0 & 0 & -2 & -3 \\ 0 & 0 & 0 & -1 \\ -2 & -3 & 2 & 3 \\ 0 & -1 & 0 & 1 \end{pmatrix}.$$

We see that $A \otimes B \neq B \otimes A$. ♣

Example 2.2. Let $\mathbf{u} = \begin{pmatrix} 1 & 0 \end{pmatrix}^T$, $\mathbf{v} = \begin{pmatrix} 0 & 1 \end{pmatrix}^T$ be the *standard basis* in $\mathbb{R}^2$ or $\mathbb{C}^2$. Then

$$\mathbf{u} \otimes \mathbf{u} = \begin{pmatrix} 1 \\ 0 \\ 0 \\ 0 \end{pmatrix}, \quad \mathbf{u} \otimes \mathbf{v} = \begin{pmatrix} 0 \\ 1 \\ 0 \\ 0 \end{pmatrix}, \quad \mathbf{v} \otimes \mathbf{u} = \begin{pmatrix} 0 \\ 0 \\ 1 \\ 0 \end{pmatrix}, \quad \mathbf{v} \otimes \mathbf{v} = \begin{pmatrix} 0 \\ 0 \\ 0 \\ 1 \end{pmatrix}.$$

We find that

$$\{ \mathbf{u} \otimes \mathbf{u}, \mathbf{u} \otimes \mathbf{v}, \mathbf{v} \otimes \mathbf{u}, \mathbf{v} \otimes \mathbf{v} \}$$

is the standard basis in $\mathbb{R}^4$ (or $\mathbb{C}^4$). ♣

Example 2.3. Consider the normalized vectors in $\mathbb{C}^2$

$$\mathbf{u} = \frac{1}{\sqrt{2}} \begin{pmatrix} 1 \\ i \end{pmatrix}, \qquad \mathbf{v} = \frac{1}{\sqrt{2}} \begin{pmatrix} 1 \\ -i \end{pmatrix}.$$

Then

$$\mathbf{u} \otimes \mathbf{u} = \frac{1}{2} \begin{pmatrix} 1 \\ i \\ i \\ -1 \end{pmatrix}, \quad \mathbf{u} \otimes \mathbf{v} = \frac{1}{2} \begin{pmatrix} 1 \\ -i \\ i \\ 1 \end{pmatrix}, \quad \mathbf{v} \otimes \mathbf{u} = \frac{1}{2} \begin{pmatrix} 1 \\ i \\ -i \\ 1 \end{pmatrix}, \quad \mathbf{v} \otimes \mathbf{v} = \frac{1}{2} \begin{pmatrix} 1 \\ -i \\ -i \\ -1 \end{pmatrix}.$$

Since $\|\mathbf{u}\| = \|\mathbf{v}\| = 1$ and $\langle \mathbf{u}, \mathbf{v} \rangle = 0$, $\{\mathbf{u}, \mathbf{v}\}$ is an orthonormal basis in the vector space $\mathbb{C}^2$. We find that $\{ \mathbf{u} \otimes \mathbf{u}, \mathbf{u} \otimes \mathbf{v}, \mathbf{v} \otimes \mathbf{u}, \mathbf{v} \otimes \mathbf{v} \}$ is an orthonormal basis in the vector space $\mathbb{C}^4$. ♣

In general we can show that if $\mathbf{v}_1, \ldots, \mathbf{v}_m$ is an orthonormal basis in $\mathbb{C}^m$ and $\mathbf{u}_1, \ldots, \mathbf{u}_n$ is an orthonormal basis in $\mathbb{C}^n$, then $\mathbf{v}_j \otimes \mathbf{u}_k$ $(j = 1, \ldots, m; k = 1, \ldots, n)$ is an orthonormal basis in the vector space $\mathbb{C}^{m \times n}$.

Example 2.4. Let

$$\mathbf{u} = \begin{pmatrix} u_1 \\ u_2 \\ u_3 \end{pmatrix} \in \mathbb{C}^3, \qquad \mathbf{v} = \begin{pmatrix} v_1 \\ v_2 \\ v_3 \end{pmatrix} \in \mathbb{C}^3.$$

Then $\mathbf{u}^* \equiv \bar{\mathbf{u}}^T = (\bar{u}_1, \bar{u}_2, \bar{u}_3)$, $\mathbf{v}^* \equiv \bar{\mathbf{v}}^T = (\bar{v}_1, \bar{v}_2, \bar{v}_3)$ and therefore

$$\mathbf{u}^* \otimes \mathbf{u} = \begin{pmatrix} \bar{u}_1 u_1 & \bar{u}_2 u_1 & \bar{u}_3 u_1 \\ \bar{u}_1 u_2 & \bar{u}_2 u_2 & \bar{u}_3 u_2 \\ \bar{u}_1 u_3 & \bar{u}_2 u_3 & \bar{u}_3 u_3 \end{pmatrix}, \qquad \mathbf{v}^* \otimes \mathbf{v} = \begin{pmatrix} \bar{v}_1 v_1 & \bar{v}_2 v_1 & \bar{v}_3 v_1 \\ \bar{v}_1 v_2 & \bar{v}_2 v_2 & \bar{v}_3 v_2 \\ \bar{v}_1 v_3 & \bar{v}_2 v_3 & \bar{v}_3 v_3 \end{pmatrix},$$

$$\mathbf{u}^* \otimes \mathbf{v} = \begin{pmatrix} \bar{u}_1 v_1 & \bar{u}_2 v_1 & \bar{u}_3 v_1 \\ \bar{u}_1 v_2 & \bar{u}_2 v_2 & \bar{u}_3 v_2 \\ \bar{u}_1 v_3 & \bar{u}_2 v_3 & \bar{u}_3 v_3 \end{pmatrix}, \qquad \mathbf{v}^* \otimes \mathbf{u} = \begin{pmatrix} \bar{v}_1 u_1 & \bar{v}_2 u_1 & \bar{v}_3 u_1 \\ \bar{v}_1 u_2 & \bar{v}_2 u_2 & \bar{v}_3 u_2 \\ \bar{v}_1 u_3 & \bar{v}_2 u_3 & \bar{v}_3 u_3 \end{pmatrix}.$$

Now

$$\mathbf{u} \otimes \mathbf{v}^* = \begin{pmatrix} u_1 \bar{v}_1 & u_1 \bar{v}_2 & u_1 \bar{v}_3 \\ u_2 \bar{v}_1 & u_2 \bar{v}_2 & u_2 \bar{v}_3 \\ u_3 \bar{v}_1 & u_3 \bar{v}_2 & u_3 \bar{v}_3 \end{pmatrix}.$$

Thus we find that $(\mathbf{u} \otimes \mathbf{v}^*)^* = \mathbf{u}^* \otimes \mathbf{v}$. Notice that $\mathbf{u}^* \otimes \mathbf{u} = \mathbf{u}\mathbf{u}^*$. ♣

Example 2.5. Let

$$\mathbf{u} = \frac{1}{\sqrt{2}} \begin{pmatrix} 1 \\ 1 \end{pmatrix}, \qquad \mathbf{v} = \frac{1}{\sqrt{2}} \begin{pmatrix} 1 \\ -1 \end{pmatrix}.$$

Obviously the set $\{\mathbf{u}, \mathbf{v}\}$ forms an orthonormal basis in $\mathbb{R}^2$. Now

$$\mathbf{u}^T \otimes \mathbf{u} = \frac{1}{2} \begin{pmatrix} 1 & 1 \\ 1 & 1 \end{pmatrix}, \qquad \mathbf{v}^T \otimes \mathbf{v} = \frac{1}{2} \begin{pmatrix} 1 & -1 \\ -1 & 1 \end{pmatrix}.$$

Consequently

$$\mathbf{u}^T \otimes \mathbf{u} + \mathbf{v}^T \otimes \mathbf{v} = \begin{pmatrix} 1 & 0 \\ 0 & 1 \end{pmatrix} = I_2.$$

We can also write $I_2 = \mathbf{u}\mathbf{u}^T + \mathbf{v}\mathbf{v}^T$. ♣

Example 2.6. Let $\mathbf{u} = \begin{pmatrix} 1 & 0 \end{pmatrix}^T$, $\mathbf{v} = \begin{pmatrix} 0 & 1 \end{pmatrix}^T$. Then

$$\mathbf{u}^T \otimes \mathbf{u} = \begin{pmatrix} 1 & 0 \\ 0 & 0 \end{pmatrix}, \quad \mathbf{v}^T \otimes \mathbf{v} = \begin{pmatrix} 0 & 0 \\ 0 & 1 \end{pmatrix} \Rightarrow \mathbf{u}^T \otimes \mathbf{u} + \mathbf{v}^T \otimes \mathbf{v} = \begin{pmatrix} 1 & 0 \\ 0 & 1 \end{pmatrix} = I_2.$$

Notice that the 2×2 unit matrix is also given by $I_2 = \mathbf{u}\mathbf{u}^T + \mathbf{v}\mathbf{v}^T$. ♣

Remark. The last two examples indicate that the unit matrix can be represented with the help of an orthonormal basis in $\mathbb{R}^2$ and the Kronecker product.

Example 2.7. Let I_n be the $n \times n$ unit matrix and let I_m be the $m \times m$ unit matrix. Then $I_n \otimes I_m$ is the $(nm) \times (mn)$ unit matrix. Obviously,

$$I_n \otimes I_m = I_m \otimes I_n = I_{m \times n}.$$

♣

Example 2.8. Let A_n be an arbitrary $n \times n$ matrix and let 0_m be the $m \times m$ zero matrix. Then $A_n \otimes 0_m = 0_{m \times n}$. ♣

Example 2.9. Given the 4×4 permutation matrix

$$A = \begin{pmatrix} 0\,0\,0\,1 \\ 0\,0\,1\,0 \\ 0\,1\,0\,0 \\ 1\,0\,0\,0 \end{pmatrix}.$$

We may ask whether A can be written as $A = B \otimes C$, where B and C are 2×2 permutation matrices. For the given A we find $B = C$ with

$$B = \begin{pmatrix} 0\,1 \\ 1\,0 \end{pmatrix}.$$

♣

Example 2.10. Let $\sigma_0 = I_2$, σ_1, σ_2, σ_3 be the Pauli spin matrices,

$$R(u, \eta) = \sum_{j=0}^{3} w_j(u, \eta) \sigma_j \otimes \sigma_j$$

and

$$w_0(u,\eta) = \sinh(u + \eta/2)\cosh(\eta/2), \quad w_1(u,\eta) = \sinh(\eta/2)\cosh(\eta/2),$$

$$w_2(u,\eta) = \sinh(\eta/2)\cosh(\eta/2), \quad w_3(u,\eta) = \sinh(\eta/2)\cosh(u + \eta/2).$$

Then one finds

$$R(u,\eta) = \begin{pmatrix} \sinh(u+\eta) & 0 & 0 & 0 \\ 0 & \sinh(u) & \sinh(\eta) & 0 \\ 0 & \sinh(\eta) & \sinh(u) & 0 \\ 0 & 0 & 0 & \sinh(u+\eta) \end{pmatrix}.$$

♣

Example 2.11. Consider the vectors (*Bell states*)

$$\frac{1}{\sqrt{2}}\begin{pmatrix}1\\0\\0\\1\end{pmatrix}, \quad \frac{1}{\sqrt{2}}\begin{pmatrix}1\\0\\0\\-1\end{pmatrix}, \quad \frac{1}{\sqrt{2}}\begin{pmatrix}0\\1\\1\\0\end{pmatrix}, \quad \frac{1}{\sqrt{2}}\begin{pmatrix}0\\1\\-1\\0\end{pmatrix}$$

in $\mathbb{C}^4$. They form an orthonormal basis in $\mathbb{C}^4$. None of them can be written in the form $\mathbf{u} \otimes \mathbf{v}$ where $\mathbf{u}, \mathbf{v} \in \mathbb{C}^2$. For example the equation

$$\frac{1}{\sqrt{2}}\begin{pmatrix}1\\0\\0\\1\end{pmatrix} = \begin{pmatrix}u_1\\u_2\end{pmatrix} \otimes \begin{pmatrix}v_1\\v_2\end{pmatrix}$$

does not admit a solution. ♣

Example 2.12. Let $\oplus$ be the direct sum. The 4×4 matrix

$$\begin{pmatrix}1&0&0&0\\0&1&0&0\\0&0&0&1\\0&0&1&0\end{pmatrix} \equiv \begin{pmatrix}1&0\\0&1\end{pmatrix} \oplus \begin{pmatrix}0&1\\1&0\end{pmatrix}$$

cannot be written as the Kronecker product of two 2×2 matrices. ♣

Let $\mathbf{x}, \mathbf{y}, \mathbf{z} \in \mathbb{R}^n$. We define a *wedge product*

$$\mathbf{x} \wedge \mathbf{y} := \mathbf{x} \otimes \mathbf{y} - \mathbf{y} \otimes \mathbf{x}.$$

Then one can easily prove that

$$(\mathbf{x} \wedge \mathbf{y}) \wedge \mathbf{z} + (\mathbf{z} \wedge \mathbf{x}) \wedge \mathbf{y} + (\mathbf{y} \wedge \mathbf{z}) \wedge \mathbf{x} = \mathbf{0}.$$

Let A be an $m \times n$ matrix and B be an $p \times q$ matrix. The *symmetric tensor product* is defined as

$$A \overset{s}{\otimes} B := \frac{1}{2}(A \otimes B + B \otimes A).$$

For example

$$A = \sigma_3 = \begin{pmatrix}1&0\\0&-1\end{pmatrix}, \quad B = \sigma_1 = \begin{pmatrix}0&1\\1&0\end{pmatrix} \Rightarrow A \overset{s}{\otimes} B = \frac{1}{2}\begin{pmatrix}0&1&1&0\\1&0&0&-1\\1&0&0&-1\\0&-1&-1&0\end{pmatrix}.$$

Given two matrices

$$A = \begin{pmatrix}\mathbf{a}_1 & \mathbf{a}_2 & \dots & \mathbf{a}_p\end{pmatrix}, \quad B = \begin{pmatrix}\mathbf{b}_1 & \mathbf{b}_2 & \dots & \mathbf{b}_p\end{pmatrix}$$

of size $m \times p$ and $n \times p$, respectively. So $\mathbf{a}_j$ are column vectors of length m and $\mathbf{b}_k$ are column vectors of length n. The columnwise *Khatri-Rao product* of A and B (denoted by $A \odot B$) is defined as the $(mn) \times p$ matrix

$$A \odot B := \left(\mathbf{a}_1 \otimes \mathbf{b}_1 \; \mathbf{a}_2 \otimes \mathbf{b}_2 \; \cdots \; \mathbf{a}_p \otimes \mathbf{b}_p \right)$$

where $\otimes$ denotes the Kronecker product. As an example consider

$$\sigma_1 = \begin{pmatrix} 0 & 1 \\ 1 & 0 \end{pmatrix}, \quad \sigma_3 = \begin{pmatrix} 1 & 0 \\ 0 & -1 \end{pmatrix} \quad \Rightarrow \quad A \odot B = \begin{pmatrix} 0 & 0 \\ 0 & -1 \\ 1 & 0 \\ 0 & 0 \end{pmatrix}.$$

Exercises. (1) An orthonormal basis in the Hilbert space $\mathbb{C}^2$ is given by

$$\mathbf{v}_1 = \frac{1}{\sqrt{2}} \begin{pmatrix} 1 \\ i \end{pmatrix}, \quad \mathbf{v}_2 = \frac{1}{\sqrt{2}} \begin{pmatrix} 1 \\ -i \end{pmatrix}.$$

Use this orthonormal basis and the Kronecker product to find an orthonormal basis in the Hilbert spaces $\mathbb{C}^4$ and $\mathbb{C}^8$.

(ii) An orthonormal basis in the Hilbert space $\mathbb{R}^3$ is given by

$$\mathbf{v}_1 = \frac{1}{\sqrt{2}} \begin{pmatrix} 1 \\ 0 \\ 1 \end{pmatrix}, \quad \mathbf{v}_2 = \begin{pmatrix} 0 \\ 1 \\ 0 \end{pmatrix}, \quad \mathbf{v}_3 = \frac{1}{\sqrt{2}} \begin{pmatrix} 1 \\ 0 \\ -1 \end{pmatrix}.$$

Use this orthonormal basis and the Kronecker product to find an orthonormal basis in the Hilbert space $\mathbb{R}^9$.

(iii) A basis in the Hilbert space $M_2(\mathbb{C})$ of 2×2 matrices is given by the *Pauli spin matrices* including the 2×2 identity matrix

$$\sigma_0 = \begin{pmatrix} 1 & 0 \\ 0 & 1 \end{pmatrix}, \quad \sigma_1 = \begin{pmatrix} 0 & 1 \\ 1 & 0 \end{pmatrix}, \quad \sigma_2 = \begin{pmatrix} 0 & -i \\ i & 0 \end{pmatrix}, \quad \sigma_3 = \begin{pmatrix} 1 & 0 \\ 0 & -1 \end{pmatrix}.$$

Use this orthonormal basis and the Kronecker product of $\sigma_j \otimes \sigma_k$ ($j, k = 0, 1, 2, 3$) to find a basis in the Hilbert space $M_4(\mathbb{C})$.

(2) Let A and B be $n \times n$ diagonal matrices. Show that $A \otimes B \neq B \otimes A$ in general. What is the condition on A and B such that $A \otimes B = B \otimes A$?

(3) Let A, B be 2×2 matrices. Find the conditions on A and B such that $A \otimes B = B \otimes A$.

(4) Let A be a 2×2 matrix and B be a 3×3 matrix. Find the conditions on A and B such that $A \otimes B = B \otimes A$.

(5) Let

$$X = \begin{pmatrix} 16 & 3 & 2 & 13 \\ 5 & 10 & 11 & 8 \\ 9 & 6 & 7 & 12 \\ 4 & 15 & 14 & 1 \end{pmatrix}.$$

X is a so-called *magic square*, since its row sums, column sums, principal diagonal sum, and principal counter diagonal are equal. Is $X \otimes X$ a magic square?

(6) A *Toeplitz matrix* is a square matrix whose elements are the same along any northwest (NW) to southeast (SE) diagonal. The Toeplitz matrix of size $2^2 \times 2^2$ is designated $T(2)$ and given by

$$T(2) = \begin{pmatrix} a_{11} & a_{12} & a_{13} & a_{14} \\ a_{21} & a_{11} & a_{12} & a_{13} \\ a_{31} & a_{21} & a_{11} & a_{12} \\ a_{41} & a_{31} & a_{21} & a_{11} \end{pmatrix}.$$

Is $T(2) \otimes T(2)$ a Toeplitz matrix?

(7) A *circulant matrix* is a square matrix whose elements in each row are obtained by a circular right shift of the elements in the preceding row. An example is

$$C(2) = \begin{pmatrix} c_{11} & c_{12} & c_{13} & c_{14} \\ c_{14} & c_{11} & c_{12} & c_{13} \\ c_{13} & c_{14} & c_{11} & c_{12} \\ c_{12} & c_{13} & c_{14} & c_{11} \end{pmatrix}.$$

Given the elements of any row, the entire matrix can be developed. Is $C(2) \otimes C(2)$ a circulant matrix?

(8) A vector $\mathbf{u}^T = (u_1, u_2, \ldots, u_n)$ is called a *probability vector* if the components are nonnegative and their sum is 1. Let $\mathbf{u}^T$ and $\mathbf{v}^T$ be probability vectors. Show that $\mathbf{u}^T \otimes \mathbf{v}^T$ is a probability vector.

(9) A square matrix $A = (a_{jk})$ is called a *stochastic matrix* if each of its rows is a probability vector, i.e. if each entry of A is nonnegative and the

sum of the entries in each row is equal to 1. Let A and B be $n \times n$ stochastic matrices. Show that AB is a stochastic matrix. Show that $A \otimes B$ is a stochastic matrix.

(10) Let A, B and C be 3×3 matrices. We set

$$D := \begin{pmatrix} A & B & 0_3 \\ C & A & B \\ 0_3 & C & A \end{pmatrix}.$$

Show that D can be written in the form $D = (I_3 \otimes A + L \otimes C + U \otimes B)$, where L and U are defined as

$$L = \begin{pmatrix} 0 & 0 & 0 \\ 1 & 0 & 0 \\ 0 & 1 & 0 \end{pmatrix}, \qquad U = \begin{pmatrix} 0 & 1 & 0 \\ 0 & 0 & 1 \\ 0 & 0 & 0 \end{pmatrix}.$$

(11) The symmetric matrix of order n

$$H = (h_{ij}), \qquad h_{ij} = \frac{1}{i+j-1}, \qquad i,j = 1,2,\ldots,n$$

is called a *Hilbert matrix*. Show that this matrix is positive definite. The number $\mathrm{cond}_2(H)$ is defined as

$$\mathrm{cond}_2(H) := \frac{\max_i |\lambda_i(H)|}{\min_i |\lambda_i(H)|}.$$

Find $\mathrm{cond}_2(H)$ for $n = 2,3,4$, and 5. Show that $H \otimes H$ is not a Hilbert matrix.

(12) Prove the identity ($\epsilon \in \mathbb{R}$)

$$(A + \epsilon \mathbf{u} \otimes \mathbf{v}^T)^{-1} \equiv A^{-1} - \frac{\epsilon}{1 + \epsilon \mathbf{v}^T A^{-1} \mathbf{u}} A^{-1} \mathbf{u} \otimes \mathbf{v}^T A^{-1}$$

where A is an invertible matrix of order n and $\mathbf{u}, \mathbf{v} \in \mathbb{R}^n$. This identity is sometimes used to find the inverse of a matrix which appears in the form of a perturbation $A + \epsilon \mathbf{u} \otimes \mathbf{v}^T$ of a matrix A whose inverse is known.

(13) Let $\mathbf{u}, \mathbf{v} \in \mathbb{R}^n$. Assume that $\|\mathbf{u}\| = 1$ and $\|\mathbf{v}\| = 1$. Is $\|\mathbf{u} \otimes \mathbf{v}\| = 1$?

(14) Let A, B be 2×2 matrices. Find the conditions on A and B such that $A \otimes I_2 \otimes B = B \otimes I_2 \otimes A$.

(15) Let $|0\rangle$, $|1\rangle$ be an orthonormal basis in the Hilbert space $\mathbb{C}^2$. Can one find a unitary 4×4 matrix such that $U(|a\rangle \otimes |b\rangle) = |b\rangle \otimes |a\rangle$ for all

$a, b \in \{0, 1\}$?

(16) Let $\mathbf{v}_1, \mathbf{v}_2 \in \mathbb{R}^2$. Find the conditions on $\mathbf{v}_1$, $\mathbf{v}_2$ such that $\mathbf{v}_1 \otimes \mathbf{v}_2 \otimes \mathbf{v}_1 = \mathbf{v}_2 \otimes \mathbf{v}_1 \otimes \mathbf{v}_2$.

(17) Consider the vector space V of $n \times n$ matrices over $\mathbb{C}$ and $A \in V$.
(i) Is the map $f(A) = A \otimes A$ linear?
(ii) Is the map $g(A) = A \otimes I_n + I_n \otimes A$ linear?

2.2 Basic Properties

Here we list some basic properties of the Kronecker product. These properties have been listed in various books on linear algebra and matrix theory (Gröbner [26], Lancaster [40], Searle [50], Horn and Johnson [32], Deif [18]). In almost all cases the proof is straightforward and is left to the reader as an exercise.

Let A be an $m \times n$ matrix, B be a $p \times q$ matrix and C be an $s \times t$ matrix. Then

$$(A \otimes B) \otimes C \equiv A \otimes (B \otimes C).$$

The proof is straightforward. The matrix $A \otimes B \otimes C$ has (mps) rows and (nqt) columns. This means the Kronecker product satisfies the *associative law*.

Let A be an $m \times n$ matrix and B be a $p \times q$ matrix. Let $c \in \mathbb{C}$. Then

$$(cA) \otimes B \equiv c(A \otimes B) \equiv A \otimes (cB).$$

Let A and B be $m \times n$ matrices and C and D be $p \times q$ matrices. Then we have

$$(A + B) \otimes (C + D) \equiv A \otimes C + A \otimes D + B \otimes C + B \otimes D.$$

This means the Kronecker product obeys the *distributive law*.

We can easily prove that the row rank and the column rank are equal to the same number. Let $r(A)$ be the rank of A and $r(B)$ be the rank of B. Then $r(A \otimes B) = r(A)r(B)$.

Example 2.13. Let

$$A = \begin{pmatrix} 0 & 1 \\ 1 & 0 \end{pmatrix}, \qquad B = \begin{pmatrix} 0 & 1 \\ 0 & 0 \end{pmatrix}.$$

Then $r(A) = 2$ and $r(B) = 1$. Therefore $r(A \otimes B) = 2$.   ♣

Let A be an $m \times n$ matrix and B be an $p \times q$ matrix. The following three properties can be easily proved

$$(A \otimes B)^T \equiv A^T \otimes B^T, \quad \overline{(A \otimes B)} \equiv \bar{A} \otimes \bar{B}, \quad (A \otimes B)^* \equiv A^* \otimes B^*.$$

Given matrices of a special type we now summarize for which matrices the Kronecker product is also of this type.

(1) If A and B are diagonal matrices, then $A \otimes B$ is a diagonal matrix. Conversely, if $A \otimes B$ is a *diagonal matrix* and $A \otimes B \neq 0$, then A and B are diagonal matrices.

(2) Let A and B be upper triangular matrices, then $A \otimes B$ is an upper *triangular matrix*. Similarly, if A and B are lower triangular, then $A \otimes B$ is a lower triangular matrix.

(3) Let A and B be normal matrices. Then $A \otimes B$ is a *normal matrix*.

(4) Let A and B be Hermitian matrices. Then $A \otimes B$ is a *Hermitian matrix*.

(5) Let A and B be *positive definite* matrices. Then $A \otimes B$ is a positive definite matrix. The same holds for positive semi-definite matrices.

(6) Let A be an *invertible* $n \times n$ matrix. Let B be an invertible $m \times m$ matrix. Then $A \otimes B$ is an invertible matrix with $(A \otimes B)^{-1} = A^{-1} \otimes B^{-1}$.

(7) Let U and V be *unitary* $n \times n$ matrices. Then $U \otimes V$ is a unitary matrix.

Note that if A and B are skew-Hermitian matrixes, then $A \otimes B$ is not skew-Hermitian in general. We have

$$(A \otimes B)^* = A^* \otimes B^* = (-A) \otimes (-B) = A \otimes B.$$

Definition 2.2. The *Kronecker powers* of an $m \times n$ matrix A are defined as

$$A^{[2]} := A \otimes A$$

and, in general

$$A^{[k+1]} := A \otimes A^{[k]}$$

where $k \in \mathbb{N}$. We find the following property $A^{[k+l]} = A^{[k]} \otimes A^{[l]}$.

Example 2.14. In genetics we find the following example (Searle [50]). Consider the n generations of random mating starting with the progeny obtained from crossing two autotetraploid plants which both have genotype AAaa. Normally the original plants would produce gametes AA, Aa and aa in the proportion 1:4:1. However suppose the proportion is

$$u : 1 - 2u : u$$

where, for example, u might take the value $(1-\alpha)/6$, for α being a measure of "diploidization" of plants: $\alpha = 0$ is the case of autotetraploids with chromosome segregation and $\alpha = 1$ is the diploid case with all gametes being Aa. What are the genotypic frequencies in the population after n generations of random mating? Let $\mathbf{u}_j$ be the vector of gametic frequencies and $\mathbf{f}_j$ the vector of genotype frequencies in the j-th generation of random mating, where $\mathbf{u}_0$ is the vector of gametic frequencies in the initial plants. Then

$$\mathbf{u}_0 = \begin{pmatrix} u \\ 1 - 2u \\ u \end{pmatrix}$$

and $\mathbf{f}_{j+1} = \mathbf{u}_j \otimes \mathbf{u}_j$ for $j = 0, 1, \ldots, n$. Furthermore, the relationship between $\mathbf{u}_j$ and $\mathbf{f}_j$ at any generation is given by $\mathbf{u}_j = B\mathbf{f}_j$, where

$$B = \begin{pmatrix} 1 & \frac{1}{2} & u & \frac{1}{2} & u & 0 & u & 0 & 0 \\ 0 & \frac{1}{2} & 1-2u & \frac{1}{2} & 1-2u & \frac{1}{2} & 1-2u & \frac{1}{2} & 0 \\ 0 & 0 & u & 0 & u & \frac{1}{2} & u & \frac{1}{2} & 1 \end{pmatrix}.$$

Thus

$$\begin{aligned}
\mathbf{f}_j &= \mathbf{u}_{j-1} \otimes \mathbf{u}_{j-1} = B\mathbf{f}_{j-1} \otimes B\mathbf{f}_{j-1} = (B \otimes B)(\mathbf{f}_{j-1} \otimes \mathbf{f}_{j-1}) \\
&= (B \otimes B)[(B \otimes B)(\mathbf{f}_{j-2} \otimes \mathbf{f}_{j-2}) \otimes (B \otimes B)(\mathbf{f}_{j-2} \otimes \mathbf{f}_{j-2})] \\
&= (B \otimes B)[(B \otimes B) \otimes (B \otimes B)][(\mathbf{f}_{j-2} \otimes \mathbf{f}_{j-2}) \otimes (\mathbf{f}_{j-2} \otimes \mathbf{f}_{j-2})].
\end{aligned}$$

It can be verified by induction that

$$\mathbf{f}_j = \otimes^2 B(\otimes^4 B)(\otimes^8 B) \cdots (\otimes^{2^{j-1}} B)(\otimes^{2^j} \mathbf{u}_0)$$

where $\otimes^n B$ denotes the Kronecker product of n B's.

♣

Exercises. (1) Let

$$A = \begin{pmatrix} 2 & 0 & 1 \\ 3 & 1 & 0 \end{pmatrix}, \qquad B = \begin{pmatrix} 1 & 2 & 2 \\ 1 & 2 & 0 \end{pmatrix}.$$

Find $r(A)$, $r(B)$ and $r(A \otimes B)$, where $r(A)$ denotes the rank of A.

(2) Let A be an $n \times n$ matrix and I_n be the $n \times n$ unit matrix. Show that in general $I_n \otimes A \neq A \otimes I_n$. What is the condition on A such that $I_n \otimes A = A \otimes I_n$?

(3) An $n \times n$ matrix A is called *skew-Hermitian* if $A^* = -A$. Let A and B be $n \times n$ skew-Hermitian matrices. Is the matrix $A \otimes B$ skew-Hermitian?

(4) Show that the Kronecker product of two Hermitian matrices is again a Hermitian matrix.

(5) Show that the Kronecker product of two positive-definite matrices is again a positive-definite matrix.

(6) An $n \times n$ matrix A is called *nilpotent* if $A^k = 0_n$ for some positive integer k. Let A and B be two $n \times n$ nilpotent matrices. Show that $A \otimes B$ is nilpotent.

(7) An $n \times n$ matrix A is called *idempotent* if $A^k = A$ for some positive integer k. Let A and B be two $n \times n$ idempotent matrices. Show that $A \otimes B$ is idempotent.

(8) The *elementary matrix* (E_{ij}) is defined as the matrix of order $m \times n$ which has 1 in the (i,j)-th position and all other elements are zero. Let (E_{ij}) and (F_{kl}) be elementary matrices of order $m \times n$ and $p \times q$, respectively. Show that $(E_{ij}) \otimes (F_{kl})$ is an elementary matrix.

(9) Let H_1, H_2 be 2×2 hermitian matrices. Find all 4×4 hermitian matrices K such that $[H_1 \otimes I_2 + I_2 \otimes H_2, K] = 0_4$.

(10) Let

$$x = \begin{pmatrix} 0 & 1 \\ 0 & 0 \end{pmatrix}, \qquad y = \begin{pmatrix} 0 & 0 \\ 1 & 0 \end{pmatrix}, \qquad h = \begin{pmatrix} 1 & 0 \\ 0 & -1 \end{pmatrix}$$

be the standard basis of the Lie algebra $s\ell(2, \mathbb{R})$. Is

$$x \otimes x, \quad x \otimes y, \quad x \otimes h, \quad y \otimes x, \quad y \otimes y, \quad y \otimes h, \quad z \otimes x, \quad z \otimes y, \quad z \otimes z$$

a basis for the Lie algebra $s\ell(4, \mathbb{R})$.

(11) Let X, Y be $n \times n$ matrices over $\mathbb{C}$.
(i) Study the map $(X, Y) \mapsto X \otimes Y - Y \otimes X$.
(ii) Study the map $(X, Y) \mapsto X \otimes Y + Y \otimes X$.
(iii) Study the map $(X, Y) \mapsto X \otimes Y + Y \otimes X - X \otimes I_n - I_n \otimes Y$.

2.3 Matrix Multiplication

In this section we describe the connection of the matrix multiplication and the Kronecker product (Gröbner [26], Lancaster [40], Brewer [12], Davis [17], Searle [50], Steeb [57], Graham [25], Gantmacher [24]).

Let A be an $m \times n$ matrix and B be an $n \times r$ matrix. Then the *matrix product* AB is defined and the matrix elements of AB are given by

$$(AB)_{kl} := \sum_{j=1}^{n} a_{kj} b_{jl}.$$

If A is of order $m \times n$, if B is of order $n \times p$, and if C is of order $p \times q$, then

$$A(BC) = (AB)C.$$

Now we can discuss the connection of Kronecker products and matrix products. Assume that the matrix A is of order $m \times n$, B of order $p \times q$, C of order $n \times r$ and D of order $q \times s$. Then by straightforward calculation we can prove that $(A \otimes B)(C \otimes D) = (AC) \otimes (BD)$.

Example 2.15. Let A and B be $n \times n$ matrices. Then

$$(A \otimes I_n)(I_n \otimes B) = A \otimes B. \qquad \clubsuit$$

Example 2.16. Let A, B be $n \times n$ matrices. Then

$$(A \otimes B \otimes I_n)(A \otimes I_n \otimes B)(I_n \otimes A \otimes B) = A^2 \otimes BA \otimes B^2$$
$$(I_n \otimes A \otimes B)(A \otimes I_n \otimes B)(A \otimes B \otimes I_n) = A^2 \otimes AB \otimes B^2. \qquad \clubsuit$$

We can prove the following theorems.

Theorem 2.1. *Let A be an invertible $m \times m$ matrix. Let B be an invertible $n \times n$ matrix. Then $A \otimes B$ is invertible and $(A \otimes B)^{-1} = A^{-1} \otimes B^{-1}$.*

Proof. Since $\det(A) \neq 0$, $\det(B) \neq 0$ and

$$\det(A \otimes B) = (\det(A))^n (\det(B))^m$$

we have $\det(A \otimes B) \neq 0$. The proof of the identity $\det(A \otimes B) = (\det(A))^n (\det(B))^m$ will be given later. Thus $(A \otimes B)^{-1}$ exists and $(A \otimes B)^{-1}(A \otimes B) = I_{mn}$ or

$$(A^{-1} \otimes B^{-1})(A \otimes B) = (A^{-1}A) \otimes (B^{-1}B) = I_{mn}. \qquad \square$$

Let A, B be $n \times n$ matrices. Let Q, R be invertible $n \times n$ matrices. Then

$$(Q^{-1} \otimes R^{-1}(A \otimes B)(Q \otimes R) = (Q^{-1} \otimes R^{-1})((AQ) \otimes (BR))$$
$$= (Q^{-1}AQ) \otimes (R^{-1}BR).$$

Theorem 2.2. *Let U and V be unitary matrices. Then $U \otimes V$ is a unitary matrix.*

Proof. Since U and V are unitary matrices we have $U^* = U^{-1}$, $V^* = V^{-1}$. Therefore

$$(U \otimes V)^* = U^* \otimes V^* = U^{-1} \otimes V^{-1} = (U \otimes V)^{-1}. \qquad \square$$

The equation $(A \otimes B)(C \otimes D) = (AC) \otimes (BD)$ can be extended as follows

$$(A_1 \otimes B_1)(A_2 \otimes B_2)(A_3 \otimes B_3) = (A_1A_2A_3) \otimes (B_1B_2B_3)$$

and

$$(A_1 \otimes A_2 \otimes A_3)(B_1 \otimes B_2 \otimes B_3) = (A_1B_1) \otimes (A_2B_2) \otimes (A_3B_3)$$

where it is assumed that the matrix products exist. The extension to r-factors is

$$(A_1 \otimes B_1)(A_2 \otimes B_2) \cdots (A_r \otimes B_r) = (A_1A_2 \cdots A_r) \otimes (B_1B_2 \cdots B_r)$$

and

$$(A_1 \otimes A_2 \otimes \cdots \otimes A_r)(B_1 \otimes B_2 \otimes \cdots \otimes B_r) = (A_1B_1) \otimes (A_2B_2) \otimes \cdots \otimes (A_rB_r).$$

Let us now study special cases of the identity

$$(A \otimes B)(C \otimes D) \equiv (AC) \otimes (BD).$$

If $A = I_n$, then the identity takes the form

$$(I_n \otimes B)(C \otimes D) = C \otimes (BD).$$

If $D = I_q$, then we find

$$(A \otimes B)(C \otimes I_q) = (AC) \otimes B.$$

Let A be an $m \times m$ matrix and B be an $n \times n$ matrix. Then

$$A \otimes B = (A \otimes I_n)(I_m \otimes B) = (I_m \otimes B)(A \otimes I_n).$$

Let

$$f(x) := \sum_{j=0}^{r} a_j x^j \equiv a_0 + a_1 x + a_2 x^2 + \cdots + a_r x^r$$

be a *polynomial*. Let A be an $n \times n$ matrix. Then

$$f(A) = \sum_{j=0}^{r} a_j A^j \equiv a_0 I_n + a_1 A + a_2 A^2 + \cdots + a_r A^r$$

where I_n is the $n \times n$ unit matrix. Now we can easily prove that

$$f(I_n \otimes A) \equiv I_n \otimes f(A), \quad f(A \otimes I_n) \equiv f(A) \otimes I_n.$$

To prove these identities one applies the equation

$$(A \otimes B)(C \otimes D) = (AC) \otimes (BD)$$

repeatedly.

Definition 2.3. Two $n \times n$ matrices A and B are called *similar* if there exists an invertible $n \times n$ matrix Q such that $B = Q^{-1}AQ$.

Theorem 2.3. *Let A and B be similar $n \times n$ matrices. Then $A \otimes A$ and $B \otimes B$ are similar and $A \otimes B$ and $B \otimes A$ are similar.*

The proof is straightforward and left to the reader.

Theorem 2.4. *Let A and B be $m \times m$ matrices and C, D be $n \times n$ matrices. Assume that $[A, B] = 0_m$, $[C, D] = 0_n$, where $[A, B] := AB - BA$ defines the commutator. Then $[A \otimes C, B \otimes D] = 0_{m \cdot n}$.*

Proof. Using $AB = BA$ and $CD = DC$ we have

$$[A \otimes C, B \otimes D] = (A \otimes C)(B \otimes D) - (B \otimes D)(A \otimes C)$$
$$= (AB) \otimes (CD) - (BA) \otimes (DC)$$
$$= (AB) \otimes (CD) - (AB) \otimes (CD) = 0_{m \cdot n}$$

where we used $(A \otimes B)(C \otimes D) = (AC) \otimes (BD)$. $\qquad\square$

From this theorem we find the corollary

Corollary 2.1. *Let A be an $m \times m$ matrix and B be an $n \times n$ matrix. I_m and I_n denote the $m \times m$ and $n \times n$ unit matrices, respectively. Then*

$$[A \otimes I_n, I_m \otimes B] = 0_{m \cdot n}$$

where $0_{m \cdot n}$ is the $(mn) \times (mn)$ zero matrix.

Proof.

$$[A \otimes I_n, I_m \otimes B] = (A \otimes I_n)(I_m \otimes B) - (I_m \otimes B)(A \otimes I_n)$$
$$= A \otimes B - A \otimes B = 0_{m \cdot n}. \qquad \square$$

Using these results we are able to prove the following theorem

Theorem 2.5. *Let A and B be $n \times n$ matrices. I_n denotes the $n \times n$ unit matrix. Then*

$$\exp(A \otimes I_n + I_n \otimes B) \equiv \exp(A) \otimes \exp(B).$$

Proof.   Since $[A \otimes I_n, I_n \otimes B] = 0_{n^2}$ we have

$$\exp(A \otimes I_n + I_n \otimes B) = \exp(A \otimes I_n)\exp(I_n \otimes B).$$

Now

$$\exp(A \otimes I_n) := \sum_{k=0}^{\infty} \frac{(A \otimes I_n)^k}{k!}, \quad \exp(I_n \otimes B) := \sum_{k=0}^{\infty} \frac{(I_n \otimes B)^k}{k!}.$$

An arbitrary term in the expansion of $\exp(A \otimes I_n)\exp(I_n \otimes B)$ is given by

$$\frac{1}{n!}\frac{1}{m!}(A \otimes I_n)^n(I_n \otimes B)^m.$$

Since

$$(A \otimes I_n)^n(I_n \otimes B)^m \equiv (A^n \otimes I_n^n)(I_n^m \otimes B^m) \equiv (A^n \otimes I_n)(I_n \otimes B^m) \equiv (A^n \otimes B^m)$$

we obtain

$$\frac{1}{n!}\frac{1}{m!}(A \otimes I_n)^n(I_n \otimes B)^m \equiv \left(\frac{1}{n!}A^n\right) \otimes \left(\frac{1}{m!}B^m\right).$$

This proves the theorem. $\qquad \square$

An extension of the formula given above is

Theorem 2.6. *Let $A_1, A_2, \ldots, A_r$ be real $n \times n$ matrices and let I_n be the $n \times n$ identity matrix. Then*

$$\exp(A_1 \otimes I_n \otimes \cdots \otimes I_n + I_n \otimes A_2 \otimes \cdots \otimes I_n + \cdots + I_n \otimes \cdots \otimes I_n \otimes A_r)$$
$$\equiv \exp(A_1) \otimes \exp(A_2) \otimes \cdots \otimes \exp(A_r).$$

The proof is left as an exercise for the reader.

We can also consider other analytic functions. For example, we have

$$\cos(I_n \otimes A) \equiv I_n \otimes \cos(A).$$

To prove this identity one uses the expansion

$$\cos(x) := \sum_{k=0}^{\infty} \frac{(-1)^k x^{2k}}{(2k)!}.$$

Analogously we can consider

$$\sin(x) := \sum_{k=0}^{\infty} \frac{(-1)^k x^{2k+1}}{(2k+1)!}.$$

Example 2.17. Let A be an $m \times m$ matrix and B be an $n \times n$ matrix. We have

$$\sin(A \otimes I_n + I_m \otimes B^T) \equiv \sin(A) \otimes \cos(B^T) + \cos(A) \otimes \sin(B^T). \quad \clubsuit$$

Definition 2.4. Let X and Y be $n \times n$ matrices. Then the *anticommutator* $[X, Y]_+$ of X and Y is defined as

$$[X, Y]_+ := XY + YX.$$

The anti-commutator plays an important role for Fermi operators.

Example 2.18. Let

$$\sigma_3 := \begin{pmatrix} 1 & 0 \\ 0 & -1 \end{pmatrix}, \qquad \sigma_+ := \begin{pmatrix} 0 & 2 \\ 0 & 0 \end{pmatrix}$$

and let I_2 be the 2×2 unit matrix. We set $A := \sigma_+ \otimes I_2$, $B := \sigma_3 \otimes \sigma_+$. Since

$$\sigma_+ \sigma_3 = \begin{pmatrix} 0 & -2 \\ 0 & 0 \end{pmatrix}, \qquad \sigma_3 \sigma_+ = \begin{pmatrix} 0 & 2 \\ 0 & 0 \end{pmatrix}$$

we find that $[A, B]_+ = 0_4$. $\quad \clubsuit$

Exercises. (1) Let $\mathbf{u}, \mathbf{v} \in \mathbb{C}^n$, where $\mathbf{u}, \mathbf{v}$ are considered as column vectors. Show that $\mathbf{u}^T \otimes \mathbf{v} = \mathbf{v}\mathbf{u}^T = \mathbf{v} \otimes \mathbf{u}^T$.

(2) Let A be an $n \times n$ Hermitian matrix. Show that $(I_n + iA)^{-1}(I_n - iA)$ is unitary. Is the matrix $(I_n + iA)^{-1} \otimes (I_n - iA)$ unitary?

(3) Let $\mathbf{u}$ and $\mathbf{v}$ be column vectors in $\mathbb{C}^n$. Let A and B be $n \times n$ matrices over $\mathbb{C}$. Show that $(\mathbf{u}^T \otimes B)(A \otimes \mathbf{v}) = (B\mathbf{v})(\mathbf{u}^T A)$.

(4) Let A be an $n \times n$ matrix over $\mathbb{C}$. Let $\mathbf{u}$ be a column vector in $\mathbb{C}^n$. Show that $(A \otimes A)(\mathbf{u} \otimes \mathbf{u}) = (A \otimes (A\mathbf{u}))\mathbf{u}$.

(5) Let A be an $n \times n$ matrix and $\mathbf{u} \in \mathbb{C}^n$. Show that $(I_n \otimes \mathbf{u})A = A \otimes \mathbf{u}$.

(6) Let A be an $n \times n$ matrix. Let $\mathbf{u}, \mathbf{v} \in \mathbb{C}^n$. Show that

$$A(\mathbf{u} \otimes \mathbf{v}^T)A = (A\mathbf{u}) \otimes (\mathbf{v}^T A).$$

(7) Let A, B be $n \times n$ matrices. Show that $(A+B)^{-1} = (I_n + A^{-1}B)^{-1}A^{-1}$, where it is assumed that the inverse matrices exist. Find

$$(A + B)^{-1} \otimes (A + B)^{-1}.$$

(8) Let A be an $n \times n$ matrix. Assume that A^{-1} exists. Let $\mathbf{u}, \mathbf{v} \in \mathbb{C}^n$. We define $\lambda := \mathbf{v}^T A^{-1}\mathbf{u}$ and assume that $|\lambda| < 1$. Show that

$$(A + \mathbf{u} \otimes \mathbf{v}^T)^{-1} = A^{-1} - \frac{(A^{-1}\mathbf{u}) \otimes (\mathbf{v}^T A^{-1})}{1 + \lambda}.$$

(9) (i) Let A be an arbitrary $n \times n$ matrix over $\mathbb{C}$. Show that

$$\exp(A \otimes I_n) \equiv \exp(A) \otimes I_n.$$

(ii) Let A be an $n \times n$ matrix and I_m be the $m \times m$ identity matrix. Show that

$$\sin(A \otimes I_m) \equiv \sin(A) \otimes I_m.$$

(iii) Let A be an $n \times n$ matrix and B be an $m \times m$ matrix. Is

$$\sin(A \otimes I_m + I_n \otimes B) \equiv (\sin(A)) \otimes (\cos(B)) + (\cos(A)) \otimes (\sin(B))?$$

(10) Let A, B be $n \times n$ matrices. Then the direct sum $A \oplus B$ is the $2n \times 2n$ matrix

$$A \oplus B = \begin{pmatrix} A & 0_n \\ 0_n & B \end{pmatrix}$$

where 0_n is the $n \times n$ zero matrix.

(i) Show that $A \oplus B$ can be written using the Kronecker product and addition of matrices.

(ii) Let C be a 2×2 matrix. Can $C \oplus C$ be written as Kronecker product of two 2×2 matrices?

(11) (i) Let A, B be $n \times n$ matrices with $A^2 = B^2 = I_n$. Show that

$$\exp(-i\omega t(A \otimes B)) = \cos(\omega t)(I_n \otimes I_n) - i\sin(\omega t)(A \otimes B).$$

Note that $(A \otimes B)^2 = I_n \otimes I_n$.

(ii) Let A, B, C be $n \times n$ matrices with $A^2 = I_n$, $B^2 = I_n$, $C^2 = I_n$. Show that

$$\exp(-i\omega t(A \otimes B \otimes C)) = \cos(\omega t)(I_n \otimes I_n \otimes I_n) - i\sin(\omega t)(A \otimes B \otimes C).$$

(12) Let $\phi \in \mathbb{R}$ and $J(\phi) = \exp(i\phi(\sigma_1 \otimes \sigma_1 \otimes \sigma_1)/2)$. Show that

$$J(\phi) = (I_2 \otimes I_2 \otimes I_2)\cos(\phi/2) + i\sigma_1 \otimes \sigma_1 \otimes \sigma_1 \sin(\phi/2).$$

Find $J(\phi = \pi/2)$.

2.4   Permutation Matrices

Permutation matrices have been introduced in section 1.13. In this section we describe the connection with the Kronecker product.

Theorem 2.7. *Let P and Q be permutation matrices of order n and m, respectively. Then $P \otimes Q$ and $Q \otimes P$ are permutation matrices of order nm.*

The proof is straightforward and is left to the reader.

Example 2.19. Let

$$P = \begin{pmatrix} 0 & 1 \\ 1 & 0 \end{pmatrix}, \qquad Q = \begin{pmatrix} 0 & 1 & 0 \\ 1 & 0 & 0 \\ 0 & 0 & 1 \end{pmatrix}$$

be two permutation matrices. Then

$$P \otimes Q = \begin{pmatrix} 0 & 0 & 0 & 0 & 1 & 0 \\ 0 & 0 & 0 & 1 & 0 & 0 \\ 0 & 0 & 0 & 0 & 0 & 1 \\ 0 & 1 & 0 & 0 & 0 & 0 \\ 1 & 0 & 0 & 0 & 0 & 0 \\ 0 & 0 & 1 & 0 & 0 & 0 \end{pmatrix}, \qquad Q \otimes P = \begin{pmatrix} 0 & 0 & 0 & 1 & 0 & 0 \\ 0 & 0 & 1 & 0 & 0 & 0 \\ 0 & 1 & 0 & 0 & 0 & 0 \\ 1 & 0 & 0 & 0 & 0 & 0 \\ 0 & 0 & 0 & 0 & 0 & 1 \\ 0 & 0 & 0 & 0 & 1 & 0 \end{pmatrix}$$

are permutation matrices. We see that $P \otimes Q \neq Q \otimes P$.   ♣

Theorem 2.8. *Let A be an $m \times n$ matrix and B be a $p \times q$ matrix, respectively. Then there exist permutation matrices P and Q such that*

$$B \otimes A = P(A \otimes B)Q.$$

Proof. Let

$$P_{p,m} := \sum_{j=1}^{m} \sum_{k=1}^{p} (\mathbf{e}_{k,p} \otimes \mathbf{e}_{j,m})(\mathbf{e}_{j,m} \otimes \mathbf{e}_{k,p})^{T}$$

and similarly

$$P_{n,q} := \sum_{j=1}^{q} \sum_{k=1}^{n} (\mathbf{e}_{k,n} \otimes \mathbf{e}_{j,q})(\mathbf{e}_{j,q} \otimes \mathbf{e}_{k,n})^{T}.$$

We find $P = P_{p,m}$ and $Q = P_{n,q}$. Thus

$$P(A \otimes B)Q = P_{p,m} \sum_{j=1}^{m} \sum_{k=1}^{n} \sum_{u=1}^{p} \sum_{v=1}^{q} (A)_{j,k}(B)_{u,v}(\mathbf{e}_{j,m}\mathbf{e}_{k,n}^{T}) \otimes (\mathbf{e}_{u,p}\mathbf{e}_{v,q}^{T})P_{n,q}$$

$$= P_{p,m} \sum_{j=1}^{m} \sum_{k=1}^{n} \sum_{u=1}^{p} \sum_{v=1}^{q} (A)_{j,k}(B)_{u,v}(\mathbf{e}_{j,m} \otimes \mathbf{e}_{u,p})(\mathbf{e}_{k,n}^{T} \otimes \mathbf{e}_{v,q}^{T})P_{n,q}$$

$$= \sum_{j=1}^{m} \sum_{k=1}^{n} \sum_{u=1}^{s} \sum_{v=1}^{t} (A)_{j,k}(B)_{u,v}P_{s,m}(\mathbf{e}_{j,m} \otimes \mathbf{e}_{u,p})(\mathbf{e}_{k,n}^{T} \otimes \mathbf{e}_{v,q}^{T})P_{n,q}$$

$$= \sum_{j=1}^{m} \sum_{k=1}^{n} \sum_{u=1}^{s} \sum_{v=1}^{t} (A)_{j,k}(B)_{u,v}(\mathbf{e}_{u,p} \otimes \mathbf{e}_{j,m})(\mathbf{e}_{v,q}^{T} \otimes \mathbf{e}_{k,n}^{T})$$

$$= \sum_{j=1}^{m} \sum_{k=1}^{n} \sum_{u=1}^{s} \sum_{v=1}^{t} (A)_{j,k}(B)_{u,v}(\mathbf{e}_{u,p}\mathbf{e}_{v,q}^{T} \otimes \mathbf{e}_{j,m}\mathbf{e}_{k,n}^{T})$$

$$= \left(\sum_{u=1}^{s} \sum_{v=1}^{t} (B)_{u,v}\mathbf{e}_{u,p}\mathbf{e}_{v,q}^{T} \right) \otimes \left(\sum_{j=1}^{m} \sum_{k=1}^{n} (A)_{j,k}\mathbf{e}_{j,m}\mathbf{e}_{k,n}^{T} \right)$$

$$= B \otimes A. \qquad \square$$

Remark. The Kronecker product $B \otimes A$ contains the same entries as $A \otimes B$ only in another order.

Example 2.20. Let

$$A = \begin{pmatrix} a_{11} & a_{12} \\ a_{21} & a_{22} \end{pmatrix}, \qquad B = \begin{pmatrix} b_{11} & b_{12} \\ b_{21} & b_{22} \end{pmatrix}.$$

Then

$$A \otimes B = \begin{pmatrix} a_{11}b_{11} & a_{11}b_{12} & a_{12}b_{11} & a_{12}b_{12} \\ a_{11}b_{21} & a_{11}b_{22} & a_{12}b_{21} & a_{12}b_{22} \\ a_{21}b_{11} & a_{21}b_{12} & a_{22}b_{11} & a_{22}b_{12} \\ a_{21}b_{21} & a_{21}b_{22} & a_{22}b_{21} & a_{22}b_{22} \end{pmatrix}$$

and

$$B \otimes A = \begin{pmatrix} b_{11}a_{11} & b_{11}a_{12} & b_{12}a_{11} & b_{12}a_{12} \\ b_{11}a_{21} & b_{11}a_{22} & b_{12}a_{21} & b_{12}a_{22} \\ b_{21}a_{11} & b_{21}a_{12} & b_{22}a_{11} & b_{22}a_{12} \\ b_{21}a_{21} & b_{21}a_{22} & b_{22}a_{21} & b_{22}a_{22} \end{pmatrix}.$$

With

$$P = Q = \begin{pmatrix} 1 & 0 & 0 & 0 \\ 0 & 0 & 1 & 0 \\ 0 & 1 & 0 & 0 \\ 0 & 0 & 0 & 1 \end{pmatrix}$$

we have $B \otimes A = P(A \otimes B)Q$. ♣

We have mentioned that the $n \times n$ permutation matrices form a finite group and the number of group elements is given by $n! \equiv 1 \cdot 2 \cdot 3 \cdot \ldots \cdot n$. The set of all 2×2 permutation matrices are given by

$$\left\{ I = \begin{pmatrix} 1 & 0 \\ 0 & 1 \end{pmatrix}, \quad P = \begin{pmatrix} 0 & 1 \\ 1 & 0 \end{pmatrix} \right\}$$

with $II = I$, $IP = P$, $PI = P$, $PP = I$. Then we obtain

$$I \otimes I = \begin{pmatrix} 1 & 0 & 0 & 0 \\ 0 & 1 & 0 & 0 \\ 0 & 0 & 1 & 0 \\ 0 & 0 & 0 & 1 \end{pmatrix}, \quad I \otimes P = \begin{pmatrix} 0 & 1 & 0 & 0 \\ 1 & 0 & 0 & 0 \\ 0 & 0 & 0 & 1 \\ 0 & 0 & 1 & 0 \end{pmatrix},$$

$$P \otimes I = \begin{pmatrix} 0 & 0 & 1 & 0 \\ 0 & 0 & 0 & 1 \\ 1 & 0 & 0 & 0 \\ 0 & 1 & 0 & 0 \end{pmatrix}, \quad P \otimes P = \begin{pmatrix} 0 & 0 & 0 & 1 \\ 0 & 0 & 1 & 0 \\ 0 & 1 & 0 & 0 \\ 1 & 0 & 0 & 0 \end{pmatrix}.$$

Since

$$(I \otimes I)(I \otimes I) = I \otimes I$$
$$(I \otimes I)(I \otimes P) = (I \otimes P)(I \otimes I) = I \otimes P$$
$$(I \otimes I)(P \otimes I) = (P \otimes I)(I \otimes I) = P \otimes I$$
$$(P \otimes P)(I \otimes I) = (I \otimes I)(P \otimes P) = P \otimes P$$
$$(P \otimes P)(I \otimes P) = (I \otimes P)(P \otimes P) = P \otimes I$$
$$(P \otimes I)(P \otimes P) = (P \otimes P)(P \otimes I) = I \otimes P$$
$$(P \otimes P)(P \otimes P) = I \otimes I$$

we find that the set
$$\{\,I \otimes I, I \otimes P, P \otimes I, P \otimes P\,\}$$
forms a group under matrix multiplication. It is a subgroup of the group of the 4×4 permutation matrices. This finite group consists of 24 matrices.

Exercises. (1) Let $\{\,P_1, \dots, P_r\,\}$ be the set of all $n \times n$ permutation matrices, i.e. $r = n!$. Show that
$$\{\,P_1 \otimes P_1, \dots, P_1 \otimes P_r, P_2 \otimes P_1, \dots, P_r \otimes P_r\,\}$$
are permutation matrices and form a group under matrix multiplication. Show that they form a subgroup of the $(n \cdot n) \times (n \cdot n)$ permutation matrices.

(2) Let P and Q be permutation matrices. Is $(P \otimes Q)(Q \otimes P)$ a permutation matrix? Is $(P \otimes P)(Q \otimes Q)$ a permutation matrix?

(3) Let P be an $n \times n$ permutation matrix with $\operatorname{tr}(P) = k$, where $k \in \mathbb{N}_0$. Let Q be an $m \times m$ with $\operatorname{tr}(Q) = l$, where $l \in \mathbb{N}_0$. Find $\operatorname{tr}(P \otimes Q)$.

(4) Let P be an $n \times n$ permutation matrix and Π be an $m \times m$ projection matrix. Is $P \otimes \Pi$ a projection matrix? Is $P \otimes \Pi \otimes P^{-1}$ a projection matrix?

(5) Consider the permutation matrix
$$P = \begin{pmatrix} 0 & 1 & 0 \\ 0 & 0 & 1 \\ 1 & 0 & 0 \end{pmatrix}.$$
Find P^2, P^3, $P \otimes P$, $(P \otimes P)^2$, and $(P \otimes P)^3$. Discuss the result. Show that P can be written as $P = M_1 + \omega^2 M_2 + \omega M_3$, where $\omega := \exp(2\pi i/3)$ and
$$M_1 := \frac{1}{3}\begin{pmatrix} 1 & 1 & 1 \\ 1 & 1 & 1 \\ 1 & 1 & 1 \end{pmatrix}, \quad M_2 := \frac{1}{3}\begin{pmatrix} 1 & \omega & \omega^2 \\ \omega^2 & 1 & \omega \\ \omega & \omega^2 & 1 \end{pmatrix}, \quad M_3 := \frac{1}{3}\begin{pmatrix} 1 & \omega^2 & \omega \\ \omega & 1 & \omega^2 \\ \omega^2 & \omega & 1 \end{pmatrix}.$$
Show that M_1, M_2 and M_3 are projection matrices.

(6) Can the permutation matrix
$$\begin{pmatrix} 0_n & I_n \\ I_n & 0_n \end{pmatrix}$$
be written as a Kronecker product of two permutation matrices?

(7) Let A be a 2×2 matrix. Can one find a 4×4 matrix P such that
$$P(I_2 \otimes A) = A \otimes I_2\,?$$

2.5   Trace and Determinant

In this section we describe the connection of the Kronecker product and trace and determinant of square matrices.

Let A be an $m \times m$ matrix and B be an $n \times n$ matrix. Then the matrix $A \otimes B$ is an $(mn) \times (mn)$ matrix. So we can calculate $\operatorname{tr}(A \otimes B)$ and we find the following theorem

Theorem 2.9. *Let A be an $m \times m$ matrix and B be an $n \times n$ matrix. Then*

$$\operatorname{tr}(A \otimes B) \equiv (\operatorname{tr}(A))(\operatorname{tr}(B)).$$

Proof.   From the definition of the Kronecker product we find

$$\operatorname{diag}(a_{jj}B) = (a_{jj}b_{11}, a_{jj}b_{22}, \dots, a_{jj}b_{nn}).$$

Therefore

$$\operatorname{tr}(A \otimes B) = \sum_{j=1}^{m} \sum_{k=1}^{n} a_{jj} b_{kk}.$$

Since $\operatorname{tr}(A) = \sum_{j=1}^{m} a_{jj}$, $\operatorname{tr}(B) = \sum_{k=1}^{n} b_{kk}$ we find the identity.   □

In statistical mechanics the following theorem plays an important role.

Theorem 2.10. *Let A, B be two arbitrary $n \times n$ matrixes. Let I_n be the $n \times n$ unit matrix. Then*

$$\operatorname{tr}(\exp(A \otimes I_n + I_n \otimes B)) \equiv (\operatorname{tr}(\exp(A)))(\operatorname{tr}(\exp(B))).$$

Proof.   Since $[A \otimes I_n, I_n \otimes B] = 0_{n^2}$ we have

$$\exp(A \otimes I_n + I_n \otimes B) = \exp(A) \otimes \exp(B).$$

Taking the trace on both sides the result follows.   □

An extension of the above formula is

$$\operatorname{tr}(\exp(A_1 \otimes I_n \otimes \cdots \otimes I_n + I_n \otimes A_2 \otimes \cdots \otimes I_n + \cdots + I_n \otimes \cdots \otimes I_n \otimes A_r))$$
$$\equiv (\operatorname{tr}(\exp(A_1)))(\operatorname{tr}(\exp(A_2))) \cdots \operatorname{tr}(\exp(A_r)).$$

Theorem 2.11. *Let A_1, A_2, B_1, B_2 be real $n \times n$ matrices and let I be the $n \times n$ identity matrix. Then*

$$\operatorname{tr}(\exp(A_1 \otimes I_n \otimes B_1 \otimes I_n + I_n \otimes A_2 \otimes I_n \otimes B_2))$$
$$\equiv (\operatorname{tr}(\exp(A_1 \otimes B_1)))(\operatorname{tr}(\exp(A_2 \otimes B_2))).$$

Proof. There is an $n^2 \times n^2$ permutation matrix P such that, for any $n \times n$ matrices M and N, $P(M \otimes N)P^T = N \otimes M$. Define an $n^4 \times n^4$ permutation matrix Q by $Q := I_n \otimes P \otimes I_n$. It follows that

$$Q(A_1 \otimes B_1 \otimes I_n \otimes I_n + I_n \otimes I_n \otimes A_2 \otimes B_2)Q^T \equiv A_1 \otimes I_n \otimes B_1 \otimes I_n + I_n \otimes A_2 \otimes I_n \otimes B_2.$$

Since $Q^T Q = I_n$ there is a similarity between the matrices

$$A_1 \otimes B_1 \otimes I_n \otimes I_n + I_n \otimes I_n \otimes A_2 \otimes B_2$$

and

$$A_1 \otimes I_n \otimes B_1 \otimes I_n + I_n \otimes A_2 \otimes I_n \otimes B_2.$$

So the two matrices have the same eigenvalues. Consequently, the same is true of the exponential function evaluated at these matrices and, in particular, they have the same trace.   □

An extension of the formula given above is

Theorem 2.12. *Let* $A_1, A_2, \ldots, A_r, B_1, B_2, \ldots, B_r$ *be real* $n \times n$ *matrices and let* I_n *be the* $n \times n$ *identity matrix. Then*

$$\begin{aligned}
\mathrm{tr}(\exp(A_1 \otimes I_n \otimes &\cdots \otimes I_n \otimes B_1 \otimes I_n \otimes \cdots \otimes I_n \\
&+ I_n \otimes A_2 \otimes \cdots \otimes I_n \otimes I_n \otimes B_2 \otimes \cdots \otimes I_n + \cdots \\
&+ I_n \otimes I_n \otimes \cdots \otimes A_r \otimes I_n \otimes I_n \otimes \cdots \otimes B_r)) \\
&= (\mathrm{tr}(\exp(A_1 \otimes B_1)))(\mathrm{tr}(\exp(A_2 \otimes B_2)))\ldots(\mathrm{tr}(\exp(A_r \otimes B_r))).
\end{aligned}$$

The proof of this theorem is completely parallel to that of theorem given above (Steeb [54]).

Let us now consider determinants.

Theorem 2.13. *Let* A *be an* $m \times m$ *matrix and* B *be an* $n \times n$ *matrix. Then*

$$\det(A \otimes B) \equiv (\det(A))^n (\det(B))^m.$$

Proof. Let I_m be the $m \times m$ unit matrix. Then

$$\det(I_m \otimes B) \equiv (\det(B))^m$$

and

$$\det(A \otimes I_n) \equiv \det(P(I_n \otimes A)P^T) \equiv (\det(P))^2 (\det(A))^n \equiv (\det(A))^n$$

where P is a permutation matrix. Now using $A \otimes B \equiv (A \otimes I_n)(I_m \otimes B)$ we obtain

$$\det(A \otimes B) \equiv \det[(A \otimes I_n)(I_m \otimes B)] \equiv \det(A \otimes I_n)\det(I_m \otimes B)$$
$$\equiv (\det(A))^n (\det(B))^m.$$ □

Another proof of this theorem with the help of the eigenvalues of $A \otimes B$ will be given in section 2.6.

Exercises. (1) Let P be an $n \times n$ permutation matrix. Show that
$$\det(P \otimes P) = 1.$$

(2) Let A be an $m \times m$ matrix and B be an $n \times n$ matrix. Show that
$$\det(\exp(A \otimes B)) \equiv \exp((\operatorname{tr}(A))(\operatorname{tr}(B))).$$
We know that $\det(e^A) \equiv \exp(\operatorname{tr}(A))$ for any $n \times n$ matrix A.

(3) Let A be an $n \times n$ idempotent matrix. Find $\det(A \otimes A)$.

(4) Let A be an $n \times n$ nilpotent matrix. Find $\det(A \otimes A)$.

(5) Let A be an $n \times n$ matrix over $\mathbb{R}$. Assume that A is skew-symmetric. Let B be an $(n+1) \times (n+1)$ skew-symmetric matrix over $\mathbb{R}$. Show that
$$\det(A \otimes B) = 0.$$

(6) Let U be a unitary matrix. Show that $\det(U \otimes U \otimes U \otimes U) = 1$.

(7) Let A be an $n \times n$ matrix with $\det(A) = 1$. Show that $\det(A^{-1} \otimes A) = 1$.

(8) Let A, B be $n \times n$ matrices. Assume that $[A, B]_+ = 0_n$. Find $\det(AB)$.

(9) Let A be an $n \times n$ matrix and I_n be the $n \times n$ identity matrix. Find
$$\operatorname{tr}(A \otimes I_n + I_n \otimes A), \qquad \operatorname{tr}(A \otimes I_n - I_n \otimes A).$$

(10) Let A, B be $n \times n$ matrices. Assume that $A \otimes B + B \otimes A = 0_{n^2}$. Find $\det(A \otimes B)$.

(11) Let A, B be $n \times n$ matrices. Show that $\operatorname{tr}(A \otimes B - B \otimes A) = 0$. Show that $\det(A \otimes B - B \otimes A) = 0$.

(12) Let A be an invertible matrix. Assume that $A = A^{-1}$. What are the possible values for $\det(A)$? What are the possible values for $\det(A \otimes A)$?

(13) Consider the vectors $\mathbf{v}$ and $\mathbf{w}$ in $\mathbb{C}^2$
$$\mathbf{v} = \begin{pmatrix} v_1 \\ v_2 \end{pmatrix}, \quad \mathbf{w} = \begin{pmatrix} w_1 \\ w_2 \end{pmatrix}.$$

Calculate $\mathbf{v} \otimes \mathbf{w}$, $\mathbf{v}^T \otimes \mathbf{w}$, $\mathbf{v} \otimes \mathbf{w}^T$, $\mathbf{v}^T \otimes \mathbf{w}^T$. Is $\mathrm{tr}(\mathbf{v}^T \otimes \mathbf{w}) = \mathrm{tr}(\mathbf{v} \otimes \mathbf{w}^T)$? Is $\det(\mathbf{v}^T \otimes \mathbf{w}) = \det(\mathbf{v} \otimes \mathbf{w}^T)$?

(14) Let A be a 2×2 matrix over $\mathbb{C}$. Find all solutions of the equation

$$\mathrm{tr}(A + A^*) = \mathrm{tr}(AA^*).$$

Find all solutions of the equation

$$\mathrm{tr}(A \otimes I_2 + I_2 \otimes A^*) = \mathrm{tr}(A \otimes A^*).$$

(15) Let $d \geq 2$ and $|0\rangle$, $|1\rangle$, ..., $|d-1\rangle$ be an orthonormal basis in the Hilbert space $\mathbb{C}^d$. Consider the Bell state in $\mathbb{C}^d$

$$|\psi\rangle = \frac{1}{\sqrt{d}} \sum_{j=0}^{d-1} |j\rangle \otimes |j\rangle.$$

For any $d \times d$ matrix over $\mathbb{C}$ we define

$$|v(A)\rangle = (A \otimes I_d)|\psi\rangle.$$

Let B be another $d \times d$ matrix over $\mathbb{C}$. Show that

$$\langle v(A)|v(B)\rangle = \frac{1}{d}\langle A, B\rangle \equiv \frac{1}{d}\mathrm{tr}(A^*B).$$

2.6   Eigenvalue Problem

Let A be an $m \times m$ matrix and let B be an $n \times n$ matrix. We now investigate the eigenvalue problem for the matrices $A \otimes B$ and $A \otimes I_n + I_m \otimes B$.

Theorem 2.14. *Let A be an $m \times m$ matrix with eigenvalue λ and the corresponding eigenvector $\mathbf{u}$. Let B be an $n \times n$ matrix with eigenvalue μ and the corresponding eigenvector $\mathbf{v}$. Then the matrix $A \otimes B$ has the eigenvalue $\lambda\mu$ with the corresponding eigenvector $\mathbf{u} \otimes \mathbf{v}$.*

Proof.   From the eigenvalue equations

$$A\mathbf{u} = \lambda\mathbf{u}, \qquad B\mathbf{v} = \mu\mathbf{v}$$

we obtain

$$(A\mathbf{u}) \otimes (B\mathbf{v}) = \lambda\mu(\mathbf{u} \otimes \mathbf{v}).$$

Since $(A\mathbf{u}) \otimes (B\mathbf{v}) \equiv (A \otimes B)(\mathbf{u} \otimes \mathbf{v})$ we arrive at

$$(A \otimes B)(\mathbf{u} \otimes \mathbf{v}) = \lambda\mu(\mathbf{u} \otimes \mathbf{v}).$$

This equation is an eigenvalue equation. Consequently, $\mathbf{u} \otimes \mathbf{v}$ is an eigenvector of $A \otimes B$ with eigenvalue $\lambda\mu$. $\square$

Example 2.21. Let

$$A = \begin{pmatrix} 0 & -i \\ i & 0 \end{pmatrix}, \quad B = \begin{pmatrix} 1 & 0 \\ 0 & -1 \end{pmatrix}.$$

Since the eigenvalues of A and B are given by $\{1, -1\}$, we find that the eigenvalues of $A \otimes B$ are given by $\{1, 1, -1, -1\}$.   ♣

Let A be an $m \times m$ matrix with m eigenvalues $\lambda_1, \ldots, \lambda_m$ and let B be an $n \times n$ matrix with n eigenvalues $\mu_1, \ldots, \mu_n$. Then

$$\det(A \otimes B) = (\det(A))^n (\det(B))^m.$$

The proof can be given using the theorem described above. First we note that $\det(A) = \lambda_1 \lambda_2 \cdots \lambda_m$ and $\det(B) = \mu_1 \mu_2 \cdots \mu_n$. We find

$$\det(A \otimes B) = \prod_{j=1}^{m} \prod_{k=1}^{n} (\lambda_j \mu_k) = \left(\prod_{j=1}^{m} \lambda_j \right)^n \left(\prod_{k=1}^{n} \mu_k \right)^m = (\det(A))^n (\det(B))^m.$$

Theorem 2.15. *Let A be an $m \times m$ matrix with eigenvalue λ and corresponding eigenvector $\mathbf{u}$. Let B be an $n \times n$ matrix with eigenvalue μ and corresponding eigenvector $\mathbf{v}$. Then*

$$A \otimes I_n + I_m \otimes B$$

has the eigenvalue $\lambda + \mu$ with corresponding eigenvector $\mathbf{u} \otimes \mathbf{v}$, where I_m is the $m \times m$ unit matrix and I_n is the $n \times n$ unit matrix.

Proof. We have

$$
\begin{aligned}
(A \otimes I_n + I_m \otimes B)(\mathbf{u} \otimes \mathbf{v}) &= (A \otimes I_n)(\mathbf{u} \otimes \mathbf{v}) + (I_m \otimes B)(\mathbf{u} \otimes \mathbf{v}) \\
&= (A\mathbf{u}) \otimes (I_n \mathbf{v}) + (I_m \mathbf{u}) \otimes (B\mathbf{v}) \\
&= (\lambda \mathbf{u}) \otimes \mathbf{v} + \mathbf{u} \otimes (\mu \mathbf{v}) \\
&= (\lambda + \mu)(\mathbf{u} \otimes \mathbf{v}).
\end{aligned}
$$

$\square$

Example 2.22. Let

$$A = \begin{pmatrix} 0 & -i \\ i & 0 \end{pmatrix}, \quad B = \begin{pmatrix} 1 & 0 \\ 0 & -1 \end{pmatrix}.$$

Since the eigenvalues of A and B are given by $\{1, -1\}$, we find that the eigenvalues of $A \otimes I_2 + I_2 \otimes B$ are given by $\{2, 0, 0, -2\}$.   ♣

The expression $A \otimes I_n + I_m \otimes B$ is sometimes called the *Kronecker sum*. The Kronecker sum plays a role when we consider the equation

$$AX + XB = C$$

where A is an $m \times m$ matrix, B is an $n \times n$ matrix, X is an $m \times n$ matrix and C is an $m \times n$ matrix. This equation can be written in the form

$$(I_n \otimes A + B^T \otimes I_m)\mathrm{vec}(X) = \mathrm{vec}(C)$$

where we used the identity $\mathrm{vec}(AYB) \equiv (B^T \otimes A)\mathrm{vec}(Y)$.

Exercises. (1) Let A and B be $n \times n$ matrices with eigenvalues $\lambda_1, \lambda_2, \ldots, \lambda_n$ and $\mu_1, \mu_2, \ldots, \mu_n$, respectively.
(i) Find the eigenvalues of $A \otimes B - B \otimes A$.
(ii) Find the eigenvalues of $A \otimes B + B \otimes A$.

(2) Let A, B, C be $n \times n$ matrices. $\lambda_1, \lambda_2, \ldots, \lambda_n$, $\mu_1, \mu_2, \ldots, \mu_n$ and $\nu_1, \nu_2, \ldots, \nu_n$, respectively. Find the eigenvalues of $A \otimes B \otimes C$.

(3) Let A be an $m \times m$ matrix with eigenvalue λ and eigenvector $\mathbf{u}$. Let B be an $n \times n$ matrix with eigenvalue μ and eigenvector $\mathbf{v}$. Show that

$$\exp(A \otimes B)(\mathbf{u} \otimes \mathbf{v}) \equiv \exp(\lambda\mu)(\mathbf{u} \otimes \mathbf{v}).$$

(4) Let A, B, C be $n \times n$ matrices. Let I_n be the $n \times n$ unit matrix. Find the eigenvalues of

$$A \otimes I_n \otimes I_n + I_n \otimes B \otimes I_n + I_n \otimes I_n \otimes C.$$

(5) Let A be an $n \times n$ matrix. Let $\lambda_1, \lambda_2, \ldots, \lambda_n$ be the eigenvalues of A. Assume that A^{-1} exists. Show that

$$\mathrm{tr}(A^{-1}) = \sum_{j=1}^{n} \lambda_j^{-1}.$$

Find the eigenvalues of $A^{-1} \otimes A$.

(6) Let A be an $n \times n$ matrix with eigenvalues $\lambda_1, \lambda_2, \ldots, \lambda_n$. Let B be an $n \times n$ matrix with eigenvalues $\mu_1, \mu_2, \ldots, \mu_n$. Let I_n be the $n \times n$ unit matrix and $\epsilon \in \mathbb{R}$. Show that

$$I_n \otimes I_n + \epsilon(A \otimes I_n + I_n \otimes B) + \epsilon^2(A \otimes B)$$

has the eigenvalues $(1 + \epsilon\lambda_j)(1 + \epsilon\mu_k)$.

(7) (i) Consider the 2×2 matrix

$$A := (u_1, u_2) \otimes \begin{pmatrix} u_1 \\ u_2 \end{pmatrix}.$$

Show that at least one eigenvalue of A is equal to zero. Hint: Calculate $\det(A)$ and use the fact that $\det(A) = \lambda_1 \lambda_2$.

(ii) Consider the 3×3 matrix

$$B := (u_1, u_2, u_3) \otimes \begin{pmatrix} u_1 \\ u_2 \\ u_3 \end{pmatrix}.$$

Show that at least one eigenvalue of B is equal to zero. Hint: Calculate $\det(B)$ and use the fact that $\det(B) = \lambda_1 \lambda_2 \lambda_3$. Generalize to the n dimensional case.

(8) Let $f : \mathbb{R}^2 \to \mathbb{R}$ designate the polynomial

$$f(x, y) = \sum_{j=0}^{p} \sum_{k=0}^{p} a_{jk} x^j y^k.$$

Let A be an $m \times m$ matrix and B be an $n \times n$ matrix. Let $f(A, B)$ designate the $mn \times mn$ matrix

$$f(A, B) := \sum_{j=0}^{p} \sum_{k=0}^{p} a_{jk} A^j \otimes B^k.$$

Show that the eigenvalues of $f(A, B)$ are given by

$$f(\lambda_r, \mu_s), \qquad r = 1, 2, \ldots, m, \qquad s = 1, 2, \ldots, n$$

where λ_r and μ_s are the eigenvalues of A and B, respectively.

(9) Let A be an $n \times n$ matrix with eigenvalues $\lambda_1, \lambda_2, \ldots, \lambda_n$. Let I_n be the $n \times n$ unit matrix. Find the eigenvalues of $I_n \otimes A$, $A \otimes I_n$, $A \otimes I_n \otimes I_n$, $I_n \otimes A \otimes I_n$ and $I_n \otimes I_n \otimes A$. Extend to n-factors.

(10) Let $a, b \in \mathbb{R}$. Show that the eigenvalues of the $n \times n$ symmetric matrix

$$A = \begin{pmatrix} a & b & 0 & \ldots & 0 \\ b & a & b & \ldots & 0 \\ 0 & b & a & \ldots & 0 \\ \vdots & & \ddots & & b \\ 0 & \ldots & & b & a \end{pmatrix}$$

are given by

$$\lambda_j = a + 2b \cos\left(\frac{\pi j}{n+1}\right)$$

where $j = 1, 2, \ldots, n$. Find the eigenvalues of $A \otimes A$, $A \otimes I_n$ and $I_n \otimes A$.

(11) (i) Find the eigenvalues and normalized eigenvectors of the 2×2 matrix

$$A(\theta) = \begin{pmatrix} \sin(\theta) & \cos(\theta) \\ -\cos(\theta) & \sin(\theta) \end{pmatrix}.$$

Find the eigenvalues and normalized eigenvectors of $A(\theta) \otimes A(\theta)$.
(ii) Find the eigenvalues and normalized eigenvectors of the 2×2 matrix

$$B(\theta, \phi) = \begin{pmatrix} \cos(\theta) & -e^{i\phi} \sin(\theta) \\ e^{i\phi} \sin(\theta) & \cos(\theta) \end{pmatrix}.$$

Find the eigenvalues and normalized eigenvectors of $B(\theta, \phi) \otimes B(\theta, \phi)$.

(12) Let A be an invertible $n \times n$ matrix. Given the eigenvalues and eigenvectors of A. What can be said about the eigenvalues and eigenvectors of $A \otimes A^{-1} + A^{-1} \otimes A$? Find $\det(A \otimes A^{-1})$ and $\operatorname{tr}(A \otimes A^{-1})$.

(13) Let A, B be $n \times n$ matrices over $\mathbb{C}$. Let $\mathbf{u}$ be an eigenvector of A and $\mathbf{v}$ be an eigenvector of B, respectively, i.e. $A\mathbf{u} = \lambda\mathbf{u}$, $B\mathbf{v} = \mu\mathbf{v}$. Find

$$\exp(A \otimes B)(\mathbf{u} \otimes \mathbf{v}), \quad \exp(A \otimes I_n + I_n \otimes B)(\mathbf{u} \otimes \mathbf{v}),$$

$$\exp(A \otimes I_n + I_n \otimes B + A \otimes B)(\mathbf{u} \otimes \mathbf{v}).$$

(14) Let A be an invertible $m \times m$ matrix and B be an $n \times n$ invertible matrix. Show that $+1$ is an eigenvalue of

$$A \otimes A^{-1} \otimes B \otimes B^{-1} \quad \text{and} \quad A \otimes B \otimes A^{-1} \otimes B^{-1}.$$

(15) Let A be an $n \times n$ matrix and B be an $m \times m$ matrix both over $\mathbb{C}$. Consider the $(n \cdot m) \times (n \cdot m)$ matrix

$$M = A \otimes I_m + I_n \otimes B + A \otimes B.$$

If λ is an eigenvalue of A with normalized eigenvector $\mathbf{v}$ and μ is an eigenvalue of B with eigenvector $\mathbf{u}$, then $\lambda + \mu + \lambda \cdot \mu$ is an eigenvalue of M with normalized eigenvector $\mathbf{v} \otimes \mathbf{u}$. Let H be a given 4×4 hermitian matrix. Can one find hermitian 2×2 matrices X and Y such that

$$H = X \otimes I_2 + I_2 \otimes Y + X \otimes Y ?$$

(16) Let σ_1, σ_2, σ_3 be the Pauli spin matrices. Show that the hermitian matrices

$$K_1 = \sigma_1 \otimes \sigma_1 + \sigma_2 \otimes \sigma_2 + \sigma_3 \otimes \sigma_3, \quad K_2 = \sigma_1 \otimes \sigma_2 + \sigma_2 \otimes \sigma_3 + \sigma_3 \otimes \sigma_1$$

admit the same eigenvalues, namely -3 (1 ×) and $+1$ (3 ×). What about the eigenvectors?

(17) Consider the Pauli spin matrices σ_1, σ_2, σ_3 and the Hadamard matrix $U_H = \frac{1}{\sqrt{2}}(\sigma_1 + \sigma_3)$ which is hermitian and unitary. Find the eigenvalues and normalized eigenvectors of the Hamilton operator

$$\hat{H} = \hbar\omega_1(U_H \otimes I_2) + \hbar\omega_2(I_2 \otimes U_H) + \hbar\omega_3(U_H \otimes U_H).$$

(18) Let $k > 0$. Consider the 2×2 matrix

$$A(x, k) = \begin{pmatrix} \cos(kx) & (1/k)\sin(kx) \\ -k\sin(kx) & \cos(kx) \end{pmatrix}.$$

Show that the eigenvalues are given by $\lambda_+ = e^{ikx}$, $\lambda_- = e^{-ikx}$. Find the normalized eigenvectors. Find the eigenvalues and normalized eigenvectors of $A(x, k) \otimes A(x, k)$. Show that

$$\lambda_+\lambda_- = \lambda_-\lambda_+ = 1.$$

Is $\frac{1}{\sqrt{2}}(0 - 1\ 1\ 0)^T$ a normalized eigenvector?

(19) Let A be an $n \times n$ hermitian matrix and v_1, v_2, ..., v_n be the normalized eigenvectors of A which should form an orthonormal basis in $\mathbb{C}^n$ and analogously u_1, u_2, ..., u_n are the normalized eigenvectors of B which should form an orthonormal basis in $\mathbb{C}^n$. The quantity

$$\max_{j,k=1,2,...,n}|\langle v_j|u_k\rangle|^2$$

plays a role for the entropic inequality. Let $A_1 = \sigma_3$, $B_1 = \sigma_1$. Find this quantity. Let $A_2 = \sigma_3 \otimes \sigma_3$, $B_2 = \sigma_1 \otimes \sigma_1$. Find this quantity. Let $A_3 = \sigma_3 \otimes \sigma_1$, $B_3 = \sigma_1 \otimes \sigma_3$. Find this quantity.

(20) Show that the 2×2 matrix

$$A(x) = \begin{pmatrix} \cos(x) & -\sin(x) \\ -\sin(x) & -\cos(x) \end{pmatrix}$$

admits the eigenvalues $\lambda_1 = +1$ and $\lambda_2 = -1$ with the corresponding normalized eigenvectors

$$v_1 = \begin{pmatrix} -\cos(x/2) \\ \sin(x/2) \end{pmatrix}, \quad v_2 = \begin{pmatrix} \sin(x/2) \\ \cos(x/2) \end{pmatrix}.$$

Note that $\sin(x) \equiv 2\sin(x/2)\cos(x/2)$, $\cos(x) \equiv \cos^2(x/2) - \sin^2(x/2)$. Find the eigenvalues and normalized eigenvectors of $A(x)$.

2.7 Projection Matrices

We introduced projection matrices in section 1.9. Here we study the Kronecker product of projection matrices.

Theorem 2.16. *Let Π be an $m \times m$ projection matrix, Π_1 be an $n \times n$ projection matrix, Π_2 be a $p \times p$ projection matrix. Let I_r be the $r \times r$ unit matrix. Then $\Pi_1 \otimes \Pi_2$ and $\Pi \otimes I_r$ are projection matrices.*

Proof.   First we notice that $(\Pi_1 \otimes \Pi_2)^* = \Pi_1^* \otimes \Pi_2^* = \Pi_1 \otimes \Pi_2$. Since

$$(\Pi_1 \otimes \Pi_2)(\Pi_1 \otimes \Pi_2) = (\Pi_1^2) \otimes (\Pi_2^2) = \Pi_1 \otimes \Pi_2$$

we find that $\Pi_1 \otimes \Pi_2$ is a projection matrix. Owing to

$$(\Pi \otimes I_r)^* = (\Pi^* \otimes I_r^*) = \Pi \otimes I_r$$

and

$$(\Pi \otimes I_r)(\Pi \otimes I_r) = (\Pi^2 \otimes I^2) = \Pi \otimes I_r$$

we find that $\Pi \otimes I_r$ is a projection matrix. $\square$

Example 2.23. Consider

$$\Pi_1 = \frac{1}{2} \begin{pmatrix} 1 & 1 \\ 1 & 1 \end{pmatrix}, \quad \Pi_2 = \begin{pmatrix} 1 & 0 \\ 0 & 0 \end{pmatrix} \Rightarrow \Pi_1 \otimes \Pi_2 = \frac{1}{2} \begin{pmatrix} 1 & 0 & 1 & 0 \\ 0 & 0 & 0 & 0 \\ 1 & 0 & 1 & 0 \\ 0 & 0 & 0 & 0 \end{pmatrix}.$$

♣

Theorem 2.17. *Let Π_1 be an $m \times m$ projection matrix, Π_2 be an $n \times n$ projection matrix. Then $I_m \otimes I_n - \Pi_1 \otimes \Pi_2$ is a projection matrix.*

Proof.   We have

$$(I_m \otimes I_n - \Pi_1 \otimes \Pi_2)^* = (I_m^* \otimes I_n^* - \Pi_1^* \otimes \Pi_2^*) = (I_m \otimes I_n - \Pi_1 \otimes \Pi_2).$$

Furthermore

$$(I_m \otimes I_n - \Pi_1 \otimes \Pi_2)^2 = (I_m^2 \otimes I_n^2 - \Pi_1 \otimes \Pi_2 - \Pi_1 \otimes \Pi_2 + \Pi_1^2 \otimes \Pi_2^2).$$

Since $\Pi_1^2 = \Pi_1$, $\Pi_2^2 = \Pi_2$, $I_m^2 = I_m$ we proved the theorem. $\square$

Exercises. (1) Let

$$\Pi_1 = \frac{1}{2} \begin{pmatrix} 1 & 1 \\ 1 & 1 \end{pmatrix}, \quad \Pi_2 = \begin{pmatrix} 0 & 0 \\ 0 & 1 \end{pmatrix}.$$

Find $\Pi_1 \otimes \Pi_2$. Is $\Pi_1 \otimes \Pi_2$ a projection matrix?

(2) Show that the matrices

$$\Pi_1 = \frac{1}{2}\begin{pmatrix} 1 & 1 \\ 1 & 1 \end{pmatrix}, \qquad \Pi_2 = \frac{1}{2}\begin{pmatrix} 1 & -1 \\ -1 & 1 \end{pmatrix}$$

are projection matrices and that $\Pi_1\Pi_2 = 0_2$. Show that

$$(\Pi_1 \otimes \Pi_2)(\Pi_2 \otimes \Pi_1) = 0_4.$$

(3) Let $\Pi_1, \ldots, \Pi_r$ be $n \times n$ projection matrices. Show that

$$\Pi_1 \otimes \Pi_2 \otimes \cdots \otimes \Pi_r$$

is a projection matrix.

(4) Let Π_1 and Π_2 be projection matrices. Find $\det(\Pi_1 \otimes \Pi_2)$. Find the eigenvalues of $\Pi_1 \otimes \Pi_2$.

(5) Let $\mathbf{u} \in \mathbb{C}^n$ with $\langle \mathbf{u}, \mathbf{u} \rangle = \mathbf{u}^*\mathbf{u} = 1$. Show that $(\mathbf{u}^* \otimes \mathbf{u})^2 = \mathbf{u}^* \otimes \mathbf{u}$.

(6) Let $\mathbf{u}, \mathbf{v} \in \mathbb{C}^n$ and $\langle \mathbf{u}, \mathbf{v} \rangle = \mathbf{v}^*\mathbf{u} = 0$. Show that $(\mathbf{u}^* \otimes \mathbf{v})^2 = 0_n$.

(7) Let Π be projection matrix and P be a permutation matrix. What is the condition on P such that $\Pi \otimes P$ is a projection matrix?

(8) Let $\mathbf{z} = \begin{pmatrix} z_0 & z_1 & \ldots & z_N \end{pmatrix}^T$ be a (column) vector in $\mathbb{C}^{N+1}$. Assume that $\mathbf{z}$ is a nonzero vector. Consider the $(N+1) \times (N+1)$ matrix

$$\Pi := I_{N+1} - \frac{1}{\mathbf{z}^*\mathbf{z}}(\mathbf{z} \otimes \mathbf{z}^*)$$

where I_{N+1} is the $(N+1) \times (N+1)$ identity matrix and

$$\mathbf{z}^* = (\overline{z}_0, \overline{z}_1, \ldots, \overline{z}_N)$$

is the transpose and complex conjugate of $\mathbf{z}$. Show that $\Pi^* = \Pi$ and $\Pi^2 = \Pi$, i.e. show that Π is a projection matrix.

2.8 Unitary, Fourier and Hadamard Matrices

In chapter 1 we introduced unitary, Fourier and Hadamard matrices. The Kronecker product of two unitary matrices is again a unitary matrix. Here we discuss Fourier matrices and its connection with the Kronecker product.

The Fourier matrices of orders 2^n may be expressed as Kronecker products. This factorization is a manifestation, essentially, of the idea known as the *Fast Fourier Transform*.

Let F'_{2^n} denote the Fourier matrices of order 2^n whose rows have been permuted according to the bit reversing permutation.

Definition 2.5. A sequence in natural order can be arranged in *bit-reversed order* as follows: for an integer expressed in binary notation, reverse the binary form and transform to decimal notation, which is then called bit-reversed notation.

Since the sequence 0, 1 is the bit reversed order of 0, 1 and 0, 2, 1, 3 is the bit reversed order of 0, 1, 2, 3 we find that the matrices F'_2 and F'_4 are given by

$$F'_2 = \frac{1}{\sqrt{2}} \begin{pmatrix} 1 & 1 \\ 1 & -1 \end{pmatrix} = F_2, \qquad F'_4 = \frac{1}{\sqrt{4}} \begin{pmatrix} 1 & 1 & 1 & 1 \\ 1 & -1 & 1 & -1 \\ 1 & i & -1 & -i \\ 1 & -i & -1 & i \end{pmatrix}.$$

We find that F'_4 can be written as

$$F'_4 = (I_2 \otimes F'_2)D_4(F'_2 \otimes I_2)$$

where D_4 is the 4×4 diagonal matrix $D_4 := \mathrm{diag}(1,1,1,i)$. Since

$$A \otimes B = P(B \otimes A)P^T$$

for some permutation matrix P that depends merely on the dimensions of A and B, we may write, for some 4×4 permutation matrix P_4

$$F'_4 = (I_2 \otimes F'_2)D_4 P_4 (I_2 \otimes F'_2)P_4$$

where $P_4^{-1} = P_4$. Similarly,

$$F'_{16} = (I_4 \otimes F'_4)D_{16}(F'_4 \otimes I_4)$$

where D_{16} is the 16×16 diagonal matrix $D_{16} := \mathrm{diag}(I, D^2, D, D^3)$ with

$$D := \mathrm{diag}(1, w, w^2, w^3), \qquad w := \exp(-2\pi i/16).$$

For an appropriate permutation matrix $P_{16} = P_{16}^{-1} = P_{16}^T$

$$F'_{16} = (I_4 \otimes F'_4)D_{16}P_{16}(I_4 \otimes F'_4)P_{16}.$$

For D_{256} we set

$$D_{256} = \mathrm{diag}(I, D^8, D^4, \dots, D^{15})$$

where the sequence $0, 8, 4, 12, 2, 10, 6, 14, 1, 9, 5, 13, 3, 11, 7, 15$ is the bit reversed order of $0, 1, 2, \ldots, 15$ and

$$D := \operatorname{diag}(1, w, \ldots, w^{15}), \qquad w := \exp(2\pi i/256).$$

Development of reduced multiplication fast Fourier transform from small N discrete Fourier transform can be accomplished using Kronecker product expansions. Let $F_L, \ldots, F_2, F_1$ be small N discrete Fourier transform algorithms with naturally ordered indices. Their dimensions are $N_L \times N_L, \ldots, N_2 \times N_2, N_1 \times N_1$, respectively. Then we can form the Kronecker products to be the $N \times N$ matrix F, where $N = N_L \cdots N_2 N_1$ and

$$F := F_L \otimes \cdots \otimes F_2 \otimes F_1.$$

For more details of this construction we refer to Elliott and Rao [20].

The *Winograd-Fourier transform* algorithm minimizes the number of multiplications, but not the number of additions, required to evaluate the reduced multiplication fast Fourier transform. The Winograd-Fourier transform algorithm (Elliott and Rao [20]) results from a Kronecker product manipulation to group input additions so that all transform multiplications follows. One writes

$$F = (S_L C_L T_L) \otimes \cdots \otimes (S_2 C_2 T_2) \otimes (S_1 C_1 T_1)$$

as

$$F = (S_L \otimes \cdots \otimes S_2 \otimes S_1)(C_L \otimes \cdots \otimes C_2 \otimes C_1)(T_L \otimes \cdots \otimes T_2 \otimes T_1)$$

where $(S_L \otimes \cdots \otimes S_2 \otimes S_1)$ describes the output additions, $(C_L \otimes \cdots \otimes C_2 \otimes C_1)$ the multiplications and $(T_L \otimes \cdots \otimes T_2 \otimes T_1)$ the input additions.

Hadamard matrices can also be constructed using the Kronecker product.

Theorem 2.18. *If A and B are Hadamard matrices of orders m and n respectively, then $A \otimes B$ is an Hadamard matrix of order mn.*

Proof. We have

$$(A \otimes B)(A \otimes B)^T = (A \otimes B)(A^T \otimes B^T) = (AA^T) \otimes (BB^T).$$

Thus

$$(A \otimes B)(A \otimes B)^T = (mI_m) \otimes (nI_n) = mn(I_m \otimes I_n) = mnI_{mn}. \qquad \square$$

Sometimes the term Hadamard matrix is limited to the matrices of order 2^n given specifically by the recursion

$$H_1 = (1), \quad H_2 = \begin{pmatrix} 1 & 1 \\ 1 & -1 \end{pmatrix}, \quad H_{2^{n+1}} = H_{2^n} \otimes H_2.$$

For example, we find

$$H_4 = \begin{pmatrix} 1 & 1 & 1 & 1 \\ 1 & -1 & 1 & -1 \\ 1 & 1 & -1 & -1 \\ 1 & -1 & -1 & 1 \end{pmatrix} = H_2 \otimes H_2.$$

These matrices have the property $H_{2^n} = H_{2^n}^T$ so that $H_{2^n}^2 = 2^n I$.

Definition 2.6. The *Walsh-Hadamard transform* is defined as

$$\hat{Z} = HZ$$

where H is an Hadamard matrix, where $Z = (z_1, z_2, \ldots, z_n)^T$.

Since H is invertible the inverse transformation can be found

$$Z = H^{-1}\hat{Z}.$$

Closely related to the Hadamard matrices are the *Haar matrices*. The Haar transform is based on the Haar functions, which are periodic, orthogonal, and complete in the Hilbert space $L_2[0,1]$ of the square integrable functions. The first two Haar matrices are given by

$$Ha(1) = \begin{pmatrix} 1 & 1 \\ 1 & -1 \end{pmatrix}, \quad Ha(2) = \begin{pmatrix} 1 & 1 & 1 & 1 \\ 1 & 1 & -1 & -1 \\ \sqrt{2}(1 & -1 & 0 & 0) \\ \sqrt{2}(0 & 0 & 1 & -1) \end{pmatrix}.$$

Higher order Haar matrices can be generated recursively using the Kronecker product as follows

$$Ha(k+1) = \begin{pmatrix} Ha(k) \otimes (1 \ 1) \\ 2^{k/2}I_{2^k} \otimes (1 \ -1) \end{pmatrix}, \quad k > 1.$$

For example we find

$$Ha(3) = \begin{pmatrix} 1 & 1 & 1 & 1 & 1 & 1 & 1 & 1 \\ 1 & 1 & 1 & 1 & -1 & -1 & -1 & -1 \\ \sqrt{2}(1 & 1 & -1 & -1 & 0 & 0 & 0 & 0) \\ \sqrt{2}(0 & 0 & 0 & 0 & 1 & 1 & -1 & -1) \\ 2 & -2 & 0 & 0 & 0 & 0 & 0 & 0 \\ 0 & 0 & 2 & -2 & 0 & 0 & 0 & 0 \\ 0 & 0 & 0 & 0 & 2 & -2 & 0 & 0 \\ 0 & 0 & 0 & 0 & 0 & 0 & 2 & -2 \end{pmatrix}.$$

The Haar transform is given by

$$\mathbf{X} = \frac{1}{N}[Ha(L)]\mathbf{x}, \qquad \mathbf{x} = [Ha(L)]^T\mathbf{X}$$

where the $N \times N$ Haar matrix $Ha(L)$ is orthogonal $Ha(L)Ha(L)^T = NI_N$. Haar matrices can be factored into sparse matrices, which lead to the fast algorithms. Based on these algorithms both the Haar transform and its inverse can be implemented in $2(N-1)$ additions or subtractions and N multiplications.

Exercises. (1) Find a unitary 4×4 matrix U such that

$$\sigma_1 \otimes \sigma_3 \equiv \begin{pmatrix} 0 & 0 & 1 & 0 \\ 0 & 0 & 0 & -1 \\ 1 & 0 & 0 & 0 \\ 0 & -1 & 0 & 0 \end{pmatrix} = U \begin{pmatrix} -1 & 0 & 0 & 0 \\ 0 & 0 & 0 & 1 \\ 0 & 0 & 1 & 0 \\ 0 & 1 & 0 & 0 \end{pmatrix} U^{-1} \equiv U \left((-1) \oplus \begin{pmatrix} 0 & 0 & 1 \\ 0 & 1 & 0 \\ 1 & 0 & 0 \end{pmatrix} \right) U^{-1}.$$

(2) Let c_1, c_2, c_3 be nonnegative real numbers.
(i) What are the conditions on c_1, c_2, c_3 such that $c_1\sigma_1 + c_2\sigma_2 + c_3\sigma_3$ is a unitary matrix?
(ii) What are the conditions on c_1, c_2, c_3 such that

$$c_1\sigma_1 \otimes \sigma_1 + c_2\sigma_2 \otimes \sigma_2 + c_3\sigma_3 \otimes \sigma_3$$

is a unitary matrix?
(iii) What are the conditions on c_1, c_2, c_3 such that

$$c_1\sigma_1 \otimes \sigma_2 + c_2\sigma_2 \otimes \sigma_3 + c_3\sigma_3 \otimes \sigma_1$$

is a unitary matrix?

2.9 Direct Sum

Let A be an $m \times n$ matrix and B be a $p \times q$ matrix. Then the *direct sum* of A and B is defined as

$$A \oplus B := \begin{pmatrix} A & 0 \\ 0 & B \end{pmatrix}.$$

All zero matrices are of appropriate order. An example is

$$(1\ 0\ 3) \oplus \begin{pmatrix} 6 & 7 \\ 8 & 9 \end{pmatrix} = \begin{pmatrix} 1 & 0 & 3 & 0 & 0 \\ 0 & 0 & 0 & 6 & 7 \\ 0 & 0 & 0 & 8 & 9 \end{pmatrix}.$$

The extension to more than two matrices is straightforward, i.e.

$$A \oplus B \oplus C = \begin{pmatrix} A & 0 & 0 \\ 0 & B & 0 \\ 0 & 0 & C \end{pmatrix}$$

and

$$\bigoplus_{j=1}^{k} A_j = \begin{pmatrix} A_1 & 0 & 0 & \cdots & 0 \\ 0 & A_2 & 0 & \cdots & 0 \\ 0 & 0 & \ddots & & \vdots \\ \vdots & & & \ddots & 0 \\ 0 & 0 & \cdots & 0 & A_k \end{pmatrix} = \text{diag}\{A_j\} \qquad \text{for } j = 1, \ldots, k.$$

The definition and its extensions apply whether or not the submatrices are the same order. All zero (null) matrices are of appropriate order.

Transposing a direct sum gives the direct sum of the transposes. The rank of a direct sum is the sum of the ranks, as is evident from the definition of rank. It is obvious that $A \oplus (-A) \neq 0$ unless A is null. Also,

$$(A \oplus B) + (C \oplus D) = (A + C) \oplus (B + D)$$

only if the necessary conditions of conformability for addition are met. Similarly,

$$(A \oplus B)(C \oplus D) = (AC) \oplus (BD)$$

provided that the matrix products exist.

The direct sum $A \oplus B$ is square only if A is $p \times q$ and B is $q \times p$. Let A and B be square matrices. Then

$$\det(A \oplus B) = \det(A)\det(B).$$

In general, we have

$$\det \left(\bigoplus_{i=1}^{N} A_i \right) = \prod_{i=1}^{N} \det(A_i), \qquad r \left(\bigoplus_{i=1}^{N} A_i \right) = \sum_{i=1}^{N} r(A_i)$$

where $r(.)$ denotes the rank. If A and B are orthogonal, then $A \oplus B$ is orthogonal. If A and B are nilpotent, then $A \oplus B$ is nilpotent. If A and B are idempotent then $A \oplus B$ is idempotent.

Let A be an $m \times m$ matrix with eigenvalues $\lambda_1, \ldots, \lambda_m$ and B be an $n \times n$ matrix with eigenvalues $\mu_1, \ldots, \mu_n$. Then the eigenvalues of $A \oplus B$ are given by $\lambda_1, \ldots, \lambda_m, \mu_1, \ldots, \mu_n$.

Theorem 2.19. *Let A, B, C be arbitrary matrices. Then*
$$(A \oplus B) \otimes C = (A \otimes C) \oplus (B \otimes C).$$
The proof is straightforward and is left as an exercise to the reader.

The Kronecker product of some matrices can be written as the direct sum of matrices. An example is
$$\begin{pmatrix} 1 & 0 \\ 0 & 1 \end{pmatrix} \otimes \begin{pmatrix} 1 & 1 \\ 1 & 1 \end{pmatrix} \equiv \begin{pmatrix} 1 & 1 \\ 1 & 1 \end{pmatrix} \oplus \begin{pmatrix} 1 & 1 \\ 1 & 1 \end{pmatrix}.$$
This plays an important role in representation theory of groups (Ludwig and Falter [43]).

The direct sum of two permutation matrices is again a permutation matrix.

The direct sum of two projection matrices is again a projection matrix.

The direct sum of two invertible matrices A and B is again an invertible matrix. We have $(A \oplus B)^{-1} = A^{-1} \oplus B^{-1}$.

Exercises. (1) Show that in general $A \otimes (B \oplus C) \neq (A \otimes B) \oplus (A \otimes C)$. Give an example.

(2) Let A be an $m \times m$ matrix with eigenvalues $\lambda_1, \lambda_2, \ldots, \lambda_m$. Let B be an $n \times n$ matrix with eigenvalues $\mu_1, \mu_2, \ldots, \mu_n$. Find the eigenvalues of $A \oplus B$.

(3) Let A be an $m \times m$ matrix with eigenvalues $\lambda_1, \lambda_2, \ldots, \lambda_m$. Let B be an $n \times n$ matrix with eigenvalues $\mu_1, \mu_2, \ldots, \mu_n$. Let C be an $r \times r$ matrix with eigenvalues $\nu_1, \nu_2, \ldots, \nu_r$. Find the eigenvalues of $(A \oplus B) \otimes C$.

(4) Let A, B, C and D be 2×2 matrices. Find the conditions on A, B, C and D such that $A \otimes B = C \oplus D$.

(5) Let A, B, C be 2×2 matrices. Let D be a 3×3 matrix and E be a 2×2 matrix. Find the conditions on A, B, C, D, E such that
$$A \oplus B \oplus C = D \otimes E.$$

(6) Let P be a permutation matrix. Is $P \oplus P$ be a permutation matrix? Is $(P \oplus P) \otimes P$ a permutation matrix?

(7) Let Π be a projection matrix. Is $\Pi \oplus \Pi$ a projection matrix? Is $(\Pi \oplus \Pi) \otimes \Pi$ a projection matrix?

(8) Let A, B be Hermitian matrices. Show that $A \oplus B$ is a Hermitian matrix.

(9) Let U_1, U_2 be unitary matrices. Is $U_1 \oplus U_2$ a unitary matrix?

(10) Let A be an $n \times n$ invertible matrix. Let B be an $m \times m$ invertible matrix. Show that $(A \oplus B)^{-1}$ exists. Find the inverse. Find the determinant of $A \oplus B$.

(11) Let A and B be binary matrices. Is $A \oplus B$ a binary matrix?

2.10   Kronecker Sum

Definition 2.7. The *Kronecker sum* $A \oplus_K B$ *of the $m \times m$ matrix A and the $n \times n$ matrix B over $\mathbb{C}$ is defined by*

$$A \oplus_K B := A \otimes I_n + I_m \otimes B.$$

Let C be $m \times m$ over $\mathbb{C}$ and D be $n \times n$ over $\mathbb{C}$. Then

$$
\begin{aligned}
(A + C) \oplus_K B &= (A + C) \otimes I_n + I_m \otimes B \\
&= A \otimes I_n + C \otimes I_n + I_m \otimes B \\
&\neq A \oplus_K B + C \oplus_K B
\end{aligned}
$$

in general (except when $B = 0_{n \times n}$), and similarly

$$A \oplus_K (B + D) \neq A \oplus_K B + A \oplus_K D$$

in general (except when $A = 0_{m \times m}$). Thus distributivity does not hold.

Let C be $m \times m$ over $\mathbb{C}$ and D be $n \times n$ over $\mathbb{C}$. Then

$$
\begin{aligned}
(A \oplus_K B)(C \oplus_K D) &= (A \otimes I_n + I_m \otimes B)(C \otimes I_n + I_m \otimes D) \\
&= (AC) \otimes I_n + A \otimes D + C \otimes B + I_m \otimes (BD) \\
&= (AC) \oplus_K (BD) + A \otimes D + C \otimes B.
\end{aligned}
$$

Applying the property of the transpose of a Kronecker product we find

$$(A \oplus_K B)^T = (A \otimes I_n + I_m \otimes B)^T = A^T \otimes I_n + I_m \otimes B^T = A^T \oplus_K B^T.$$

Theorem 2.20. *Let* $\mathbf{x}_\lambda$ *be an eigenvector of* A *corresponding to the eigenvalue* λ *and* $\mathbf{y}_\mu$ *be an eigenvector of* B *corresponding to the eigenvalue* μ. *Then* $\lambda + \mu$ *is an eigenvalue of* $A \oplus_K B$ *corresponding to the eigenvector* $\mathbf{x}_\lambda \otimes \mathbf{y}_\mu$.

Proof. Let $\mathbf{x}_\lambda$ be an eigenvector of A corresponding to the eigenvalue λ and $\mathbf{y}_\mu$ be an eigenvector of B corresponding to the eigenvalue μ. Then

$$\begin{aligned}
(A \oplus_K B)(\mathbf{x}_\lambda \otimes \mathbf{y}_\mu) &= (A \otimes I_n)(\mathbf{x}_\lambda \otimes \mathbf{y}_\mu) + (I_m \otimes B)(\mathbf{x}_\lambda \otimes \mathbf{y}_\mu) \\
&= (A\mathbf{x}_\lambda) \otimes (\mathbf{y}_\mu) + (\mathbf{x}_\lambda) \otimes (B\mathbf{y}_\mu) \\
&= (\lambda\mathbf{x}_\mu) \otimes (\mathbf{y}_\mu) + (\mathbf{x}_\mu) \otimes (\mu\mathbf{y}_\mu) \\
&= (\lambda + \mu)(\mathbf{x}_\lambda \otimes \mathbf{y}_\mu).
\end{aligned}$$

Thus $\lambda + \mu$ is an eigenvalue of $A \oplus_K B$ corresponding to the eigenvector $\mathbf{x}_\lambda \otimes \mathbf{y}_\mu$. $\qquad\square$

Applying the property of the trace of a Kronecker product we find

$$\mathrm{tr}(A \oplus_K B) = \mathrm{tr}(A \otimes I_n + I_m \otimes B) = \mathrm{tr}(A \otimes I_n) + \mathrm{tr}(I_m \otimes B) = n\,\mathrm{tr}(A) + m\,\mathrm{tr}(B).$$

Let A and B be square matrices ($m = n$ and $s = t$). Since the determinant of $A \oplus_k B$ is the product of the eigenvalues of $A \oplus_k B$, and the eigenvalues of $A \oplus_k B$ are all the sums $\lambda + \mu$ of eigenvalues λ of A and μ of B we have

$$\det(A \oplus_K B) = \prod_\lambda \left(\prod_\mu (\lambda + \mu) \right).$$

In general we do not find any additional structure.

Suppose A and B are normal, then A has a spectral decomposition

$$A = V_A D_A V_A^*$$

and B has a spectral decomposition $B = V_B D_B V_B^*$. It follows that

$$\begin{aligned}
A \oplus_K B &= (V_A D_A V_A^*) \otimes I_n + I_m \otimes (V_B D_B V_B^*) \\
&= (V_A D_A V_A^*) \otimes (V_B I_n V_B^*) + (V_A I_m V_A^*) \otimes (V_B D_B V_B^*) \\
&= (V_A \otimes V_B)(D_A \otimes I_n + I_m \otimes D_B)(V_A^* \otimes V_B^*) \\
&= (V_A \otimes V_B)(D_A \oplus_K D_B)(V_A \otimes V_B)^*
\end{aligned}$$

is clearly a spectral decomposition of $A \oplus_K B$.

Example 2.24. Let

$$\sigma_1 = \begin{pmatrix} 0 & 1 \\ 1 & 0 \end{pmatrix}, \qquad \sigma_2 = \begin{pmatrix} 0 & -i \\ i & 0 \end{pmatrix}.$$

Then

$$\sigma_1 \oplus_K \sigma_2 = \sigma_1 \otimes I_2 + I_2 \otimes \sigma_2 = \begin{pmatrix} 0 & -i & 1 & 0 \\ i & 0 & 0 & 1 \\ 1 & 0 & 0 & -i \\ 0 & 1 & i & 0 \end{pmatrix}.$$

Note that this matrix is not invertible although σ_1 and σ_2 are.   ♣

2.11   Matrix Decompositions

Suppose A and B are normal, then A has a spectral decomposition $A = V_A D_A V_A^*$ and B has a spectral decomposition $B = V_B D_B V_B^*$. It follows that

$$A \otimes B = (V_A D_A V_A^*) \otimes (V_B D_B V_B^*) = (V_A \otimes V_B)(D_A \otimes D_B)(V_A^* \otimes V_B^*)$$
$$= (V_A \otimes V_B)(D_A \otimes D_B)(V_A \otimes V_B)^*$$

is clearly a spectral decomposition of $A \otimes B$.

Suppose A and B are square, then A has a polar decomposition $A = U_A H_A$ and B has a polar decomposition $B = U_B H_B$. It follows that

$$A \otimes B = (U_A H_A) \otimes (U_B H_B) = (U_A \otimes U_B)(H_A \otimes H_B)$$

is clearly a polar decomposition of $A \otimes B$.

For any matrices A and B, A has a singular value decomposition $A = U_A \Sigma_A V_A^*$ and B has a singular value decomposition $B = U_B \Sigma_B V_B^*$. It follows that

$$A \otimes B = (U_A \Sigma_A V_A^*) \otimes (V_B \Sigma_B V_B^*) = (U_A \otimes U_B)(\Sigma_A \otimes \Sigma_B)(V_A^* \otimes V_B^*)$$
$$= (U_A \otimes U_B)(\Sigma_A \otimes \Sigma_B)(V_A \otimes V_B)^*.$$

If $\Sigma_A \otimes \Sigma_B$ has decreasing values on the diagonal then we have a singular value decomposition. This is not the case in general.

In the following we do not insist that the singular values are arranged in nondecreasing order for the singular value decomposition.

Here we consider $2n \times 2n$ matrices acting on $\mathbb{C}^n \otimes \mathbb{C}^2$. Let U be a $2n \times 2n$ unitary matrix acting on $\mathbb{C}^n \otimes \mathbb{C}^2$. We can write U in the form

$$U = \begin{pmatrix} 1 & 0 \\ 0 & 0 \end{pmatrix} \otimes Q_0 + \begin{pmatrix} 0 & 1 \\ 0 & 0 \end{pmatrix} \otimes Q_1 + \begin{pmatrix} 0 & 0 \\ 1 & 0 \end{pmatrix} \otimes Q_2 + \begin{pmatrix} 0 & 0 \\ 0 & 1 \end{pmatrix} \otimes Q_3$$

where Q_0, Q_1, Q_2 and Q_3 are $n \times n$ matrices (some of which may be the zero matrix). Since U is a unitary matrix we have

$$UU^* = \begin{pmatrix} 1 & 0 \\ 0 & 0 \end{pmatrix} \otimes (Q_0 Q_0^* + Q_1 Q_1^*) + \begin{pmatrix} 0 & 1 \\ 0 & 0 \end{pmatrix} \otimes (Q_0 Q_2^* + Q_1 Q_3^*)$$

$$+ \begin{pmatrix} 0 & 0 \\ 0 & 1 \end{pmatrix} \otimes (Q_2 Q_2^* + Q_3 Q_3^*) + \begin{pmatrix} 0 & 0 \\ 1 & 0 \end{pmatrix} \otimes (Q_2 Q_0^* + Q_3 Q_1^*)$$

$$= I_n \otimes I_2 = U^* U.$$

From the condition $U^* U = UU^* = I_n \otimes I_2$ we find the conditions on the matrices Q_j

$$Q_0 Q_0^* + Q_1 Q_1^* = I_n, \qquad Q_0^* Q_0 + Q_2^* Q_2 = I_n,$$
$$Q_2 Q_2^* + Q_3 Q_3^* = I_n, \qquad Q_1^* Q_1 + Q_3^* Q_3 = I_n,$$
$$Q_2 Q_0^* + Q_3 Q_1^* = 0_n, \qquad Q_1^* Q_0 + Q_3^* Q_2 = 0_n$$

where 0_n denotes the $n \times n$ zero matrix. Note that Q_0 is square so that in the singular value decomposition $Q_0 = U_0 \Sigma_0 V_0^*$ we find $\Sigma_0^* = \Sigma_0$. Applying the singular value decomposition $Q_0 = U_0 \Sigma_0 V_0^*$ we obtain for the first equation above

$$U_0 \Sigma_0^2 U_0^* + Q_1 Q_1^* = I_n$$

where Σ_0 is an $n \times n$ diagonal matrix. Thus $\Sigma_0^2 + U_0^* Q_1 Q_1^* U_0 = I_n$ and we find that $Q_1 Q_1^*$ is diagonalized by U_0^*. It follows that we can find a singular value decomposition $Q_1 = U_0 \Sigma_1 V_1^*$ of Q_1 where $\Sigma_0^2 + \Sigma_1^2 = I_n$. Similarly

$$V_0 \Sigma_0^2 V_0^* + Q_2^* Q_2 = I_n \Rightarrow \Sigma_0^2 + V_0^* Q_2^* Q_2 V_0 = I_n$$

and we find that $Q_2^* Q_2$ is diagonalized by V_0^*. It follows that we can find a singular value decomposition $Q_2 = U_2 \Sigma_2 V_0^*$ of Q_2 where $\Sigma_0^2 + \Sigma_2^2 = I_n$. Thus $\Sigma_1 = \Sigma_2$ (since they are diagonal matrices with nonnegative entries). We follow the same procedure with Q_2 and Q_3, except for one sign change to ensure all equations are satisfied. From

$$Q_1^* Q_1 + V_1 \Sigma_3^2 V_1^* = I_n \Rightarrow V_1^* Q_1^* Q_1 V_1 + \Sigma_3^2 = I_n$$

we find that Q_3 has a singular value decomposition $Q_3 = U_1 \Sigma_3 V_1^*$ where

$$\Sigma_1^2 + \Sigma_3^2 = I_n.$$

Then

$$Q_2 Q_2^* + Q_3 Q_3^* = Q_2 Q_2^* + U_1 \Sigma_3^2 U_1^* = I_n$$

or equivalently

$$U_1^* Q_2 Q_2^* U_1 + \Sigma_3^2 = (-U_1)^* Q_2 Q_2^* (-U_1) + \Sigma_3^2 = I_n$$

so that $Q_2 Q_2^*$ is diagonalized by $(-U_1)^*$. It follows that Q_2 ha a singular value decomposition $(U_2 = -U_1)$ $Q_2 = (-U_1)\Sigma_3 V_0^*$ where

$$\Sigma_2^2 + \Sigma_3^2 = I_n \ \Rightarrow \ \Sigma_3 = \Sigma_0.$$

To summarize, we can construct singular value decompositions for the matrices Q_0, Q_1, Q_2 and Q_3 such that

$$Q_0 = U_0 \Sigma_0 V_0^*, \quad Q_1 = U_0 \Sigma_1 V_1^*, \quad Q_2 = (-U_1)\Sigma_1 V_0^*, \quad Q_3 = U_1 \Sigma_0 V_1^*.$$

For the last two equations we have

$$Q_2 Q_0^* + Q_3 Q_1^* = -U_1 \Sigma_1 \Sigma_0 U_0^* + U_1 \Sigma_0 \Sigma_1 U_0^* = 0_n$$

since $\Sigma_1 \Sigma_0 = \Sigma_0 \Sigma_1$ (diagonal matrices commute) and similarly

$$Q_1^* Q_0 + Q_3^* Q_2 = V_1 \Sigma_1 \Sigma_0 V_0^* - V_1 \Sigma_0 \Sigma_1 V_0^* = 0_n.$$

Since $\Sigma_0^2 + \Sigma_1^2 = I_n$ we have $\Sigma_1^2 = I_n - \Sigma_0^2$. Furthermore, the entries of Σ_0 and Σ_1 are real valued and thus each pair of diagonal entries correspond to the identity $\cos^2(\theta) + \sin^2(\theta) = 1$ for some $\theta \in \mathbb{R}$. Consequently we write

$$\Sigma_0 = C := \mathrm{diag}(\cos(\theta_1), \cos(\theta_2), \ldots, \cos(\theta_n)),$$
$$\Sigma_1 = S := \mathrm{diag}(\sin(\theta_1), \sin(\theta_2), \ldots, \sin(\theta_n))$$

so that

$$Q_0 = U_0 C V_0^*, \quad Q_1 = U_0 S V_1^*, \quad Q_2 = (-U_1)S V_0^*, \quad Q_3 = U_1 C V_1^*.$$

Lastly we note that

$$\begin{pmatrix} 1 & 0 \\ 0 & 0 \end{pmatrix} \equiv \begin{pmatrix} 1 & 0 \\ 0 & 0 \end{pmatrix}\begin{pmatrix} 1 & 0 \\ 0 & 0 \end{pmatrix}\begin{pmatrix} 1 & 0 \\ 0 & 0 \end{pmatrix}, \qquad \begin{pmatrix} 0 & 1 \\ 0 & 0 \end{pmatrix} \equiv \begin{pmatrix} 1 & 0 \\ 0 & 0 \end{pmatrix}\begin{pmatrix} 0 & 1 \\ 0 & 0 \end{pmatrix}\begin{pmatrix} 0 & 0 \\ 0 & 1 \end{pmatrix},$$

$$\begin{pmatrix} 0 & 0 \\ 1 & 0 \end{pmatrix} \equiv \begin{pmatrix} 0 & 0 \\ 0 & 1 \end{pmatrix}\begin{pmatrix} 0 & 0 \\ 1 & 0 \end{pmatrix}\begin{pmatrix} 1 & 0 \\ 0 & 0 \end{pmatrix}, \qquad \begin{pmatrix} 0 & 0 \\ 0 & 1 \end{pmatrix} \equiv \begin{pmatrix} 0 & 0 \\ 0 & 1 \end{pmatrix}\begin{pmatrix} 0 & 0 \\ 0 & 1 \end{pmatrix}\begin{pmatrix} 0 & 0 \\ 0 & 1 \end{pmatrix}$$

which yields the Cosine-Sine decomposition

$$U = \begin{pmatrix} 1 & 0 \\ 0 & 0 \end{pmatrix} \otimes U_0 C V_0^* + \begin{pmatrix} 0 & 1 \\ 0 & 0 \end{pmatrix} \otimes U_0 S V_1^*$$
$$- \begin{pmatrix} 0 & 0 \\ 1 & 0 \end{pmatrix} \otimes U_1 S V_0^* + \begin{pmatrix} 0 & 0 \\ 0 & 1 \end{pmatrix} \otimes U_1 C V_1^*$$
$$= \begin{pmatrix} 1 & 0 \\ 0 & 0 \end{pmatrix} \begin{pmatrix} 1 & 0 \\ 0 & 0 \end{pmatrix} \begin{pmatrix} 1 & 0 \\ 0 & 0 \end{pmatrix} \otimes U_0 C V_0^* + \begin{pmatrix} 1 & 0 \\ 0 & 0 \end{pmatrix} \begin{pmatrix} 0 & 1 \\ 0 & 0 \end{pmatrix} \begin{pmatrix} 0 & 0 \\ 0 & 1 \end{pmatrix} \otimes U_0 S V_1^*$$
$$- \begin{pmatrix} 0 & 0 \\ 0 & 1 \end{pmatrix} \begin{pmatrix} 0 & 0 \\ 1 & 0 \end{pmatrix} \begin{pmatrix} 1 & 0 \\ 0 & 0 \end{pmatrix} \otimes U_1 S V_0^* + \begin{pmatrix} 0 & 0 \\ 0 & 1 \end{pmatrix} \begin{pmatrix} 0 & 0 \\ 0 & 1 \end{pmatrix} \begin{pmatrix} 0 & 0 \\ 0 & 1 \end{pmatrix} \otimes U_1 C V_1^*$$
$$= \left[\begin{pmatrix} 1 & 0 \\ 0 & 0 \end{pmatrix} \otimes U_0 + \begin{pmatrix} 0 & 0 \\ 0 & 1 \end{pmatrix} \otimes U_1 \right] \left[\begin{pmatrix} 1 & 0 \\ 0 & 1 \end{pmatrix} \otimes C + \begin{pmatrix} 0 & 1 \\ -1 & 0 \end{pmatrix} \otimes S \right]$$
$$\times \left[\begin{pmatrix} 1 & 0 \\ 0 & 0 \end{pmatrix} \otimes V_0^* + \begin{pmatrix} 0 & 0 \\ 0 & 1 \end{pmatrix} \otimes V_1^* \right]$$
$$= \begin{pmatrix} U_0 & 0_n \\ 0_n & U_1 \end{pmatrix} \begin{pmatrix} C & S \\ -S & C \end{pmatrix} \begin{pmatrix} V_0 & 0_n \\ 0_n & V_1 \end{pmatrix}^*.$$

This is a *Cosine-sine decomposition* (Stewart [65], van Loan [69]) of U. Each of the three matrices appearing in the decomposition are unitary.

We compute a Cosine-Sine decomposition of the symmetric 4×4 matrix

$$\frac{1}{2} \begin{pmatrix} 1 & 1 & 1 & 1 \\ 1 & 1 & -1 & -1 \\ 1 & -1 & 1 & -1 \\ 1 & -1 & -1 & 1 \end{pmatrix}.$$

The four quadrants and their singular value decompositions (without the ordering of the singular values) are given by

$$Q_0 = \frac{1}{2} \begin{pmatrix} 1 & 1 \\ 1 & 1 \end{pmatrix} = \frac{1}{\sqrt{2}} \begin{pmatrix} 1 & 1 \\ 1 & -1 \end{pmatrix} \begin{pmatrix} 1 & 0 \\ 0 & 0 \end{pmatrix} \frac{1}{\sqrt{2}} \begin{pmatrix} 1 & -1 \\ 1 & 1 \end{pmatrix}^*$$
$$Q_1 = \frac{1}{2} \begin{pmatrix} 1 & 1 \\ -1 & -1 \end{pmatrix} = \frac{1}{\sqrt{2}} \begin{pmatrix} 1 & 1 \\ 1 & -1 \end{pmatrix} \begin{pmatrix} 0 & 0 \\ 0 & 1 \end{pmatrix} \frac{1}{\sqrt{2}} \begin{pmatrix} 1 & 1 \\ -1 & 1 \end{pmatrix}^*$$
$$Q_2 = \frac{1}{2} \begin{pmatrix} 1 & -1 \\ 1 & -1 \end{pmatrix} = -\frac{1}{\sqrt{2}} \begin{pmatrix} 1 & 1 \\ -1 & 1 \end{pmatrix} \begin{pmatrix} 0 & 0 \\ 0 & 1 \end{pmatrix} \frac{1}{\sqrt{2}} \begin{pmatrix} 1 & -1 \\ 1 & 1 \end{pmatrix}^*$$
$$Q_3 = \frac{1}{2} \begin{pmatrix} 1 & -1 \\ -1 & 1 \end{pmatrix} = \frac{1}{\sqrt{2}} \begin{pmatrix} 1 & 1 \\ -1 & 1 \end{pmatrix} \begin{pmatrix} 1 & 0 \\ 0 & 0 \end{pmatrix} \frac{1}{\sqrt{2}} \begin{pmatrix} 1 & 1 \\ -1 & 1 \end{pmatrix}^*.$$

Of course other decompositions are possible. Thus we find

$$\frac{1}{2}\begin{pmatrix}1 & 1 & 1 & 1\\ 1 & 1 & -1 & -1\\ 1 & -1 & 1 & -1\\ 1 & -1 & -1 & 1\end{pmatrix} = \frac{1}{\sqrt{2}}\begin{pmatrix}1 & 1 & 0 & 0\\ 1 & -1 & 0 & 0\\ 0 & 0 & 1 & 1\\ 0 & 0 & -1 & 1\end{pmatrix}\begin{pmatrix}1 & 0 & 0 & 0\\ 0 & 0 & 0 & 1\\ 0 & 0 & 1 & 0\\ 0 & -1 & 0 & 0\end{pmatrix}\frac{1}{\sqrt{2}}\begin{pmatrix}1 & -1 & 0 & 0\\ 1 & 1 & 0 & 0\\ 0 & 0 & 1 & 1\\ 0 & 0 & -1 & 1\end{pmatrix}^*.$$

Definition 2.8. Let M be an $(ms) \times (nt)$ matrix over $\mathbb{C}$. The *Schmidt rank* of M is the minimum number r such that M can be written in the form

$$M = \sum_{j=1}^{r} A_j \otimes B_j$$

for some $m \times n$ matrices $A_1, \ldots, A_r$ and $s \times t$ matrices $B_1, \ldots, B_r$. Here r depends on M, m, n, s and t.

Let the matrices E_j $(j = 1, \ldots, mn)$ be a basis for the $m \times n$ matrices and let F_k $(k = 1, \ldots, st)$ be a basis for the $s \times t$ matrices. It follows that M can be written in the form

$$M = \sum_{j=1}^{mn} \sum_{k=1}^{st} c_{jk} E_j \otimes F_k.$$

For $c_{jk} \in \mathbb{C}$, where $j = 1, \ldots, mn$ and $k = 1, \ldots, st$. It follows that the Schmidt rank r above is the rank of the matrix $C := (c_{jk})$, i.e. $r = r(C)$. Let $C = U\Sigma V^*$ be a singular value decomposition of C, where U is an $(mn) \times (mn)$ unitary matrix, V is an $(st) \times (st)$ unitary matrix and Σ is a $(mn) \times (st)$ matrix with only the first $r(C)$ diagonal entries nonzero (i.e. the singular values nonzero $\sigma_1, \sigma_2, \ldots, \sigma_r$ of C). Since the values c_{jk} are scalars, we find

$$M = \sum_{j=1}^{mn} \sum_{k=1}^{st} c_{jk} \otimes E_j \otimes F_k = \sum_{j=1}^{mn} \sum_{k=1}^{st} (e_{j,mn}^T C e_{k,st}) \otimes E_j \otimes F_k$$

$$= \sum_{j=1}^{mn} \sum_{k=1}^{st} (e_{j,mn}^T U\Sigma V^* e_{k,st}) \otimes E_j \otimes F_k$$

$$= \sum_{j=1}^{mn} \sum_{k=1}^{st} ((e_{j,mn}^T U) \otimes E_j \otimes I_s)(\Sigma \otimes I_n \otimes I_s)((V^* e_{k,st}) \otimes I_n \otimes F_k)$$

$$= \left(\sum_{j=1}^{mn} (e_{j,mn}^T U) \otimes E_j \otimes I_s\right)(\Sigma \otimes I_n \otimes I_s)\left(\sum_{k=1}^{st} (V^* e_{k,st}) \otimes I_n \otimes F_k\right).$$

Let

$$U = \sum_{p=1}^{mn} \mathbf{u}_p \mathbf{e}_{p,mn}^T, \qquad V = \sum_{q=1}^{st} \mathbf{v}_q \mathbf{e}_{q,st}^T$$

where $\mathbf{u}_1, \ldots, \mathbf{u}_{mn}$ are the columns of U and $\mathbf{v}_1, \ldots, \mathbf{v}_{st}$ are the columns of V. Using these expressions for U and V and the fact that

$$\mathbf{e}_{p,mn}^T \Sigma \mathbf{e}_{q,st} = \begin{cases} \sigma_p & p = q, \, p \in \{1, 2, \ldots, r\} \\ 0 & \text{otherwise} \end{cases}$$

we have

$$M = \left(\sum_{j,p=1}^{mn} (\mathbf{e}_{j,mn}^T \mathbf{u}_p \mathbf{e}_{p,mn}^T) \otimes E_j \otimes I_s \right)$$

$$\times (\Sigma \otimes I_n \otimes I_s) \left(\sum_{k,q=1}^{st} (\mathbf{e}_{q,st} \mathbf{v}_q^* \mathbf{e}_{k,st}) \otimes I_n \otimes F_k \right)$$

$$= \sum_{p=1}^{r} \sigma_p \left(\sum_{j=1}^{mn} (\mathbf{e}_{j,mn}^T \mathbf{u}_p) \otimes E_j \otimes I_s \right) \left(\sum_{k=1}^{st} (\mathbf{v}_p^* \mathbf{e}_{k,st}) \otimes I_n \otimes F_k \right)$$

$$= \sum_{p=1}^{r} \sigma_p \left(\sum_{j=1}^{mn} (\mathbf{e}_{j,mn}^T \mathbf{u}_p) E_j \right) \otimes \left(\sum_{k=1}^{st} (\mathbf{v}_p^* \mathbf{e}_{k,st}) F_k \right).$$

The above decomposition is called a *Schmidt decomposition* of M. The terms

$$\left(\sum_{j=1}^{mn} (\mathbf{e}_{j,mn}^T \mathbf{u}_p) E_j \right), \qquad \left(\sum_{k=1}^{st} (\mathbf{v}_p^* \mathbf{e}_{k,st}) F_k \right)$$

are never zero by construction.

Example 2.25. Let M be the 4×1 matrix

$$M = \frac{1}{\sqrt{2}} \begin{pmatrix} 1 \\ 0 \\ 0 \\ 1 \end{pmatrix}$$

where we have chosen $m = s = 2$ and $n = t = 1$. We write M in terms of the standard basis ($E_1 = F_1 = \mathbf{e}_{1,2}$ and $E_2 = F_2 = \mathbf{e}_{2,2}$)

$$M = \frac{1}{\sqrt{2}} \mathbf{e}_{1,2} \otimes \mathbf{e}_{1,2} + 0 \, \mathbf{e}_{1,2} \otimes \mathbf{e}_{2,2} + 0 \, \mathbf{e}_{2,2} \otimes \mathbf{e}_{1,2} + \frac{1}{\sqrt{2}} \mathbf{e}_{2,2} \otimes \mathbf{e}_{2,2}.$$

Thus we use

$$C = \frac{1}{\sqrt{2}} I_2$$

with the trivial singular value decomposition $U = V = I_2$ and $\Sigma = C$. Hence $\sigma_1 = \sigma_2 = \frac{1}{\sqrt{2}}$, $\mathbf{u}_1 = \mathbf{v}_1 = \mathbf{e}_{1,2}$, $\mathbf{u}_2 = \mathbf{v}_2 = \mathbf{e}_{2,2}$ and

$$\left(\sum_{j=1}^{2} (\mathbf{e}_{j,2}^T \mathbf{u}_1) E_j \right) = \mathbf{e}_{1,2}, \qquad \left(\sum_{j=1}^{2} (\mathbf{e}_{j,2}^T \mathbf{u}_2) E_j \right) = \mathbf{e}_{2,2},$$

$$\left(\sum_{k=1}^{2} (\mathbf{v}_1^* \mathbf{e}_{k,2}) F_k \right) = \mathbf{e}_{1,2}, \qquad \left(\sum_{k=1}^{2} (\mathbf{v}_2^* \mathbf{e}_{k,2}) F_k \right) = \mathbf{e}_{2,2}.$$

Hence the Schmidt rank is 2 and we have the Schmidt decomposition

$$M = \frac{1}{\sqrt{2}} \mathbf{e}_{1,2} \otimes \mathbf{e}_{1,2} + \frac{1}{\sqrt{2}} \mathbf{e}_{2,2} \otimes \mathbf{e}_{2,2}.$$

♣

2.12   Vec Operator and Sylvester Equation

In section 1.17 we have introduced the vec operator. Here we investigate the connection of the vec operator with the Kronecker product. A comprehensive discussion is given by Graham [25], Barnett [3] and Henderson and Searle [30].

The following theorems give useful properties of the vec operator. It is convenient to work in terms of the elementary vector $\mathbf{e}_{j,n}$, the j-th column of a unit matrix, i.e.

$$\mathbf{e}_{1,n} = \begin{pmatrix} 1 \\ 0 \\ \vdots \\ 0 \end{pmatrix}, \quad \mathbf{e}_{2,n} = \begin{pmatrix} 0 \\ 1 \\ \vdots \\ 0 \end{pmatrix}, \quad \dots, \quad \mathbf{e}_{n,n} = \begin{pmatrix} 0 \\ 0 \\ \vdots \\ 1 \end{pmatrix}.$$

The Kronecker product allows us to give a formal algebraic definition for the vec operator. The vec operator applied to the $m \times n$ matrix A is defined as the $mn \times 1$ matrix

$$\mathrm{vec}\,(A) := \sum_{j=1}^{n} \mathbf{e}_{j,n} \otimes (A\mathbf{e}_{j,n}) = (I_n \otimes A) \sum_{j=1}^{n} \mathbf{e}_{j,n} \otimes \mathbf{e}_{j,n}.$$

This operation is invertible, so that we can define $\text{vec}_{m \times n}^{-1}(\mathbf{x})$ by

$$\text{vec}_{m \times n}^{-1}(\mathbf{x}) = \sum_{i=1}^{m} \sum_{j=1}^{n} \left((\mathbf{e}_{j,n} \otimes \mathbf{e}_{i,m})^* \mathbf{x} \right) \mathbf{e}_{i,m} \otimes \mathbf{e}_{j,n}^*$$

where $\mathbf{x}$ is an $mn \times 1$ matrix.

Theorem 2.21. *Let A, B, C be matrices such that the matrix product ABC exists. Then*

$$\text{vec}(ABC) \equiv (C^T \otimes A)\text{vec}(B).$$

Proof. The j-th column of ABC, and hence the j-th subvector of $\text{vec}(ABC)$ is, for C having r rows,

$$ABC\mathbf{e}_{j,n} = AB \sum_i \mathbf{e}_{i,n} \mathbf{e}_{i,n}^T C \mathbf{e}_{j,n} = \sum_i c_{ij} AB\mathbf{e}_{i,n}$$

$$= (c_{1j}A \; c_{2j}A \cdots c_{rj}A) \begin{pmatrix} B\mathbf{e}_{1,n} \\ B\mathbf{e}_{2,n} \\ \vdots \\ B\mathbf{e}_{r,n} \end{pmatrix}$$

which is the j-th subvector of $(C^T \otimes A)\text{vec}(B)$. $\qquad\square$

Corollary 2.2. *Let A, B be $n \times n$ matrices. Then*

$$\text{vec}(AB) = (I_n \otimes A)\text{vec}(B)$$
$$\text{vec}(AB) = (B^T \otimes A)\text{vec}(I_n)$$
$$\text{vec}(AB) = \sum_{k=1}^{n} (B^T)_k \otimes A_k$$

where A_k denotes the k-th column of the matrix A.

The proof is straightforward and left as an exercise to the reader.

Theorem 2.22. *Let A, B, C, D be $n \times n$ matrices. Then*

$$\text{tr}(AD^T BDC) = (\text{vec}(D))^T (CA \otimes B^T)\text{vec}(D).$$

Proof.

$$\text{tr}(AD^T BDC) = \text{tr}(D^T BDCA) = (\text{vec}(D))^T \text{vec}(BDCA)$$
$$= (\text{vec}(D))^T (A^T C^T \otimes B)\text{vec}(D)$$

and, being a scalar, this equals its own transpose. $\qquad\square$

Consider the following equation (*Sylvester equation*)

$$AX + XB = C$$

where A is an $n \times n$ matrix, B an $m \times m$ matrix and X an $n \times m$ matrix. Assume that A, B and C are given. The task is to find the matrix X. To solve this problem we apply the vec operator to the left and right-hand side of the equation given above. Thus the equation can be written in the form

$$(I_m \otimes A + B^T \otimes I_n)\text{vec}(X) = \text{vec}(C).$$

This can easily be seen by applying the vec operator to both sides of the Sylvester equation. Thus the equation $AX + XB = C$ has a unique solution if and only if the matrix $(I_m \otimes A + B^T \otimes I_n)$ is nonsingular. It follows that all the eigenvalues must be nonzero. The eigenvalues are of the form $\lambda_i + \mu_j$, where λ_i are the eigenvalues of A and μ_j are the eigenvalues of B. Thus the equation has a unique solution if and only if

$$\lambda_i + \mu_j \neq 0, \qquad \text{for all } i, j.$$

We showed that $AX + XB = C$ has a unique solution if and only if A and $-B$ have no common eigenvalue.

For example consider

$$A = B = \begin{pmatrix} 0 & 1 \\ 1 & 0 \end{pmatrix}.$$

Then the eigenvalues of A and $-B$ are the same, namely ± 1. Thus the equation $AX + XB = C$ does not have a unique solution.

Let A be an $m \times n$ matrix. Obviously $\text{vec}(A^T)$ is just a permutation of $\text{vec}(A)$. The permutation (Hardy and Steeb [29])

$$P_{m,n} := \sum_{j=1}^{m} \sum_{k=1}^{n} (\mathbf{e}_{k,n} \otimes \mathbf{e}_{j,m})(\mathbf{e}_{j,m} \otimes \mathbf{e}_{k,n})^T$$

provides $P_{n,m}\text{vec}(A) = \text{vec}(A^T)$, $P_{m,n}\text{vec}(A^T) = \text{vec}(A)$.

Example 2.26. Consider the problem of finding a matrix A that implements the transforms

$$A\mathbf{x}_j = \mathbf{b}_j$$

where $j = 1, \ldots, k$, $\mathbf{x}_1, \ldots \mathbf{x}_k \in \mathbb{C}^m \setminus \{\mathbf{0}\}$ and $\mathbf{b}_1, \ldots, \mathbf{b}_k \in \mathbb{C}^n$. We write

$$AX = B, \qquad X = (\mathbf{x}_1, \ldots, \mathbf{x}_k), \qquad B = (\mathbf{b}_1, \ldots, \mathbf{b}_k).$$

Thus

$$\text{vec}(AX) = \text{vec}(I_m AX) = (X^T \otimes I_m)\text{vec}(A) = \text{vec}(B).$$

Solutions to this equation can be found using the pseudo inverse of the matrix $X^T \otimes I_m$. ♣

Example 2.27. We construct the 2×2 matrix A such that

$$A\begin{pmatrix} 1 \\ 0 \end{pmatrix} = \frac{1}{\sqrt{2}}\begin{pmatrix} 1 \\ 1 \end{pmatrix}, \qquad A\begin{pmatrix} 0 \\ 1 \end{pmatrix} = \frac{1}{\sqrt{2}}\begin{pmatrix} 1 \\ -1 \end{pmatrix}.$$

We solve for A from

$$A\begin{pmatrix} 1 & 0 \\ 0 & 1 \end{pmatrix} = \frac{1}{\sqrt{2}}\begin{pmatrix} 1 & 1 \\ 1 & -1 \end{pmatrix}.$$

Using the vec operator yields

$$\left[\begin{pmatrix} 1 & 0 \\ 0 & 1 \end{pmatrix}^T \otimes \begin{pmatrix} 1 & 0 \\ 0 & 1 \end{pmatrix}\right] \text{vec}(A) = \frac{1}{\sqrt{2}}\begin{pmatrix} 1 \\ 1 \\ 1 \\ -1 \end{pmatrix}.$$

We find the *Hadamard matrix*

$$A = \frac{1}{\sqrt{2}}\begin{pmatrix} 1 & 1 \\ 1 & -1 \end{pmatrix} \equiv \frac{1}{\sqrt{2}}(\sigma_1 + \sigma_3).$$

♣

Exercises. (1) Let L, X, N, Y be $n \times n$ matrices. Assume that

$$LX + XN = Y.$$

Show that $(I \otimes L + N^T \otimes I)\text{vec}(X) = \text{vec}(Y)$.

(2) Let $\mathbf{u}, \mathbf{v} \in \mathbb{C}^n$. Show that $\text{vec}(\mathbf{u}\mathbf{v}^T) = \mathbf{v} \otimes \mathbf{u}$.

(3) Show that $\text{vec}(A \otimes \mathbf{u}) = (\text{vec}(A)) \otimes \mathbf{u}$.

(4) Let

$$A = \begin{pmatrix} 1 & -1 \\ 0 & 2 \end{pmatrix}, \qquad B = \begin{pmatrix} 2 & -1 \\ 1 & 1 \end{pmatrix}, \qquad C = \begin{pmatrix} 1 & 3 \\ 2 & 2 \end{pmatrix}.$$

Find the solution to $AX + XB = C$.

(5) Let A, B, C, and X be $n \times n$ matrices over $\mathbb{R}$. Solve the linear matrix differential equation $dX/dt = AX + XB$, $X(t = 0) = C$ using the vec operator. As an example consider the matrices given by (4). Hint: The solution is given by $X(t) = \exp(tA)C\exp(tB)$.

(6) Let A, B and X be $n \times n$ matrices over $\mathbb{R}$. Show that the differential equation $dX/dt = AXB$, $X(t = 0) = X(0)$ can be transformed into

$$\frac{d}{dt}\text{vec}(X) = (A \otimes B^T)\text{vec}(X).$$

Show that the general solution is given by

$$\text{vec}(X(t)) = \exp(t(A \otimes B^T))\text{vec}(X(0)).$$

(7) Find the condition for the equation $AX - XA = \mu X$, $\mu \in \mathbb{C}$ to have a nontrivial solution.

(8) Find a 4×4 unitary matrix U such that

$$\sigma_1 \otimes \sigma_3 \equiv \begin{pmatrix} 0 & 0 & 1 & 0 \\ 0 & 0 & 0 & -1 \\ 1 & 0 & 0 & 0 \\ 0 & -1 & 0 & 0 \end{pmatrix} = U \begin{pmatrix} -1 & 0 & 0 & 0 \\ 0 & 0 & 0 & 1 \\ 0 & 0 & 1 & 0 \\ 0 & 1 & 0 & 0 \end{pmatrix} U^{-1}.$$

Apply the vec operator.

(9) Let $X, Y \in \mathbb{C}^{n \times n}$ and $[X, Y]_+$ denotes the anti-commutator. Show that

$$\text{vec}([X, Y]_+) \equiv (X \otimes I_n + I_n \otimes X)\text{vec}(Y).$$

2.13   Groups

In section 1.21 we introduced groups and some of their properties. Here we study the connection with the Kronecker product. In the following we consider groups which are given as $n \times n$ matrices with nonzero determinant and the matrix multiplication is the group multiplication. We start with an example.

Example 2.28. Let

$$e = \begin{pmatrix} 1 & 0 \\ 0 & 1 \end{pmatrix}, \qquad a = \begin{pmatrix} 0 & 1 \\ 1 & 0 \end{pmatrix}.$$

Then $ee = e$, $ea = a$, $ae = a$, $aa = e$. Obviously $\{e, a\}$ forms a group under matrix multiplication. The group is abelian. Now we can ask whether

$$\{e \otimes e, \ e \otimes a, \ a \otimes e, \ a \otimes a\}$$

form a group under matrix multiplication. We find

$$(e \otimes e)(e \otimes e) = e \otimes e, \qquad (e \otimes a)(e \otimes e) = e \otimes a,$$

$$(e \otimes e)(e \otimes a) = e \otimes a, \qquad (e \otimes a)(e \otimes a) = e \otimes e,$$

$$(a \otimes e)(e \otimes a) = a \otimes a, \qquad (e \otimes a)(a \otimes e) = a \otimes a,$$

$$(a \otimes e)(a \otimes e) = e \otimes e, \qquad (a \otimes a)(a \otimes a) = e \otimes e$$

and

$$(e \otimes e)^{-1} = e \otimes e, \quad (e \otimes a)^{-1} = e \otimes a, \quad (a \otimes e)^{-1} = a \otimes e, \quad (a \otimes a)^{-1} = a \otimes a.$$

Thus, the set $\{e \otimes e, e \otimes a, a \otimes e, a \otimes a\}$ forms a group under matrix multiplication. ♣

We consider now the general case. Let G be a group represented as $n \times n$ matrices and let G' be a group represented by $m \times m$ matrices. If $g_1, g_2 \in G$ and $g_1', g_2' \in G'$, then we have

$$(g_1 \otimes g_1')(g_2 \otimes g_2') = (g_1 g_2) \otimes (g_1' g_2').$$

We find that all pairs $g \otimes g'$ with $g \in G$ and $g' \in G'$ form a group. The identity element is $e \otimes e'$, where e is the identity element of G (i.e. the $n \times n$ unit matrix) and e' is the identity element of G' (i.e the $m \times m$ unit matrix). The inverse of $g \otimes g'$ is given by

$$(g \otimes g')^{-1} = g^{-1} \otimes g'^{-1}.$$

Since $(g_1 g_2)g_3 = g_1(g_2 g_3)$, $(g_1' g_2')g_3' = g_1'(g_2' g_3')$ we have

$$[(g_1 \otimes g_1')(g_2 \otimes g_2')] (g_3 \otimes g_3') = (g_1 g_2 \otimes g_1' g_2')(g_3 \otimes g_3')$$
$$= ((g_1 g_2)g_3 \otimes (g_1' g_2')g_3')$$
$$= (g_1(g_2 g_3) \otimes g_1'(g_2' g_3'))$$
$$= (g_1 \otimes g_1') [(g_2 \otimes g_2')(g_3 \otimes g_3')].$$

The above consideration can be formulated in group theory (Miller [44]) in a more abstract way.

Let G and G' be two groups.

Definition 2.9. The *direct product* $G \times G'$ is the group consisting of all ordered pairs (g, g') with $g \in G$ and $g' \in G'$. The product of two group elements is given by

$$(g_1, g_1')(g_2, g_2') := (g_1 g_2, g_1' g_2').$$

Obviously, $G \times G'$ is a group with identity element (e, e'), where e, e' are the identity elements of G, G', respectively. Indeed

$$(g, g')^{-1} = (g^{-1}, g'^{-1})$$

and the associative law is trivial to verify. The subgroup

$$G \times \{e'\} = \{(g, e') : g \in G\}$$

of $G \times G'$ is isomorphic to G with the isomorphism given by $(g, e') \leftrightarrow g$. Similarly the subgroup $\{e\} \times G'$ is isomorphic to G'. Since

$$(g, e')(e, g') = (e, g')(g, e') = (g, g')$$

it follows that (1) the elements of $G \times \{e'\}$ commute with the elements of $\{e\} \times G'$ and (2) every element of $G \times G'$ can be written uniquely as a product of an element in $g \times \{e'\}$ and an element in $\{e\} \times G'$.

Example 2.29. Let

$$e = \begin{pmatrix} 1 & 0 \\ 0 & 1 \end{pmatrix}, \qquad a = \begin{pmatrix} 0 & 1 \\ 1 & 0 \end{pmatrix}.$$

Then $ee = e$, $ea = a$, $ae = a$, $aa = e$. Obviously the set $\{e, a\}$ forms a group under matrix multiplication. Now we can ask whether

$$\{e \otimes e, a \otimes a\}$$

forms a group under matrix multiplication. We find

$$(e \otimes e)(e \otimes e) = e \otimes e, \qquad (e \otimes e)(a \otimes a) = a \otimes a$$

$$(a \otimes a)(e \otimes e) = a \otimes a, \qquad (a \otimes a)(a \otimes a) = e \otimes e$$

and $(e \otimes e)^{-1} = e \otimes e$, $(a \otimes a)^{-1} = a \otimes a$. The associative law is obviously satisfied. Thus the set $\{e \otimes e, a \otimes a\}$ forms a group under matrix multiplication. ♣

Obviously, the group $\{e \otimes e, a \otimes a\}$ is a subgroup of the commutative group $\{e \otimes e, e \otimes a, a \otimes e, a \otimes a\}$.

The above consideration can be formulated in group theory in a more abstract way.

Let G be a group. Let $g_j \in G$. We consider now the set of all ordered pairs (g_j, g_j). We define

$$(g_j, g_j)(g_k, g_k) := (g_j g_k, g_j g_k).$$

With this definition the set $\{(g_j, g_j)\}$ forms a group with the group multiplication defined above.

Exercises. (1) Let

$$M = \frac{1}{\sqrt{2}} \begin{pmatrix} 1 & i & 0 & 0 \\ 0 & 0 & i & 1 \\ 0 & 0 & i & -1 \\ 1 & -i & 0 & 0 \end{pmatrix}.$$

Show that the matrix is unitary. Consider the unitary matrices

$$U_H = \frac{1}{\sqrt{2}} \begin{pmatrix} 1 & 1 \\ 1 & -1 \end{pmatrix}, \quad U_S = \begin{pmatrix} 1 & 0 \\ 0 & i \end{pmatrix}, \quad U_{CNOT2} = \begin{pmatrix} 1 & 0 & 0 & 0 \\ 0 & 0 & 0 & 1 \\ 0 & 0 & 1 & 0 \\ 0 & 1 & 0 & 0 \end{pmatrix}.$$

Show that the matrix M can be written as

$$M = U_{CNOT2}(I_2 \otimes U_H)(U_S \otimes U_S).$$

Let $SO(4)$ be the special orthogonal Lie group and $SU(2)$ be the special unitary Lie group. Show that for every real orthogonal matrix $U \in SO(4)$, the matrix MUM^{-1} is the Kronecker product of two 2-dimensional special unitary matrices, i.e. $MUM^{-1} \in SU(2) \otimes SU(2)$.

(2) The Pauli spin matrices σ_1, σ_2, σ_3 are not elements of the Lie group $SU(2)$, but elements of the Lie group $U(2)$. The matrices $\tau_1 = i\sigma_1$, $\tau_2 = i\sigma_2$, $\tau_3 = i\sigma_3$ are elements of $SU(2)$. Find the group generated by

$$\tau_1 = i\sigma_1 = \begin{pmatrix} 0 & i \\ i & 0 \end{pmatrix}, \quad \tau_2 = i\sigma_2 = \begin{pmatrix} 0 & 1 \\ -1 & 0 \end{pmatrix}, \quad \tau_3 = i\sigma_3 = \begin{pmatrix} i & 0 \\ 0 & -i \end{pmatrix}$$

under matrix multiplication. Find the group generated by

$$\tau_1 \otimes \tau_1, \quad \tau_2 \otimes \tau_2, \quad \tau_3 \otimes \tau_3$$

under matrix multiplication. Find the group generated by

$$\tau_1 \otimes \tau_2, \quad \tau_2 \otimes \tau_3, \quad \tau_3 \otimes \tau_1.$$

2.14 Group Representation Theory

Let V be the vector space $\mathbb{C}^n$. Let $GL(n, \mathbb{C})$ be the set of all invertible $n \times n$ matrices over $\mathbb{C}$. Thus $GL(n, \mathbb{C})$ is a group under matrix multiplication.

For a more detailed discussion of group representation theory we refer to Miller [44], Ludwig and Falter [43] and Fulton and Harris [23].

Definition 2.10. An n dimensional *matrix representation* of G is a homomorphism

$$T : g \longrightarrow T(g)$$

of G into $GL(n, \mathbb{C})$.

As a consequence of this definition we find that

$$T(g_1)T(g_2) = T(g_1 g_2), \quad T(g)^{-1} = T(g^{-1}), \quad T(e) = I_n, \quad g_1, g_2, g \in G$$

where e is the identity element of G and I_n is the unit matrix in $\mathbb{C}^n$.

Example 2.30. Consider the group $G = \{\, e, a \,\}$ with $ee = e$, $ea = a$, $ae = a$, $aa = e$. Then $e \mapsto 1$, $a \mapsto 1$ is a one dimensional representation. Another one dimensional representation is $e \mapsto 1$, $a \mapsto -1$. A two dimensional representation is given by

$$e \mapsto \begin{pmatrix} 1 & 0 \\ 0 & 1 \end{pmatrix}, \qquad a \mapsto \begin{pmatrix} 0 & 1 \\ 1 & 0 \end{pmatrix}.$$

♣

Example 2.31. Consider the group $(\{a, b, c\}, \cdot)$ given by the group table

·	a	b	c
a	a	b	c
b	b	c	a
c	c	a	b

We list three representations, others are possible. Noticing that a is the group identity, $b^3 = a$ and $c^2 = b$ we have three choices for the one dimensional representation of b namely 1, $e^{i2\pi/3}$, $e^{i4\pi/3}$. Choosing $b \to e^{i2\pi/3}$ yields $c \to e^{i\pi/3}$, but then $b^2 \to e^{i4\pi/3}$ which is not the chosen representation of c. Here we choose $b \to e^{i4\pi/3}$.

$$T_1 : \{a, b, c\} \to GL(\mathbb{C}, 1), \quad T_1(a) = 1, \quad T_1(b) = e^{i4\pi/3}, \quad T_1(c) = e^{i2\pi/3}.$$

Since $e^{i\theta}$ can be interpreted as a rotation on the unit circle around the origin in the complex plane we choose the 2×2 representation $T_2 : \{a, b, c\} \to GL(\mathbb{C}, 2)$

$$T_2(a) = \begin{pmatrix} 1 & 0 \\ 0 & 1 \end{pmatrix}, \quad T_2(b) = \begin{pmatrix} \cos(4\pi/3) & \sin(4\pi/3) \\ -\sin(4\pi/3) & \cos(4\pi/3) \end{pmatrix},$$

$$T_2(c) = \begin{pmatrix} \cos(2\pi/3) & \sin(2\pi/3) \\ -\sin(2\pi/3) & \cos(2\pi/3) \end{pmatrix}$$

i.e.

$$T_2(a) = \begin{pmatrix} 1 & 0 \\ 0 & 1 \end{pmatrix}, \quad T_2(b) = \frac{1}{2}\begin{pmatrix} -1 & -\sqrt{3} \\ \sqrt{3} & -1 \end{pmatrix}, \quad T_2(c) = \frac{1}{2}\begin{pmatrix} -1 & \sqrt{3} \\ -\sqrt{3} & -1 \end{pmatrix}.$$

For a 3×3 matrix representation we can use permutation matrices for the representation: $T_3 : \{a, b, c\} \to GL(\mathbb{C}, 3)$

$$T_3(a) = \begin{pmatrix} 1 & 0 & 0 \\ 0 & 1 & 0 \\ 0 & 0 & 1 \end{pmatrix}, \quad T_3(b) = \begin{pmatrix} 0 & 1 & 0 \\ 0 & 0 & 1 \\ 1 & 0 & 0 \end{pmatrix}, \quad T_3(c) = \begin{pmatrix} 0 & 0 & 1 \\ 1 & 0 & 0 \\ 0 & 1 & 0 \end{pmatrix}.$$

♣

Definition 2.11. The *character* of $T(g)$ is defined by

$$\chi(g) := \text{tr}(T(g))$$

where tr denotes the trace.

Any group representation T of G with representation space V defines many matrix representations. For, if $\{\mathbf{v}_1, \ldots, \mathbf{v}_n\}$ is a basis of $\mathbb{C}^n$, the matrices $T(g) = (T(g)_{kj})$ defined by

$$T(g)\mathbf{v}_k = \sum_{j=1}^{n} T(g)_{jk}\mathbf{v}_j, \quad 1 \le k \le n$$

form an n dimensional matrix representation of G. Every choice of a basis for V yields a new matrix representation of G defined by T. However, any two such matrix representations T, T' are equivalent in the sense that there exists a matrix $S \in GL(n, \mathbb{C})$ such that

$$T'(g) = ST(g)S^{-1}$$

for all $g \in G$. If T, T' correspond to the bases $\{\mathbf{v}_i\}$, $\{\mathbf{v}'_i\}$ respectively, then for S we can take the matrix (S_{ji}) defined by

$$\mathbf{v}_i = \sum_{j=1}^{n} S_{ji}\mathbf{v}'_j, \quad i = 1, \ldots, n.$$

In order to determine all possible representations of a group G it is enough to find one representation T in each equivalence class.

Let T be a finite dimensional representation of a finite group G acting on the complex vector space $V = \mathbb{C}^n$.

Definition 2.12. A subspace W of V is *invariant* under T if

$$T(g)\mathbf{w} \in W$$

for every $g \in G, \mathbf{w} \in W$.

If W is invariant under T we can define a representation $T' = T|W$ of G on W by

$$T'(g)\mathbf{w} := T(g)\mathbf{w}, \qquad \mathbf{w} \in W.$$

This representation is called the restriction of T to W. If T is a unitary matrix so is T'.

Definition 2.13. The representation T is *reducible* if there is a proper subspace W of V which is invariant under T. Otherwise, T is *irreducible*.

A representation is irreducible if the only invariant subspaces of V are $\{\mathbf{0}\}$ and V itself. One dimensional and zero dimensional representations are necessarily irreducible.

Suppose T is reducible and W is a proper invariant subspace of V. If $\dim(W) = k$ and $\dim(V) = n$ we can find a basis $\mathbf{v}_1, \ldots, \mathbf{v}_n$ for V such that $\mathbf{v}_1, \ldots, \mathbf{v}_k$, $1 \le k \le n$, form a basis for W. Then the matrices $T(g)$ with respect to this basis take the form

$$\begin{pmatrix} T'(g) & * \\ 0 & T''(g) \end{pmatrix}.$$

The $k \times k$ matrices $T'(g)$ and the $(n-k) \times (n-k)$ matrices $T''(g)$ separately define matrix representations of G. In particular $T'(g)$ is the matrix of the representation $T'(g)$, with respect to the basis $\mathbf{v}_1, \ldots, \mathbf{v}_k$ of W. Here 0 is the $(n-k)k$ zero matrix.

Every reducible representation can be decomposed into irreducible representations in an almost unique manner. Thus the problem of constructing all representations of G simplifies to the problem of constructing all irreducible representations.

Definition 2.14. Let V and V' be vector spaces of dimensions n, m, respectively. Let $\{\mathbf{u}_j\}$, $\{\mathbf{u}'_k\}$ be bases for these vector spaces. We define $V \otimes V'$, the direct product (also called tensor product) of V and V', as the $n \times m$ dimensional vector space with basis

$$\{\, \mathbf{u}_j \otimes \mathbf{u}'_k \,\} \qquad 1 \le j \le n, \qquad 1 \le k \le m.$$

If $V = \mathbb{C}^n$ and $V' = \mathbb{C}^m$ then $\mathbf{u}_j \otimes \mathbf{u}'_k$ would be the Kronecker product. Thus, any $\mathbf{w} \in V \otimes V'$ can be written uniquely in the form

$$\mathbf{w} = \sum_{j=1}^{n} \sum_{k=1}^{m} \alpha_{jk} \mathbf{u}_j \otimes \mathbf{u}'_k.$$

If

$$\mathbf{u} = \sum_{j=1}^{n} \alpha_j \mathbf{u}_j, \qquad \mathbf{u}' = \sum_{k=1}^{m} \beta_k \mathbf{u}'_k$$

we define the vector $\mathbf{u} \otimes \mathbf{u}' \in V \otimes V'$ by

$$\mathbf{u} \otimes \mathbf{u}' := \sum_{j=1}^{n} \sum_{k=1}^{m} \alpha_j \beta_k \mathbf{u}_j \otimes \mathbf{u}'_k.$$

Definition 2.15. If T_1 and T_2 are matrix representation of the groups G_1 and G_2, respectively, we can define a matrix representation T of the direct product group $G_1 \times G_2$ on $V_1 \otimes V_2$ by

$$(T(g_1 g_2))(\mathbf{u}_1 \otimes \mathbf{u}_2) := (T_1(g_1)\mathbf{u}_1) \otimes (T_2(g_2)\mathbf{u}_2)$$

where $g_1 \in G_1$, $g_2 \in G_2$, $\mathbf{u}_1 \in V_1$, $\mathbf{u}_2 \in V_2$.

If T_1 is n_1 dimensional and T_2 is n_2 dimensional, then T is $n_1 n_2$ dimensional. Furthermore, we find that the character χ of T is given by

$$\chi(g_1 g_2) = \chi_1(g_1)\chi_2(g_2)$$

where χ_j is the character of T_j. This is due to the fact that

$$\text{tr}(A \otimes B) \equiv \text{tr}(A)\text{tr}(B)$$

where A is an $m \times m$ matrix and B is an $n \times n$ matrix.

Theorem 2.23. *The matrix representation T is irreducible if and only if T_1 and T_2 are irreducible.*

For the proof we refer to the books of Miller [44] and Ludwig and Falter [43].

Exercises. (1) Show that

$$e \mapsto \begin{pmatrix} 1 & 0 & 0 \\ 0 & 1 & 0 \\ 0 & 0 & 1 \end{pmatrix}, \qquad a \mapsto \begin{pmatrix} 0 & 0 & 1 \\ 0 & 1 & 0 \\ 1 & 0 & 0 \end{pmatrix}$$

is a three dimensional reducible representation of the group $G = \{e, a\}$.

(2) Consider the group $G = \{e, a\}$ with $ee = e$, $ea = a$, $ae = a$, $aa = e$. Show that the representations $e \mapsto 1$, $a \mapsto 1$ and $e \mapsto 1$, $a \mapsto -1$ are irreducible. Show that the representations

$$e \mapsto \begin{pmatrix} 1 & 0 \\ 0 & 1 \end{pmatrix}, \qquad a \mapsto \begin{pmatrix} 1 & 0 \\ 0 & -1 \end{pmatrix}$$

and

$$e \mapsto \begin{pmatrix} 1 & 0 \\ 0 & 1 \end{pmatrix}, \qquad a \mapsto \begin{pmatrix} 0 & 1 \\ 1 & 0 \end{pmatrix}$$

are reducible.

(3) Show that the set $\{ +1, -1, +i, -i \}$ forms a group under multiplication. Is the group Abelian? Find a two dimensional representation. Find all irreducible representations. We recall that the number of irreducible representations is equal to the number of conjugacy classes. Thus first find all the conjugacy classes.

(4) Consider the group G_1 of all 2×2 permutation matrices and the group G_2 of all 3×3 permutation matrices. Obviously $V_1 = \mathbb{R}^2$ and $V_2 = \mathbb{R}^3$. Find reducible and irreducible matrix representations of the direct product group $G_1 \times G_2$ on $V_1 \otimes V_2$.

(5) Let $(G, \cdot)$ be an arbitrary finite group with n elements, i.e.

$$G = \{g_1, \ldots, g_n\}.$$

Let $\phi : G \to \{1, 2, \ldots, n\}$ be the one-to-one function $\phi(g_j) = j$. Define $T : G \to GL(n, \mathbb{C})$ by

$$T(g) \begin{pmatrix} x_{\phi(g_1)} \\ x_{\phi(g_2)} \\ \vdots \\ x_{\phi(g_n)} \end{pmatrix} = \begin{pmatrix} x_{\phi(g \cdot g_1)} \\ x_{\phi(g \cdot g_2)} \\ \vdots \\ x_{\phi(g \cdot g_n)} \end{pmatrix}, \qquad \forall x_{\phi(g_1)}, x_{\phi(g_2)}, \ldots, x_{\phi(g_n)} \in \mathbb{C}.$$

Does T provide a matrix representation of the group $(G, \cdot)$? Let

$$G = \{I_2, \sigma_1, i\sigma_2, \sigma_3, -I_2, -\sigma_1, -i\sigma_2, -\sigma_3\}$$

and $\cdot$ denote matrix multiplication. Find $T(G)$.

2.15   Commutators and Anti-Commutators

In this section we consider commutators and anticommutators of direct sums, Kronecker sums and Kronecker products.

First we consider the direct sum.

Theorem 2.24. *Let A and C be $m \times m$ matrices and let B and D be $n \times n$ matrices. Then*

$$[A \oplus B, C \oplus D] = [A, C] \oplus [B, D].$$

Proof. Straightforward calculation yields

$$[A \oplus B, C \oplus D] = (AC) \oplus (BD) - (CA) \oplus (DB) = [A, C] \oplus [B, D]$$

where we used the fact that $A \oplus B + C \oplus D = (A + C) \oplus (B + D)$. □

Theorem 2.25. *Let A and C be $m \times m$ matrices and let B and D be $n \times n$ matrices. Then*

$$[A \oplus B, C \oplus D]_+ = [A, C]_+ \oplus [B, D]_+.$$

A similar theorem holds for the commutator of Kronecker sums. However, we do not have an analogous result for the anticommutator.

Theorem 2.26. *Let A and C be $m \times m$ matrices and let B and D be $n \times n$ matrices. Then*

$$[A \oplus_K B, C \oplus_K D] = [A, C] \oplus_K [B, D].$$

Proof. Since

$$[A \otimes I_n, I_m \otimes D] = A \otimes D - A \otimes D = 0_{m+n}$$

where 0_{m+n} is the $(m + n) \times (m + n)$ zero matrix, we find

$$\begin{aligned}
[A \oplus_K B, C \oplus_K D] &= [A \otimes I_n + I_m \otimes B, C \otimes I_n + I_m \otimes D] \\
&= [A \otimes I_m, C \otimes I_n] + [A \otimes I_n, I_m \otimes D] \\
&\quad + [I_m \otimes B, C \otimes I_n] + [I_m \otimes B, I_m \otimes D] \\
&= [A, C] \otimes I_n + I_m \otimes [B, D] = [A, C] \oplus_K [B, D].
\end{aligned}$$
□

The commutator and anticommutator of Kronecker products can be expressed in terms of commutators and anticommutators as follows.

Theorem 2.27. *Let A and C be $m \times m$ matrices and let B and D be $n \times n$ matrices. Then*

$$[A \otimes B, C \otimes D] \equiv \frac{1}{2} \left([A, C] \otimes [B, D]_+ + [A, C]_+ \otimes [B, D] \right)$$

$$[A \otimes B, C \otimes D]_+ \equiv \frac{1}{2} \left([A, C] \otimes [B, D] + [A, C]_+ \otimes [B, D]_+ \right).$$

The proof is straightforward.

Consider the Pauli spin matrices σ_1, σ_2, σ_3. The commutators and anti-commutators are given by

$$
\begin{array}{lll}
[\sigma_1, \sigma_1] = 0_2, & [\sigma_1, \sigma_2] = 2i\sigma_3, & [\sigma_1, \sigma_3] = -2i\sigma_2, \\
[\sigma_2, \sigma_2] = 0_2, & [\sigma_2, \sigma_3] = 2i\sigma_1, & [\sigma_3, \sigma_3] = 0_2, \\
[\sigma_1, \sigma_1]_+ = 2I_2, & [\sigma_1, \sigma_2]_+ = 0_2, & [\sigma_1, \sigma_3]_+ = 0_2, \\
[\sigma_2, \sigma_2]_+ = 2I_2, & [\sigma_2, \sigma_3]_+ = 0_2, & [\sigma_3, \sigma_3]_+ = 2I_2.
\end{array}
$$

It follows that

$$
[\sigma_1 \otimes \sigma_2, \sigma_2 \otimes \sigma_3] = \frac{1}{2} \left([\sigma_1, \sigma_2] \otimes [\sigma_2, \sigma_3]_+ + [\sigma_1, \sigma_2]_+ \otimes [\sigma_2, \sigma_3] \right) = 0_4
$$

and $[\sigma_1 \otimes \sigma_2, \sigma_2 \otimes \sigma_3]_+ = -2\sigma_3 \otimes \sigma_1$.

Theorem 2.28. *Let A and B be $n \times n$ matrices. If $A \otimes B = B \otimes A$, then $[A, B] = 0_n$.*

Proof. If $A \otimes B = B \otimes A$, then $A = 0_n$ (and therefore $[A, B] = 0_n$ trivially) or $A \neq 0_n$. If $A \neq 0_n$, then there exists $(A)_{jk} \neq 0$ and

$$
(A)_{jk} B = (B)_{jk} A
$$

so that

$$
B = \frac{(B)_{jk}}{(A)_{jk}} A
$$

and

$$
[A, B] = A \frac{(B)_{jk}}{(A)_{jk}} A - \frac{(B)_{jk}}{(A)_{jk}} A^2 = \frac{(B)_{jk}}{(A)_{jk}} (A^2 - A^2) = 0_n. \qquad \square
$$

2.16   Inversion of Partitioned Matrices

Partitioned matrices where the elements in each block satisfy some particular pattern, arise widely in systems theory. Block Toeplitz, block Hankel and block circulant matrices play a central role in linear control theory, signal processing and statistics (Davis [17]). Any partitioned matrix, each of whose blocks has the same pattern amongst its elements, is similar to a matrix in which this pattern occurs in block form. This transformation is advantageous for the purpose of inversion. A simple derivation (Barnett [4]) is given using the Kronecker matrix product.

Suppose that A is an $mn \times mn$ matrix partitioned into blocks A_{ij}, each of which is $n \times n$ and has the same pattern amongst its elements. We show that inversion of A is equivalent to inverting a matrix B in which the pattern appears amongst the blocks. For example, if each A_{ij} has Toeplitz form, then B is a block Toeplitz matrix, and can be efficiently inverted using a standard algorithm.

Let X be an $M \times N$ matrix having columns $\mathbf{x}_1, \ldots, \mathbf{x}_M$, and define an $MN \times 1$ column vector, $\mathrm{vec}(X)$ by

$$\mathrm{vec}(X) := \begin{pmatrix} \mathbf{x}_1 \\ \mathbf{x}_2 \\ \vdots \\ \mathbf{x}_M \end{pmatrix}.$$

The $MN \times MN$ vec-permutation matrix $P_{M,N}$ is then defined by

$$\mathrm{vec}(X) = P_{M,N}\mathrm{vec}(X^T)$$

where the superscript T denotes transpose. It follows that

$$P_{N,M}P_{M,N} = I_{MN}$$

where I_{MN} is the unit matrix of order MN. The key property of the vec-permutation matrix is that it reverses the order of a Kronecker product $X \otimes Y$ (here Y is $P \times R$), namely

$$P_{M,P}(X \otimes Y)P_{R,N} = Y \otimes X.$$

In the following we assume that $m = 2$, so that

$$A := \begin{pmatrix} A_{11} & A_{12} \\ A_{21} & A_{22} \end{pmatrix}$$

and let each block be an $n \times n$ Toeplitz matrix, i.e.

$$A_{ij} = (a_{rs}^{(ij)}) = (a_{r-s}^{(ij)}), \qquad i,j = 1,2; \qquad r,s = 1,2,\ldots,n.$$

The matrix A can be written as

$$A = B_0 \otimes I_n + B_{-1} \otimes K_1 + B_{-2} \otimes K_2 + \cdots + B_{1-n} \otimes K_{n-1}$$
$$+ B_1 \otimes K_1^T + B_2 \otimes K_2^T + \cdots + B_{n-1} \otimes K_{n-1}^T$$

where

$$B_k := \begin{pmatrix} a_k^{(11)} & a_k^{(12)} \\ a_k^{(21)} & a_k^{(22)} \end{pmatrix}$$

for $k = 1 - n, 2 - n, \ldots, n - 2, n - 1$ and I_n is the $n \times n$ unit matrix. Here, K_i denotes an $n \times n$ matrix all of whose elements are zero except for a line of ones parallel to, and above, the principal diagonal, from entry $(1, i + l)$ to entry $(n - 1, n)$. Applying

$$P_{N,P}(X \otimes Y)P_{R,M} = Y \otimes X$$

with $N = M = 2$, $P = R = n$ yields

$$I_{2,n}AI_{n,2} = I_n \otimes B_0 + K_1 \otimes B_{-1} + \cdots + K_{n-1} \otimes B_{1-n}$$
$$+ K_1^T \otimes B_1 + \cdots + K_{n-1}^T \otimes B_{n-1}.$$

Thus the right-hand side is a $2n \times 2n$ matrix B partitioned into 2×2 blocks

$$B_{rs} = B_{r-s}, \qquad r, s = 1, \ldots, n.$$

Therefore, the $2n \times 2n$ matrix A having Toeplitz blocks is similar to this block Toeplitz matrix B. The result can be extended (Barnett [4]).

2.17   Nearest Kronecker Product

This section describes the nearest Kronecker product problem and its solution for the case when we use the Frobenius norm (van Loan and Pitsianis [71]).

We consider first the general case. Let M be an $ms \times nt$ matrix over $\mathbb{C}$. It is not generally true that $m \times n$ and $s \times t$ matrices A and B exist such that $M = A \otimes B$, except of course if $m = n = 1$ or $s = t = 1$.

Example 2.32. Let $m = s = 2$, $n = t = 1$ and $M = (1, 0, 0, 1)^T$. We try to solve for $A = \begin{pmatrix} a_1 & a_2 \end{pmatrix}^T$ and $B = \begin{pmatrix} b_1 & b_2 \end{pmatrix}^T$ such that

$$M = \begin{pmatrix} 1 \\ 0 \\ 0 \\ 1 \end{pmatrix} = A \otimes B = \begin{pmatrix} a_1b_1 \\ a_1b_2 \\ a_2b_1 \\ a_2b_2 \end{pmatrix}.$$

The product of the first and fourth entries provides $1 \cdot 1 = a_1b_1 = a_2b_2$, while the product of the second and third entries provides $0 \cdot 0 = a_1b_2 = a_2b_1$ which are contradictory. ♣

Example 2.33. Let $m = s = 2$, $n = t = 2$. We try to solve for

$$A = \begin{pmatrix} a_1 & a_2 \\ a_3 & a_4 \end{pmatrix}, \qquad B = \begin{pmatrix} b_1 & b_2 \\ b_3 & b_4 \end{pmatrix}$$

such that

$$M = \begin{pmatrix} 1 & 0 & 0 & 0 \\ 0 & 1 & 0 & 0 \\ 0 & 0 & 0 & 1 \\ 0 & 0 & 1 & 0 \end{pmatrix} = A \otimes B = \begin{pmatrix} a_1b_1 & a_1b_2 & a_2b_1 & a_2b_2 \\ a_1b_3 & a_1b_4 & a_2b_3 & a_2b_4 \\ a_3b_1 & a_3b_2 & a_4b_1 & a_4b_2 \\ a_3b_3 & a_3b_4 & a_4b_3 & a_4b_4 \end{pmatrix}.$$

Thus we have $a_1b_1 = 1$ so that $a_1 \neq 0$, $a_1b_3 = 0$ so that $b_3 = 0$ and consequently $a_4b_3 \neq 1$. Thus the equation has no solution for A and B. ♣

Thus we consider the *nearest Kronecker product* problem, i.e. to find $m \times n$ and $s \times t$ matrices A and B such that

$$\| M - A \otimes B \|$$

is a minimum and $\| \cdot \|$ denotes some norm. Different norms lead to different A and B. In this section we consider the Frobenius norm $\| \cdot \|_F$.

Using the fact that

$$\| A \|_F = \sqrt{\sum_{j=1}^{m} \sum_{k=1}^{n} |(A)_{j,k}|^2}$$

is independent of the order in which the sum over the elements of A is taken, we find

$$\| A \|_F = \| R(A) \|_F$$

where $R(A)$ is a matrix with mn entries, where each entry of A appears exactly once in $R(A)$. Specifically let $R = R_{m \times n, s \times t}$ be the *reshaping operator*

$$R_{m \times n, s \times t}(A \otimes B) = (\mathrm{vec}(A))(\mathrm{vec}(B))^T$$

for all $m \times n$ matrices A and $s \times t$ matrices B. Clearly $A \otimes B$ and $(\mathrm{vec}(A))(\mathrm{vec}(B))^T$ have the same entries in a different matrix configuration.

Example 2.34. We find $\mathbf{a} \in \mathbb{R}^2$ and $\mathbf{b} \in \mathbb{R}^2$ which minimizes

$$\| (1\ 0\ 0\ 1)^T - \mathbf{a} \otimes \mathbf{b} \|^2, \quad \mathbf{a} = \begin{pmatrix} a_1 \\ a_2 \end{pmatrix}, \quad \mathbf{b} = \begin{pmatrix} b_1 \\ b_2 \end{pmatrix}$$

where $\| \mathbf{x} \| = \sqrt{\mathbf{x}^T \mathbf{x}}$ denotes the Euclidean norm for $\mathbf{x} \in \mathbb{R}^4$. Since

$$\mathbf{a} \otimes \mathbf{b} = \begin{pmatrix} a_1b_1 \\ a_1b_2 \\ a_2b_1 \\ a_2b_2 \end{pmatrix}$$

we obtain

$$f(a_1, a_2, b_1, b_2) = \| \begin{pmatrix} 1 & 0 & 0 & 1 \end{pmatrix}^T - \mathbf{a} \otimes \mathbf{b} \|^2$$
$$= (a_1 b_1 - 1)^2 + (a_1 b_2)^2 + (a_2 b_1)^2 + (a_2 b_2 - 1)^2.$$

To minimize we have to solve the system of four nonlinear equations

$$\frac{\partial f}{\partial a_1} = 2(a_1 b_1 - 1)b_1 + 2a_1 b_2^2 = 0, \qquad \frac{\partial f}{\partial a_2} = 2a_2 b_1^2 + 2(a_2 b_2 - 1)b_2 = 0,$$

$$\frac{\partial f}{\partial b_1} = 2(a_1 b_1 - 1)a_1 + 2a_2^2 b_1 = 0, \qquad \frac{\partial f}{\partial b_2} = 2a_1^2 b_2 + 2(a_2 b_2 - 1)a_2 = 0.$$

These equations can be written as

$$b_1 = a_1(b_1^2 + b_2^2), \quad b_2 = a_2(b_1^2 + b_2^2), \quad a_1 = b_1(a_1^2 + a_2^2), \quad a_2 = b_2(a_1^2 + a_2^2).$$

If $b_1^2 + b_2^2 = 0$, then $a_1 = a_2 = b_1 = b_2 = 0$ and $f(0,0,0,0) = 2$. Now consider $b_1^2 + b_2^2 \neq 0$. Then

$$a_1 = \frac{b_1}{b_1^2 + b_2^2}, \qquad a_2 = \frac{b_2}{b_1^2 + b_2^2}.$$

For these values we find

$$f(b_1/(b_1^2 + b_2^2), b_2/(b_1^2 + b_2^2), b_1, b_2) = 1.$$

To determine whether this is the global minimum we first note that

$$f(a_1, a_2, b_1, b_2) = (\mathbf{a}^T \mathbf{a})(\mathbf{b}^T \mathbf{b}) - 2\mathbf{a}^T \mathbf{b} + 2.$$

The *Cauchy-Schwarz inequality* provides $(\mathbf{a}^T \mathbf{a})(\mathbf{b}^T \mathbf{b}) \geq (\mathbf{a}^T \mathbf{b})^2$. Consequently

$$f(a_1, a_2, b_1, b_2) \geq (\mathbf{a}^T \mathbf{b})^2 - 2\mathbf{a}^T \mathbf{b} + 2 = (\mathbf{a}^T \mathbf{b} - 1)^2 + 1 \geq 1.$$

Hence the global minimum is found at

$$\mathbf{a} = \frac{1}{b_1^2 + b_2^2} \mathbf{b}, \quad \mathbf{b} \in \mathbb{R}^2 \setminus \left\{ \begin{pmatrix} 0 \\ 0 \end{pmatrix} \right\}. \qquad \clubsuit$$

Example 2.35. Consider

$$A = \begin{pmatrix} a_1 & a_2 \\ a_3 & a_4 \end{pmatrix}, \quad B = \begin{pmatrix} b_1 & b_2 \\ b_3 & b_4 \\ b_5 & b_6 \end{pmatrix} \Rightarrow A \otimes B = \begin{pmatrix} a_1 b_1 & a_1 b_2 & a_2 b_1 & a_2 b_2 \\ a_1 b_3 & a_1 b_4 & a_2 b_3 & a_2 b_4 \\ a_1 b_5 & a_1 b_6 & a_2 b_5 & a_2 b_6 \\ a_3 b_1 & a_3 b_2 & a_4 b_1 & a_4 b_2 \\ a_3 b_3 & a_3 b_4 & a_4 b_3 & a_4 b_4 \\ a_3 b_5 & a_3 b_6 & a_4 b_5 & a_4 b_6 \end{pmatrix}$$

and

$$(\operatorname{vec}(A))(\operatorname{vec}(B))^T = \begin{pmatrix} a_1 \\ a_3 \\ a_2 \\ a_4 \end{pmatrix} \begin{pmatrix} b_1 \\ b_3 \\ b_5 \\ b_2 \\ b_4 \\ b_6 \end{pmatrix}^T = \begin{pmatrix} a_1b_1 & a_1b_3 & a_1b_5 & a_1b_2 & a_1b_4 & a_1b_6 \\ a_3b_1 & a_3b_3 & a_3b_5 & a_3b_2 & a_3b_4 & a_3b_6 \\ a_2b_1 & a_2b_3 & a_2b_5 & a_2b_2 & a_2b_4 & a_2b_6 \\ a_4b_1 & a_4b_3 & a_4b_5 & a_4b_2 & a_4b_4 & a_4b_6 \end{pmatrix}.$$

Matching entries from the first matrix with the second allows us to calculate the reshaping of a matrix

$$R_{2\times2,2\times3} \begin{pmatrix} 1 & 2 & 3 & 4 \\ 5 & 6 & 7 & 8 \\ 9 & 10 & 11 & 12 \\ 13 & 14 & 15 & 16 \\ 17 & 18 & 19 & 20 \\ 21 & 22 & 23 & 24 \end{pmatrix} = \begin{pmatrix} 1 & 5 & 9 & 2 & 6 & 10 \\ 13 & 17 & 21 & 14 & 18 & 22 \\ 3 & 7 & 11 & 4 & 8 & 12 \\ 15 & 19 & 23 & 16 & 20 & 24 \end{pmatrix}.$$

A more concise way to obtain the reshaping operator is as follows
$$R_{m\times n,s\times t}(A \otimes B)$$

$$= \operatorname{vec}(A)(\operatorname{vec}(B))^T = \left(\sum_{j=1}^{n} \mathbf{e}_{j,n} \otimes A\mathbf{e}_{j,n} \right) (\operatorname{vec}(B))^T$$

$$= \left(\sum_{j=1}^{n} \sum_{k=1}^{m} \mathbf{e}_{j,n} \otimes (\mathbf{e}_{k,m}^T A\mathbf{e}_{j,n})\mathbf{e}_{k,m} \right) (\operatorname{vec}(B))^T$$

$$= \sum_{j=1}^{n} \sum_{k=1}^{m} \mathbf{e}_{j,n} \otimes \mathbf{e}_{k,m} \left(\operatorname{vec}\left(\mathbf{e}_{k,m}^T A\mathbf{e}_{j,n}B \right) \right)^T$$

$$= \sum_{j=1}^{n} \sum_{k=1}^{m} \mathbf{e}_{j,n} \otimes \mathbf{e}_{k,m} \left(\operatorname{vec}\left([\mathbf{e}_{k,m}^T \otimes I_s]A \otimes B[\mathbf{e}_{j,n} \otimes I_t] \right) \right)^T.$$

Thus we take the vec (transposed) of each sub matrix of $A \otimes B$ with the same dimensions as B, in row then column order, to form the reshaped matrix

$$R_{2\times2,2\times3} \left(\begin{array}{cc|cc} 1 & 2 & 3 & 4 \\ 5 & 6 & 7 & 8 \\ 9 & 10 & 11 & 12 \\ \hline 13 & 14 & 15 & 16 \\ 17 & 18 & 19 & 20 \\ 21 & 22 & 23 & 24 \end{array} \right) = \begin{pmatrix} 1 & 5 & 9 & 2 & 6 & 10 \\ 13 & 17 & 21 & 14 & 18 & 22 \\ 3 & 7 & 11 & 4 & 8 & 12 \\ 15 & 19 & 23 & 16 & 20 & 24 \end{pmatrix}.$$

Now applying the reshaping operator to the minimization problem we find

$$\|M - A \otimes B\|_F = \|R_{m \times n, s \times t}(M - A \otimes B)\|_F$$
$$= \|R_{m \times n, s \times t}(M) - R_{m \times n, s \times t}(A \otimes B)\|_F$$
$$= \|R_{m \times n, s \times t}(M) - \text{vec}(A)(\text{vec}(B))^T\|_F.$$

Notice that $R_{m \times n, s \times t}(M)$ is $mn \times st$ while M is $ms \times nt$. Clearly $\text{vec}(A)(\text{vec}(B))^T$ has rank 1. Thus we need to find a rank-1 approximation of $R_{m \times n, s \times t}(M)$. Let $R_{m \times n, s \times t}(M) = U\Sigma V^*$ be a singular value decomposition of $R_{m \times n, s \times t}(M)$, then the minimum is found when

$$\text{vec}(A)(\text{vec}(B))^T = U(\sigma_1 \mathbf{e}_{1,mn} \mathbf{e}_{1,st}^T)V^* = (\sigma_1 U\mathbf{e}_{1,mn})(V\mathbf{e}_{1,st})^*$$

where σ_1 is the largest singular value of $R_{m \times n, s \times t}(M)$. Thus we may set (amongst other choices) $\text{vec}(A) = \sigma_1 U\mathbf{e}_{1,mn}$ and $\text{vec}(B) = \overline{V\mathbf{e}_{1,st}}$.

Next we consider the case when A is known, and B has to be determined. Given $A \neq 0_{m \times n}$, then

$$(B)_{ij} = \frac{\text{tr}(A^* \tilde{M}_{ij})}{\|A\|_F^2}$$

minimizes $\|M - A \otimes B\|_F^2$, where $\tilde{M}_{ij} := (I_m \otimes \mathbf{e}_{i,s})^* M (I_n \otimes \mathbf{e}_{j,t})$. Note that if A is the $m \times n$ zero matrix then B is arbitrary. First note that

$$\sum_{i=1}^{s} \sum_{j=1}^{t} \tilde{M}_{ij} \otimes \mathbf{e}_{i,s} \mathbf{e}_{j,t}^* = \sum_{i=1}^{s} \sum_{j=1}^{t} \left[(I_m \otimes \mathbf{e}_{i,s})^* M (I_n \otimes \mathbf{e}_{j,t}) \right] \otimes \mathbf{e}_{i,s} \mathbf{e}_{j,t}^*$$

$$= \sum_{i=1}^{s} \sum_{j=1}^{t} \left(I_m \otimes \mathbf{e}_{i,s} \mathbf{e}_{i,s}^* \right) M \left(I_n \otimes \mathbf{e}_{j,t} \mathbf{e}_{j,t}^* \right)$$

$$= \left(I_m \otimes \sum_{i=1}^{s} \mathbf{e}_{i,s} \mathbf{e}_{i,s}^* \right) M \left(I_n \otimes \sum_{j=1}^{t} \mathbf{e}_{j,t} \mathbf{e}_{j,t}^* \right)$$

$$= (I_m \otimes I_s) M (I_n \otimes I_t) = I_{ms} M I_{nt} = M.$$

Since B is a critical point of $\|M - A \otimes B\|_F^2$, we find the critical points of $\|M - A \otimes B\|_F^2$. Expressing the entries of B in terms of their real and imaginary parts $(B)_{ij} = \Re(B)_{ij} + \Im(B)_{ij}\sqrt{-1}$ we obtain

$$
\frac{\partial}{\partial\Re(B)_{ij}}\|M - A \otimes B\|_F^2
$$

$$
= \frac{\partial}{\partial\Re(B)_{ij}}\mathrm{tr}([M - A \otimes B]^*[M - A \otimes B])
$$

$$
= \frac{\partial}{\partial\Re(B)_{ij}}\mathrm{tr}(M^*M - [A \otimes B]^*M - M^*[A \otimes B] + (A^*A) \otimes (B^*B))
$$

$$
= -\frac{\partial}{\partial\Re(B)_{ij}}\Big[\mathrm{tr}([A \otimes B]^*M) + \mathrm{tr}(M^*[A \otimes B])\Big] + \mathrm{tr}(A^*A)\frac{\partial\mathrm{tr}(B^*B)}{\partial\Re(B)_{ij}}
$$

$$
= -\frac{\partial}{\partial\Re(B)_{ij}}2\Re\Big[\mathrm{tr}(M^*[A \otimes B])\Big] + 2\|A\|_F^2\Re(B)_{ij} = 0
$$

and

$$
\frac{\partial}{\partial\Im(B)_{ij}}\|M - A \otimes B\|_F^2
$$

$$
= \frac{\partial}{\partial\Im(B)_{ij}}\mathrm{tr}([M - A \otimes B]^*[M - A \otimes B])
$$

$$
= \frac{\partial}{\partial\Im(B)_{ij}}\mathrm{tr}(M^*M - [A \otimes B]^*M - M^*[A \otimes B] + (A^*A) \otimes (B^*B))
$$

$$
= -\frac{\partial}{\partial\Im(B)_{ij}}\Big[\mathrm{tr}([A \otimes B]^*M) + \mathrm{tr}(M^*[A \otimes B])\Big] + \mathrm{tr}(A^*A)\frac{\partial\mathrm{tr}(B^*B)}{\partial\Im(B)_{ij}}
$$

$$
= -\frac{\partial}{\partial\Im(B)_{ij}}2\Re\Big[\mathrm{tr}(M^*[A \otimes B])\Big] + 2\|A\|_F^2\Im(B)_{ij} = 0
$$

since $\mathrm{tr}(X) = \overline{\mathrm{tr}(X^*)} \Rightarrow \mathrm{tr}(X) + \mathrm{tr}(X^*) = \mathrm{tr}(X) + \overline{\mathrm{tr}(X)} = 2\Re\Big[\mathrm{tr}(X)\Big]$ for any square matrix X. It follows that

$$
\Re(B)_{ij} = \frac{1}{\|A\|_F^2}\Re\left[\frac{\partial}{\partial\Re(B)_{ij}}\mathrm{tr}(M^*[A \otimes B])\right]
$$

$$
\Im(B)_{ij} = \frac{1}{\|A\|_F^2}\Re\left[\frac{\partial}{\partial\Im(B)_{ij}}\mathrm{tr}(M^*[A \otimes B])\right].
$$

Now

$$\text{tr}(M^*[A \otimes B]) = \text{tr}\left(\sum_{i=1}^{s}\sum_{j=1}^{t}\left(\tilde{M}_{ij} \otimes \mathbf{e}_{i,s}\mathbf{e}_{j,t}^*\right)^*[A \otimes B]\right)$$

$$= \text{tr}\left(\sum_{i=1}^{s}\sum_{j=1}^{t}\left(\tilde{M}_{ij}^*A\right) \otimes (\mathbf{e}_{j,t}\mathbf{e}_{i,s}B)\right)$$

$$= \sum_{u=1}^{n}\sum_{v=1}^{t}\sum_{i=1}^{s}\sum_{j=1}^{t}\left(\mathbf{e}_{u,n}^*\tilde{M}_{ij}^*A\mathbf{e}_{u,n}\right) \otimes \left(\mathbf{e}_{v,t}^*\mathbf{e}_{j,t}\mathbf{e}_{i,s}B\mathbf{e}_{v,t}\right)$$

$$= \sum_{i=1}^{s}\sum_{j=1}^{t}\sum_{u=1}^{n}\left(\mathbf{e}_{u,n}^*\tilde{M}_{ij}^*A\mathbf{e}_{u,n}\right) \otimes (\mathbf{e}_{i,s}B\mathbf{e}_{j,t})$$

$$= \sum_{i=1}^{s}\sum_{j=1}^{t}\text{tr}\left(\tilde{M}_{ij}^*A\right)(B)_{ij}$$

$$= \sum_{i=1}^{s}\sum_{j=1}^{t}\text{tr}\left(\tilde{M}_{ij}^*A\right)\left[\Re(B)_{ij} + \Im(B)_{ij}\sqrt{-1}\right].$$

Thus we have

$$\Re(B)_{ij} = \frac{1}{\|A\|_F^2}\Re\left[\text{tr}\left(\tilde{M}_{ij}^*A\right)\right]$$

$$\Im(B)_{ij} = \frac{1}{\|A\|_F^2}\Re\left[\text{tr}\left(\tilde{M}_{ij}^*A\right)\sqrt{-1}\right] = -\frac{1}{\|A\|_F^2}\Im\left[\text{tr}\left(\tilde{M}_{ij}^*A\right)\right]$$

or equivalently

$$\Re(B)_{ij} = \frac{1}{\|A\|_F^2}\Re\left[\text{tr}\left(A^*\tilde{M}_{ij}\right)\right], \qquad \Im(B)_{ij} = \frac{1}{\|A\|_F^2}\Im\left[\text{tr}\left(A^*\tilde{M}_{ij}\right)\right]$$

so that

$$(B)_{ij} = \Re(B)_{ij} + \Im(B)_{ij}\sqrt{-1} = \frac{\text{tr}\left(A^*\tilde{M}_{ij}\right)}{\|A\|_F^2}.$$

This yields a minimum since

$$\frac{\partial^2}{\partial\Re(B)_{ij}^2}\|M - A \otimes B\|_F^2 = \frac{\partial^2}{\partial\Im(B)_{ij}^2}\|M - A \otimes B\|_F^2 = 2\|A\|_F^2$$

and for $u \neq i$ or $v \neq j$

$$\frac{\partial^2}{\partial\Re(B)_{ij}\partial\Re(B)_{uv}}\|M - A \otimes B\|_F^2 = \frac{\partial^2}{\partial\Im(B)_{ij}\partial\Im(B)_{uv}}\|M - A \otimes B\|_F^2 = 0$$

and for all i, j, u and v

$$\frac{\partial^2}{\partial \Re(B)_{ij} \partial \Im(B)_{uv}} \| M - A \otimes B \|_F^2 = 0$$

so that the Hessian matrix is diagonal with all eigenvalues $2\|A\|_F^2 > 0$, i.e. we have a minimum. Now suppose $B \neq 0_{s \times t}$ is known, then

$$(A)_{ij} = \frac{\mathrm{tr}(B^* \check{M}_{ij})}{\|B\|_F^2}$$

minimizes $\| M - A \otimes B \|_F^2$, where

$$\check{M}_{ij} := (\mathbf{e}_{i,m} \otimes I_s)^* M (\mathbf{e}_{j,n} \otimes I_t).$$

Note that if B is the $s \times t$ zero matrix then A is arbitrary. The derivation of this formula is very similar to the derivation in the previous section and is left as an exercise.

Exercise. Let M be a 4×4 hermitian matrix. Find 2×2 matrices A and B such that $\| M - (A \otimes I_2 + I_2 \otimes B + A \otimes B) \|$ is a minimum.

2.18 Gâteaux Derivative and Matrices

Let X and Y be Banach space and $f : X \to Y$. A function f is said to be *Gâteaux differentiable* at $x \in X$ if there exists a bounded linear operator $T_x \in \mathcal{B}(X, Y)$ such that for all $v \in X$

$$\lim_{\epsilon \to 0} \frac{f(x + \epsilon v) - f(x)}{\epsilon} = T_x v.$$

The operator T_x is called the Gâteaux derivative of f at x.

If for some fixed $v \in X$ the limits

$$\delta_v f(x) := \frac{d}{d\epsilon} f(x + \epsilon v) \Big|_{\epsilon=0}$$

exists, we say that f has a directional derivative at x in the direction v. Consequently the map f is Gâteaux differentiable at x if and only if all the directional derivatives $\delta_v f(x)$ exist and form a bounded linear operator $Df(x) : v \mapsto \delta_v f(x)$.

Let V_1 and V_2 be locally convex topological vector spaces (examples are Banach spaces and Hilbert spaces). Let U be an open subset of V_1 and

$f : V_1 \to V_2$. The Gâteaux differential $df(u; v_1)$ of f at $u \in U$ in the direction $v_1 \in C_1$ is defined as

$$df(u; v_1) = \frac{d}{d\epsilon} f(u + \epsilon v_1) \Big|_{\epsilon=0}.$$

If the limit exists for all $v_1 \in V_1$, then f is Gâteaux differentiable at u.

Example 2.36. Let $V_1 = V_2 = V$ be the Hilbert space of $n \times n$ matrices over $\mathbb{C}$ and $A \in V$. Consider $f(A) = A^2$. Then

$$\frac{d}{d\epsilon} f(A + \epsilon B) \Big|_{\epsilon=0} = \frac{d}{d\epsilon}(A + \epsilon B)(A + \epsilon B) \Big|_{\epsilon=0} = AB + BA \equiv [A, B]_+ \quad \clubsuit$$

Example 2.37. Let V_1 be the Hilbert space of $n \times n$ matrices over $\mathbb{C}$, V_2 be the Hilbert space of $n^2 \times n^2$ matrices over $\mathbb{C}$. Consider $f(A) = A \otimes A$. Then

$$\frac{d}{d\epsilon} f(A + \epsilon B) \Big|_{\epsilon=0} = \frac{d}{d\epsilon}((A + \epsilon B) \otimes (A + \epsilon B)) \Big|_{\epsilon=0} = A \otimes B + B \otimes A. \quad \clubsuit$$

Exercises. (1) Let A be an $n \times n$ matrix over $\mathbb{C}$. Find the Gâteaux derivative of $f(A) = A^3$ and $g(A) = A \otimes A \otimes A$.

(2) Let A be an $n \times n$ matrix over $\mathbb{C}$. Find the Gâteaux derivative of $f(A) = A \otimes I_n$. Find the Gâteaux derivative of $f(A) = A \oplus A$, where $\oplus$ denotes the direct sum.

(3) Consider the map $d : \mathcal{H} \to \mathbb{C}$, $d(A) := \det(A)$. Find

$$\frac{d}{d\epsilon} d(A + \epsilon B) \Big|_{\epsilon=0}.$$

(4) Let A be a fixed element of the Lie group $SL(2, \mathbb{R})$. Consider the vector space of all $\mathbb{R}^{2 \times 2}$ of all 2×2 matrices over $\mathbb{R}$. Consider the map $f : \mathbb{R}^{2 \times 2} \to \mathbb{R}$, $f(X) = \text{tr}(X^T A X)$. Show that

$$\frac{d}{d\epsilon} f(X + \epsilon Y) \Big|_{\epsilon=0} = \text{tr}(Y^T A X + X^T A Y).$$

(5) Let A, B be $n \times n$ matrices over $\mathbb{C}$. Find

$$\frac{d}{d\epsilon}(e^{A+\epsilon B}) \Big|_{\epsilon=0}, \qquad \frac{d}{d\epsilon} \det(e^{A+\epsilon B}) \Big|_{\epsilon=0}.$$

(6) Consider the map $f : \mathcal{H} \to \mathcal{H} \otimes \mathcal{H}$, $f(A) = A \otimes I_n + I_n \otimes A$. Find

$$\frac{d}{d\epsilon} f(A + \epsilon B) \Big|_{\epsilon=0}.$$

Chapter 3

Applications

3.1 Trace and Partial Trace

Let $\mathcal{H}$ be the finite dimensional Hilbert space $\mathbb{C}^n$ with an orthonormal basis $\{ |\phi_j\rangle : j = 1, 2, \ldots, n \}$. Let A be a linear operator ($n \times n$ matrix) acting in this Hilbert space. Then the *trace* of A is defined as

$$\operatorname{tr}(A) := \sum_{j=1}^{n} \langle \phi_j | A | \phi_j \rangle.$$

The trace is independent of the chosen orthonormal basis. For the trace we have cyclic invariance. Let A, B, C be $n \times n$ matrices over $\mathbb{C}$. Then

$$\operatorname{tr}(AB) = \operatorname{tr}(BA)$$

and (*cyclic invariance*)

$$\operatorname{tr}(ABC) = \operatorname{tr}(CAB) = \operatorname{tr}(BCA).$$

The trace of an $n \times n$ matrix A is the sum of the eigenvalues counting multiplicities. The eigenvalues of A can be reconstructed from

$$\operatorname{tr}(A) = \lambda_1 + \lambda_2 + \cdots + \lambda_n$$
$$\operatorname{tr}(A^2) = \lambda_1^2 + \lambda_2^2 + \cdots + \lambda_n^2$$
$$\vdots$$
$$\operatorname{tr}(A^n) = \lambda_1^n + \lambda_2^n + \cdots + \lambda_n^n.$$

If $|\psi\rangle$ is a normalized state in $\mathbb{C}^n$, then

$$\operatorname{tr}(|\psi\rangle\langle\psi|) = 1.$$

For any $n \times n$ matrix A over $\mathbb{C}$ we have the identity

$$\det(\exp(A)) \equiv \exp(\operatorname{tr}(A)).$$

179

Let A be an $n \times n$ matrix over $\mathbb{C}$ and B be an $m \times m$ matrix over $\mathbb{C}$. Then

$$\text{tr}(A \otimes B) = \text{tr}(A)\text{tr}(B).$$

The calculation of the *partial trace* plays a central role in quantum computing. Suppose a finite dimensional quantum system S_{AB} is a system composed of two subsystems S_A and S_B. The finite dimensional Hilbert space $\mathcal{H}$ of S_{AB} is given by the tensor product of the individual Hilbert spaces $\mathcal{H}_A \otimes \mathcal{H}_B$. Let $N_A := \dim(\mathcal{H}_A)$ and $N_B := \dim(\mathcal{H}_B)$. Let ρ_{AB} be the density matrix of S_{AB}. Using the partial trace we can define the density operators ρ_A and ρ_B in the subspaces $\mathcal{H}_A$ and $\mathcal{H}_B$ as follows

$$\rho_A := \text{tr}_B(\rho_{AB}) \equiv \sum_{j=1}^{N_B}(I_A \otimes \langle \phi_j|)\rho_{AB}(I_A \otimes |\phi_j\rangle)$$

and

$$\rho_B := \text{tr}_A(\rho_{AB}) \equiv \sum_{j=1}^{N_A}(\langle \psi_j| \otimes I_B)\rho_{AB}(|\psi_j\rangle \otimes I_B))$$

where I_A is the identity operator in $\mathcal{H}_A$, I_B is the identity operator in $\mathcal{H}_B$ and

$$|\phi_j\rangle, \quad (j = 1, 2, \ldots, N_B)$$

is an orthonormal basis in $\mathcal{H}_B$ and

$$|\psi_j\rangle, \quad (j = 1, 2, \ldots, N_A)$$

is an orthonormal basis in $\mathcal{H}_A$. For example we could select the standard bases in the two finite dimensional Hilbert spaces $\mathcal{H}_A$ and $\mathcal{H}_B$.

The partial trace can also be calculated as follows. Consider a bipartite state

$$|\psi\rangle = \sum_{j=0}^{n-1}\sum_{k=0}^{n-1} c_{jk}|jk\rangle \equiv \sum_{j=0}^{n-1}\sum_{k=0}^{n-1} c_{jk}|j\rangle \otimes |k\rangle, \qquad \sum_{j=0}^{n-1}\sum_{k=0}^{n-1} c_{jk}c_{jk}^* = 1$$

in the finite-dimensional Hilbert space $\mathcal{H} = \mathbb{C}^n \otimes \mathbb{C}^n$. We can define the $n \times n$ matrix

$$\Lambda_{jk} := c_{jk}, \qquad j, k = 0, 1, \ldots, n - 1.$$

Then we have (prove it)

$$\rho_A = \text{tr}_B(\rho) = \text{tr}_B(|\psi\rangle\langle\psi|) = \Lambda\Lambda^*.$$

Exercises. (1) Let A be a normal 3×3 matrix over $\mathbb{R}$ with
$$\operatorname{tr}(A) = 0, \quad \operatorname{tr}(A^2) = 0, \quad \operatorname{tr}(A^3) = 3.$$
Can the matrix be reconstructed from this information? First find the eigenvalues of A.

(2) Consider the Hilbert space $\mathcal{H} = \mathcal{H}_1 \otimes \mathcal{H}_2 = \mathbb{C}^4$ with $\mathcal{H}_1 = \mathcal{H}_2 = \mathbb{C}^2$. Consider the density matrix (pure state)
$$\rho = |\psi\rangle\langle\psi| = \frac{1}{2} \begin{pmatrix} 1 & 0 & 0 & 1 \\ 0 & 0 & 0 & 0 \\ 0 & 0 & 0 & 0 \\ 1 & 0 & 0 & 1 \end{pmatrix}.$$

Show that
$$\operatorname{tr}_{\mathcal{H}_1}(|\psi\rangle\langle\psi|) = \frac{1}{2}I_2, \quad \operatorname{tr}_{\mathcal{H}_2}(|\psi\rangle\langle\psi|) = \frac{1}{2}I_2.$$

3.2 Pauli Spin Matrices

In this section we introduce the Pauli spin matrices and give some application in connection with the Kronecker product. In the first application we calculate the eigenvalues and eigenvectors of the two-point Heisenberg model. In the second application we show that the representation of the Clifford algebra can be expressed with the help of the Kronecker product.

The *Pauli spin matrices* are defined by
$$\sigma_1 := \begin{pmatrix} 0 & 1 \\ 1 & 0 \end{pmatrix}, \qquad \sigma_2 := \begin{pmatrix} 0 & -i \\ i & 0 \end{pmatrix}, \qquad \sigma_3 := \begin{pmatrix} 1 & 0 \\ 0 & -1 \end{pmatrix}.$$
Instead of σ_1, σ_2 and σ_3 one also writes sometimes σ_x, σ_y and σ_z. The matrices σ_+ and σ_- are defined as
$$\sigma_+ := \sigma_1 + i\sigma_2 \equiv \begin{pmatrix} 0 & 2 \\ 0 & 0 \end{pmatrix}, \qquad \sigma_- := \sigma_1 - i\sigma_2 \equiv \begin{pmatrix} 0 & 0 \\ 2 & 0 \end{pmatrix}.$$
First we list the properties of the Pauli spin matrices. We find
$$(\sigma_1)^2 = (\sigma_2)^2 = (\sigma_3)^2 = \begin{pmatrix} 1 & 0 \\ 0 & 1 \end{pmatrix} \equiv I_2$$
$\operatorname{tr}(\sigma_1) = 0$, $\operatorname{tr}(\sigma_2) = 0$, $\operatorname{tr}(\sigma_3) = 0$ and
$$\sigma_1\sigma_2 = i\sigma_3, \qquad \sigma_2\sigma_1 = -i\sigma_3$$
$$\sigma_2\sigma_3 = i\sigma_1, \qquad \sigma_3\sigma_2 = -i\sigma_1$$
$$\sigma_3\sigma_1 = i\sigma_2, \qquad \sigma_1\sigma_3 = -i\sigma_2.$$

The Pauli spin matrices are hermitian and unitary, respectively. This means $(\sigma_1)^* = \sigma_1$, $(\sigma_2)^* = \sigma_2$, $(\sigma_3)^* = \sigma_3$ and

$$(\sigma_1)^* = (\sigma_1)^{-1}, \qquad (\sigma_2)^* = (\sigma_2)^{-1}, \qquad (\sigma_3)^* = (\sigma_3)^{-1}.$$

Therefore the eigenvalues of σ_1, σ_2 and σ_3 are given by $\{+1, -1\}$. The normalized eigenvectors of σ_1 are

$$\frac{1}{\sqrt{2}} \begin{pmatrix} 1 \\ 1 \end{pmatrix}, \qquad \frac{1}{\sqrt{2}} \begin{pmatrix} 1 \\ -1 \end{pmatrix}.$$

The normalized eigenvectors of σ_2 are given by

$$\frac{1}{\sqrt{2}} \begin{pmatrix} 1 \\ i \end{pmatrix}, \qquad \frac{1}{\sqrt{2}} \begin{pmatrix} 1 \\ -i \end{pmatrix}.$$

For σ_3 we obviously find the normalized eigenvectors

$$\begin{pmatrix} 1 \\ 0 \end{pmatrix}, \qquad \begin{pmatrix} 0 \\ 1 \end{pmatrix}.$$

The *commutator* of the Pauli spin matrices gives

$$[\sigma_1, \sigma_2] = 2i\sigma_3, \qquad [\sigma_3, \sigma_1] = 2i\sigma_2, \qquad [\sigma_2, \sigma_3] = 2i\sigma_1.$$

The Pauli spin matrices form a basis of a semisimple Lie algebra under the commutator. The *anticommutator* of the Pauli spin matrices vanishes, i.e.

$$[\sigma_1, \sigma_2]_+ = 0_2, \qquad [\sigma_3, \sigma_1]_+ = 0_2, \qquad [\sigma_2, \sigma_3]_+ = 0_2.$$

Let $z \in \mathbb{C}$. For the exponential map of the Pauli spin matrices we find

$$\exp(z\sigma_1) = I_2 \cosh(z) + \sigma_1 \sinh(z)$$
$$\exp(z\sigma_2) = I_2 \cosh(z) + \sigma_2 \sinh(z)$$
$$\exp(z\sigma_3) = I_2 \cosh(z) + \sigma_3 \sinh(z).$$

Furthermore we have

$$\exp(-z\sigma_1)\sigma_2 \exp(z\sigma_1) = \sigma_2 \cosh(2z) - i\sigma_3 \sinh(2z)$$
$$\exp(-z\sigma_2)\sigma_3 \exp(z\sigma_2) = \sigma_3 \cosh(2z) - i\sigma_1 \sinh(2z)$$
$$\exp(-z\sigma_3)\sigma_1 \exp(z\sigma_3) = \sigma_1 \cosh(2z) - i\sigma_2 \sinh(2z).$$

The Pauli spin matrices σ_1, σ_2, σ_3 and I_2 form an orthogonal basis in the Hilbert space of the 2×2 matrices over $\mathbb{C}$ with the scalar product

$$\langle A, B \rangle := \operatorname{tr}(AB^*).$$

The *spin matrices* for spin-$\frac{1}{2}$ are defined by $S_1 := \frac{1}{2}\sigma_1$, $S_2 := \frac{1}{2}\sigma_2$, $S_3 := \frac{1}{2}\sigma_3$.

Note that the matrices

$$U_{12} = \frac{1}{\sqrt{2}}(\sigma_1 + \sigma_2), \quad U_{13} = \frac{1}{\sqrt{2}}(\sigma_1 + \sigma_3), \quad U_{23} = \frac{1}{\sqrt{2}}(\sigma_2 + \sigma_3)$$

are also hermitian and unitary. Hence $U_{12}^2 = U_{13}^2 = U_{23}^2 = I_2$ and the eigenvalues are $+1$ and -1. Furthermore the matrix

$$U_{123} = \frac{1}{\sqrt{3}}(\sigma_1 + \sigma_2 + \sigma_3)$$

is hermitian and unitary with the eigenvalues $+1$ and -1.

Definition 3.1. Let $j = 1, 2, \ldots, N$. We define

$$\sigma_{\alpha,j} := I_2 \otimes \cdots \otimes I_2 \otimes \sigma_\alpha \otimes I_2 \otimes \cdots \otimes I_2$$

where I_2 is the 2×2 unit matrix, $\alpha = 1, 2, 3$ and σ_α is the α-th Pauli matrix in the j-th location. Thus $\sigma_{\alpha,j}$ is a $2^N \times 2^N$ matrix. Analogously, we define

$$S_{\alpha,j} := I_2 \otimes \cdots \otimes I_2 \otimes S_\alpha \otimes I_2 \otimes \cdots \otimes I_2.$$

In the following we set $\mathbf{S}_j := (S_{1,j}, S_{2,j}, S_{3,j})^T$.

Example 3.1. We calculate the eigenvalues and eigenvectors for the two-point *Heisenberg model*. The model is given by the Hamilton operator

$$\hat{H} = J \sum_{j=1}^{2} \mathbf{S}_j \cdot \mathbf{S}_{j+1}$$

where J is the so-called exchange constant ($J > 0$ or $J < 0$) and $\cdot$ denotes the scalar product. We impose *cyclic boundary conditions*, i.e. $\mathbf{S}_3 \equiv \mathbf{S}_1$. It follows that

$$\hat{H} = J(\mathbf{S}_1 \cdot \mathbf{S}_2 + \mathbf{S}_2 \cdot \mathbf{S}_3) \equiv J(\mathbf{S}_1 \cdot \mathbf{S}_2 + \mathbf{S}_2 \cdot \mathbf{S}_1).$$

Therefore

$$\hat{H} = J(S_{1,1}S_{1,2} + S_{2,1}S_{2,2} + S_{3,1}S_{3,2} + S_{1,2}S_{1,1} + S_{2,2}S_{2,1} + S_{3,2}S_{3,1}).$$

Since $S_{1,1} = S_1 \otimes I_2$, $S_{1,2} = I_2 \otimes S_1$ etc. where I_2 is the 2×2 unit matrix, it follows that

$$\hat{H} = J[(S_1 \otimes I_2)(I_2 \otimes S_1) + (S_2 \otimes I_2)(I_2 \otimes S_2) + (S_3 \otimes I_2)(I_2 \otimes S_3)$$
$$+ (I_2 \otimes S_1)(S_1 \otimes I_2) + (I_2 \otimes S_2)(S_2 \otimes I_2) + (I_2 \otimes S_3)(S_3 \otimes I_2)].$$

Therefore

$$\hat{H} = 2J[(S_1 \otimes S_1) + (S_2 \otimes S_2) + (S_3 \otimes S_3)].$$

Since $S_1 := \frac{1}{2}\sigma_1$, $S_2 := \frac{1}{2}\sigma_2$, $S_3 := \frac{1}{2}\sigma_3$ we obtain

$$S_1 \otimes S_1 = \frac{1}{4} \begin{pmatrix} 0 & 1 \\ 1 & 0 \end{pmatrix} \otimes \begin{pmatrix} 0 & 1 \\ 1 & 0 \end{pmatrix} = \frac{1}{4} \begin{pmatrix} 0 & 0 & 0 & 1 \\ 0 & 0 & 1 & 0 \\ 0 & 1 & 0 & 0 \\ 1 & 0 & 0 & 0 \end{pmatrix}$$

etc.. Then the Hamilton operator $\hat{H}$ is given by the 4×4 symmetric matrix

$$\hat{H} = \frac{J}{2} \begin{pmatrix} 1 & 0 & 0 & 0 \\ 0 & -1 & 2 & 0 \\ 0 & 2 & -1 & 0 \\ 0 & 0 & 0 & 1 \end{pmatrix} \equiv \frac{J}{2} \left[(1) \oplus \begin{pmatrix} -1 & 2 \\ 2 & -1 \end{pmatrix} \oplus (1) \right]$$

where $\oplus$ denotes the direct sum. The eigenvalues and eigenvectors can now easily be calculated. We define

$$| \uparrow \rangle := \begin{pmatrix} 1 \\ 0 \end{pmatrix} \text{ spin up,} \qquad | \downarrow \rangle := \begin{pmatrix} 0 \\ 1 \end{pmatrix} \text{ spin down.}$$

Then we define

$$| \uparrow\uparrow \rangle := | \uparrow \rangle \otimes | \uparrow \rangle, \quad | \uparrow\downarrow \rangle := | \uparrow \rangle \otimes | \downarrow \rangle,$$

$$| \downarrow\uparrow \rangle := | \downarrow \rangle \otimes | \uparrow \rangle, \quad | \downarrow\downarrow \rangle := | \downarrow \rangle \otimes | \downarrow \rangle.$$

Consequently,

$$| \uparrow\uparrow \rangle = \begin{pmatrix} 1 \\ 0 \\ 0 \\ 0 \end{pmatrix}, \quad | \uparrow\downarrow \rangle = \begin{pmatrix} 0 \\ 1 \\ 0 \\ 0 \end{pmatrix}, \quad | \downarrow\uparrow \rangle = \begin{pmatrix} 0 \\ 0 \\ 1 \\ 0 \end{pmatrix}, \quad | \downarrow\downarrow \rangle = \begin{pmatrix} 0 \\ 0 \\ 0 \\ 1 \end{pmatrix}.$$

One sees at once that $| \uparrow\uparrow \rangle$ and $| \downarrow\downarrow \rangle$ are eigenvectors of the Hamilton operator with eigenvalues $J/2$ and $J/2$, respectively. This means the eigenvalue $J/2$ is degenerate. The eigenvalues of the matrix

$$\frac{J}{2} \begin{pmatrix} -1 & 2 \\ 2 & -1 \end{pmatrix}$$

are given by $\{ J/2, -3J/2 \}$. The corresponding normalized eigenvectors of $\hat{H}$ are given by the *entangled states*

$$\frac{1}{\sqrt{2}} (| \uparrow\downarrow \rangle + | \downarrow\uparrow \rangle), \qquad \frac{1}{\sqrt{2}} (| \uparrow\downarrow \rangle - | \downarrow\uparrow \rangle).$$

Thus the eigenvalue $J/2$ is three times degenerate. If $J > 0$, then $-3J/2$ is the ground state energy and the ground state is nondegenerate. If $J < 0$, then $J/2$ is the ground state energy.   ♣

Example 3.2. Within the representation of the *Clifford algebra* the Pauli spin matrices and Kronecker product play the central role (Steiner [64], Fletcher [22]). For every pair of nonnegative integers (p, q) we denote by $C(p, q)$ the Clifford algebra of the nondegenerate real bilinear symmetric from of rank $p + q$ and signature $p - q$. This is a 2^{p+q} dimensional real algebra with generators $e_1, \ldots, e_{p+q}$ satisfying the following relations

$$e_i^2 = 1 \qquad \text{for } i = 1, \ldots, p$$
$$e_i^2 = -1 \qquad \text{for } i = p+1, \ldots, p+q$$
$$e_i e_j = -e_j e_i \text{ for } i \neq j.$$

We can now give a description of these representations. We express the image of the generators e_i of the Clifford algebras as a Kronecker product of real 2×2 matrices

$$\sigma_0 = \begin{pmatrix} 1 & 0 \\ 0 & 1 \end{pmatrix}, \quad \sigma_1 = \begin{pmatrix} 0 & 1 \\ 1 & 0 \end{pmatrix}, \quad \rho_2 = \begin{pmatrix} 0 & -1 \\ 1 & 0 \end{pmatrix}, \quad \sigma_3 = \begin{pmatrix} 1 & 0 \\ 0 & -1 \end{pmatrix}$$

where $\rho_2 := -i\sigma_2$. We find that $\rho_2^2 = -I_2$, where $\sigma_0 = I_2$ is the 2×2 identity matrix. We obtain the irreducible representation of $C(1, 1)$

$$\alpha : C(1, 1) \rightarrow M_2(\mathbb{R})$$

$$f_1 \mapsto \sigma_1, \qquad f_2 \mapsto \rho_2$$

where $M_2(\mathbb{R})$ is the four dimensional vector space of the 2×2 matrices over $\mathbb{R}$ and f_1, f_2 are the two generators of the presentation given above. Owing to the modulo-8 periodicity in the structure of the algebras, the basic representations are those of $C(0, 8)$ and $C(8, 0)$. We write $g_1, \ldots, g_8$ and $h_1, \ldots, h_8$, respectively. The quantities $g_1, \ldots, g_8$ and $h_1, \ldots, h_8$ are 16×16 matrices (with trace equal to 0) over the real numbers $\mathbb{R}$ and given by the Kronecker product of 2×2 matrices

$$\beta : \quad C(0, 8) \rightarrow M_{16}(\mathbb{R})$$
$$g_1 \mapsto \rho_2 \otimes I_2 \otimes I_2 \otimes I_2$$
$$g_2 \mapsto \sigma_3 \otimes \rho_2 \otimes I_2 \otimes I_2$$
$$g_3 \mapsto \sigma_1 \otimes \rho_2 \otimes \sigma_1 \otimes I_2$$
$$g_4 \mapsto \sigma_3 \otimes \sigma_3 \otimes \rho_2 \otimes I_2$$
$$g_5 \mapsto \sigma_1 \otimes I_2 \otimes \rho_2 \otimes \sigma_1$$
$$g_6 \mapsto \sigma_3 \otimes \sigma_1 \otimes \rho_2 \otimes \sigma_1$$
$$g_7 \mapsto \sigma_1 \otimes \rho_2 \otimes \sigma_3 \otimes \sigma_1$$
$$g_8 \mapsto \sigma_3 \otimes \sigma_3 \otimes \sigma_3 \otimes \rho_2$$

and

$$\tau \ : \ C(8,0) \to M_{16}(\mathbb{R})$$
$$h_1 \mapsto \sigma_1 \otimes I_2 \otimes I_2 \otimes I_2$$
$$h_2 \mapsto \sigma_3 \otimes \sigma_1 \otimes I_2 \otimes I_2$$
$$h_3 \mapsto \sigma_3 \otimes \sigma_3 \otimes \sigma_1 \otimes I_2$$
$$h_4 \mapsto \rho_2 \otimes \sigma_1 \otimes \rho_2 \otimes I_2$$
$$h_5 \mapsto \sigma_3 \otimes \sigma_3 \otimes \sigma_4 \otimes \sigma_1$$
$$h_6 \mapsto \rho_2 \otimes I_2 \otimes \sigma_1 \otimes \rho_2$$
$$h_7 \mapsto \sigma_3 \otimes \rho_2 \otimes \sigma_1 \otimes \rho_2$$
$$h_8 \mapsto \rho_2 \otimes \sigma_1 \otimes \sigma_3 \otimes \rho_2.$$

We have $g_j^2 \mapsto -I_{16}$ for $j = 1, 2, \ldots, 8$. Analogously, we have $h_j^2 \mapsto I_{16}$ for $j = 1, 2, \ldots, 8$. The representations of $C(0,n)$ and $C(n,0)$ with $n < 8$ can easily be found from those of $C(0,8)$ and $C(8,0)$, respectively. One has to consider the first n generators and take the first factors in their images, in order to obtain representations of dimension $d(0,n)$ and $d(n,0)$, respectively.   ♣

Exercises. (1) Show that any 2×2 matrix over $\mathbb{C}$ can be written as a linear combination of the 2×2 unit matrix I_2 and the Pauli spin matrices.

(2) Show that σ_1, σ_2 and σ_3 form a basis of a semisimple Lie algebra under the commutator.

(3) Let $\mathbf{x} = (x_1 \, x_2 \, x_3)$ and σ_1, σ_2, σ_3 be the Pauli spin matrices. Find the condition on x_1, x_2, x_3 such that

$$(\mathbf{x} \cdot \boldsymbol{\sigma})^2 \equiv (x_1 \sigma_1 + x_2 \sigma_2 + x_3 \sigma_3)^2 = I_2.$$

(4) Let I_2 be the 2×2 unit matrix. Do the 4×4 matrices

$$\{ \sigma_1 \otimes I_2, \ I_2 \otimes \sigma_1, \ \sigma_2 \otimes I_2, \ I_2 \otimes \sigma_2, \ \sigma_3 \otimes I_2, \ I_2 \otimes \sigma_3 \}$$

form a basis of a Lie algebra under the commutator?

(5) Let $\epsilon \in \mathbb{R}$. Show that

$$\exp(\epsilon i \sigma_2) = \begin{pmatrix} \cos(\epsilon) & \sin(\epsilon) \\ -\sin(\epsilon) & \cos(\epsilon) \end{pmatrix}.$$

Find $\exp(\epsilon i \sigma_2 \otimes \sigma_2)$ and $\exp(\epsilon i \sigma_2 \otimes \sigma_3)$.

(6) Show that $\exp(\sigma_1 + \sigma_2) \neq \exp(\sigma_1)\exp(\sigma_2)$.

(7) Show that the matrices $\sigma_1 \otimes I_2$, $I_2 \otimes \sigma_2$, $\sigma_1 \otimes \sigma_2$ commute with each other. Then show that

$$\exp(\sigma_1 \otimes I_2 + I_2 \otimes \sigma_2 + \sigma_1 \otimes \sigma_2) \equiv \exp(\sigma_1 \otimes I_2)\exp(I_2 \otimes \sigma_2)\exp(\sigma_1 \otimes \sigma_2).$$

(8) Find all the eigenvalues and eigenvectors of the 16×16 matrices

$$\sigma_1 \otimes \sigma_1 \otimes \sigma_1 \otimes \sigma_1, \quad \sigma_2 \otimes \sigma_2 \otimes \sigma_2 \otimes \sigma_2, \quad \sigma_3 \otimes \sigma_3 \otimes \sigma_3 \otimes \sigma_3.$$

(9) Let

$$\hat{H} = J\sum_{j=1}^{3} \mathbf{S}_j \cdot \mathbf{S}_{j+1} = J(\mathbf{S}_1 \cdot \mathbf{S}_2 + \mathbf{S}_2 \cdot \mathbf{S}_3 + \mathbf{S}_3 \cdot \mathbf{S}_1)$$

with $\mathbf{S}_4 = \mathbf{S}_1$, i. e. we have cyclic boundary conditions. Here

$$S_{1,1} = S_1 \otimes I_2 \otimes I_2, \qquad S_{1,2} = I_2 \otimes S_1 \otimes I_2, \qquad S_{1,3} = I_2 \otimes I_2 \otimes S_1$$

and I_2 is the 2×2 unit matrix. Find the eigenvalues of $\hat{H}$.

(10) Let $\hat{H} = J(\mathbf{S}_1 \cdot \mathbf{S}_2 + \mathbf{S}_2 \cdot \mathbf{S}_3)$. Find the eigenvalues and eigenvectors of $\hat{H}$. Here we have open end boundary conditions.

(11) The following hermitian and unitary 4×4 matrices

$$A_1 = \begin{pmatrix} 0&0&0&1 \\ 0&0&1&0 \\ 0&1&0&0 \\ 1&0&0&0 \end{pmatrix}, \quad A_2 = \begin{pmatrix} 0&0&0&-i \\ 0&0&i&0 \\ 0&-i&0&0 \\ i&0&0&0 \end{pmatrix}, \quad A_3 = \begin{pmatrix} 0&0&1&0 \\ 0&0&0&-1 \\ 1&0&0&0 \\ 0&-1&0&0 \end{pmatrix}$$

play an important role in the *Dirac theory* of the electron. Show that the matrices A_1, A_2 and A_3 can be written as Kronecker products of the Pauli spin matrices.

(12) For describing spin-1 particles (for example the *photon*) the following 3×3 matrices are important

$$A_1 = \begin{pmatrix} 0&0&0 \\ 0&0&-i \\ 0&i&0 \end{pmatrix}, \quad A_2 = \begin{pmatrix} 0&0&i \\ 0&0&0 \\ -i&0&0 \end{pmatrix}, \quad A_3 = \begin{pmatrix} 0&-i&0 \\ i&0&0 \\ 0&0&0 \end{pmatrix}.$$

Find the eigenvalues and eigenvectors of A_1, A_2 and A_3. Find the eigenvalues and eigenvectors of $A_1 \otimes A_1$, $A_2 \otimes A_2$ and $A_3 \otimes A_3$.

(13) The *graviton* is a spin-2 particle and described by the 5×5 matrices

$$S_1 = \begin{pmatrix} 0 & 1 & 0 & 0 & 0 \\ 1 & 0 & \sqrt{6}/2 & 0 & 0 \\ 0 & \sqrt{6}/2 & 0 & \sqrt{6}/2 & 0 \\ 0 & 0 & \sqrt{6}/2 & 0 & 1 \\ 0 & 0 & 0 & 1 & 0 \end{pmatrix}, \quad S_2 = i \begin{pmatrix} 0 & -1 & 0 & 0 & 0 \\ 1 & 0 & -\sqrt{6}/2 & 0 & 0 \\ 0 & \sqrt{6}/2 & 0 & -\sqrt{6}/2 & 0 \\ 0 & 0 & \sqrt{6}/2 & 0 & -1 \\ 0 & 0 & 0 & 1 & 0 \end{pmatrix},$$

$$S_3 = \begin{pmatrix} 2 & 0 & 0 & 0 & 0 \\ 0 & 1 & 0 & 0 & 0 \\ 0 & 0 & 0 & 0 & 0 \\ 0 & 0 & 0 & -1 & 0 \\ 0 & 0 & 0 & 0 & -2 \end{pmatrix}$$

which are hermitian. Find the eigenvalues of S_1, S_2, S_3 and $S_1 \otimes S_2 \otimes S_3$.

(14) Find the spectrum of the Hamilton operator

$$\hat{H} = \hbar\omega_1(I_2 \otimes I_2) + \hbar\omega_2(I_2 \otimes \sigma_3 + \sigma_3 \otimes I_2) + \hbar\omega_3(\sigma_1 \otimes \sigma_1).$$

3.3 Spin Coherent States

Let S_1, S_2, S_3 be the spin matrices for spin

$$s = \frac{1}{2}, \ 1, \ \frac{3}{2}, \ 2, \ldots \ .$$

The matrices are $(2s+1) \times (2s+1)$ hermitian matrices (S_1 and S_3 are real symmetric) with trace equal to 0 satisfying the commutation relations

$$[S_1, S_2] = iS_3, \quad [S_2, S_3] = iS_1, \quad [S_3, S_1] = iS_2. \tag{1}$$

The eigenvalues of S_1, S_2, S_3 are $s, s-1, \ldots, -s$ for a given s. It is known that

$$S_1^2 + S_2^2 + S_3^2 = s(s+1)I_{2s+1} \tag{2}$$

where I_{2s+1} is the $(2s+1) \times (2s+1)$ identity matrix. Furthermore we have

$$\text{tr}(S_j^2) = \frac{1}{3}s(s+1)(2s+1) \tag{3}$$

and

$$\text{tr}(S_j S_k) = 0 \quad \text{for } j \neq k \quad \text{and } j, k = 1, 2, 3. \tag{4}$$

In general we have that $\text{tr}(S_j^n) = 0$ if n is odd and $\text{tr}(S_j^n) = \frac{1}{3}s(s+1)(2s+1)$ if n is even. From eq.(4) it follows that $\text{tr}((S_j S_k) \otimes (S_\ell S_m)) = 0$ if $j \neq k$

and $\ell \neq m$. Note that for spin-$\frac{1}{2}$ we have $S_j = \frac{1}{2}\sigma_j$ $(j = 1, 2, 3)$, where σ_1, σ_2, σ_3 are the Pauli spin matrices.

One defines

$$S_+ := S_1 + iS_2, \qquad S_- := S_1 - iS_2.$$

Let $|0\rangle$ be the ground state in Hilbert space $\mathbb{C}^{2s+1}$, i.e.

$$|0\rangle = \begin{pmatrix} 1 & 0 & 0 & \cdots & 0 \end{pmatrix}^T.$$

Let $z \in \mathbb{C}$. The normalized *coherent spin state* $|z\rangle$ is defined as

$$|z\rangle = \frac{1}{(1 + |z|^2)^s} \exp(zS_-)|0\rangle.$$

Then we find for the scalar product $z_1, z_2 \in \mathbb{C}$

$$\langle z_1|z_2\rangle = \frac{(1 + \overline{z}_1 z_2)^{2s}}{(1 + |z_1|^2)^s (1 + |z_2|^2)^s}$$

and the *completeness relation* is given by

$$\int_{\mathbb{C}} dz\, m(|z|^2)|z\rangle\langle z| = I$$

where I is the $(2s + 1) \times (2s + 1)$ identity matrix and the measure is

$$m(w) = \frac{1 + 2s}{\pi(1 + w)^2}.$$

It can be shown that

$$\langle z_1|S_3|z_2\rangle = s\frac{1 - \overline{z}_1 z_2}{1 + \overline{z}_1 z_2}\langle z_1|z_2\rangle$$

$$\langle z_1|S_+|z_2\rangle = 2s\frac{z_2}{1 + \overline{z}_1 z_2}\langle z_1|z_2\rangle$$

$$\langle z_1|S_-|z_2\rangle = 2s\frac{\overline{z}_1}{1 + \overline{z}_1 z_2}\langle z_1|z_2\rangle.$$

As an example consider spin $s = 1/2$ and the spin matrices

$$S_1 = \frac{1}{2}\begin{pmatrix} 0 & 1 \\ 1 & 0 \end{pmatrix}, \quad S_2 = \frac{1}{2}\begin{pmatrix} 0 & -i \\ i & 0 \end{pmatrix}, \quad S_3 = \frac{1}{2}\begin{pmatrix} 1 & 0 \\ 0 & -1 \end{pmatrix}.$$

Consequently

$$S_+ = \begin{pmatrix} 0 & 1 \\ 0 & 0 \end{pmatrix}, \quad S_- = \begin{pmatrix} 0 & 0 \\ 1 & 0 \end{pmatrix}.$$

The ground state is

$$|0\rangle = \begin{pmatrix} 1 \\ 0 \end{pmatrix}.$$

Then with $\exp(zS_-) = I_2 + zS_-$ we obtain

$$|z\rangle = \frac{1}{(1 + |z|^2)^s} \exp(zS_-)|0\rangle = \frac{1}{(1 + |z|^2)^{1/2}} \begin{pmatrix} 1 \\ z \end{pmatrix}.$$

Consider spin $s = 1$ with the corresponding spin matrices

$$S_1 = \frac{1}{\sqrt{2}} \begin{pmatrix} 0 & 1 & 0 \\ 1 & 0 & 1 \\ 0 & 1 & 0 \end{pmatrix}, \quad S_2 = \frac{i}{\sqrt{2}} \begin{pmatrix} 0 & -1 & 0 \\ 1 & 0 & -1 \\ 0 & 1 & 0 \end{pmatrix}, \quad S_3 = \begin{pmatrix} 1 & 0 & 0 \\ 0 & 0 & 0 \\ 0 & 0 & -1 \end{pmatrix}.$$

Consequently

$$S_+ = \begin{pmatrix} 0 & \sqrt{2} & 0 \\ 0 & 0 & \sqrt{2} \\ 0 & 0 & 0 \end{pmatrix}, \quad S_- = \begin{pmatrix} 0 & 0 & 0 \\ \sqrt{2} & 0 & 0 \\ 0 & \sqrt{2} & 0 \end{pmatrix}$$

with $(S_+)^3 = (S_-)^3 = 0_3$. Then

$$\exp(zS_-) = \begin{pmatrix} 1 & 0 & 0 \\ \sqrt{2}z & 1 & 0 \\ z^2 & \sqrt{2}z & 1 \end{pmatrix}$$

and

$$|z\rangle = \exp(zS_-) \begin{pmatrix} 1 \\ 0 \\ 0 \end{pmatrix} = \frac{1}{1 + |z|^2} \begin{pmatrix} 1 \\ \sqrt{2}z \\ z^2 \end{pmatrix}.$$

Exercises. Study the case for spin $s = 3/2$ and spin $s = 2$.

3.4   Pauli Group, Clifford Groups and Bell Group

The n-qubit *Pauli group* is defined by

$$\mathcal{P}_n := \{ I_2, \sigma_1, \sigma_2, \sigma_3 \}^{\otimes n} \otimes \{ \pm 1, \pm i \}$$

where σ_1, σ_2, σ_3 are the 2×2 Pauli matrices and I_2 is the 2×2 identity matrix. The dimension of the Hilbert space under consideration is

$$\dim(\mathcal{H}) = 2^n.$$

Thus each element of the Pauli group $\mathcal{P}_n$ is (up to an overall phase ± 1, $\pm i$) a Kronecker product of Pauli matrices and 2×2 identity matrices acting on n qubits. The order of the Pauli group is 2^{2n+2}. Thus for $n = 1$ we have the order 16 and for $n = 2$ we have the order 64. The center of $\mathcal{P}_1$ is

$$Z(\mathcal{P}_1) = \{ \pm I_2, \pm i I_2 \}.$$

The two-qubit Pauli group may be generated by

$$P_2 = \langle \sigma_1 \otimes \sigma_1, \sigma_3 \otimes \sigma_3, \sigma_1 \otimes \sigma_2, \sigma_2 \otimes \sigma_3, \sigma_3 \otimes \sigma_1 \rangle.$$

The n-qubit *Clifford group* $\mathcal{C}_n$ is the normalizer of the Pauli group. A $2^n \times 2^n$ unitary matrix U acting on n qubits is an element of the Clifford group iff

$$UMU^* \in P_n \quad \text{for each} \quad M \in P_n.$$

This means the unitary matrix U acting by conjugation takes a Kronecker product of Pauli matrices to Kronecker product of Pauli matrices. An element of the Clifford group is defined as this action by conjugation, so that the overall phase of the unitary matrix U is not relevant. In other words the Clifford group is the group of all matrices that leave the Pauli group invariant.

Example 3.3. The *Hadamard gate*

$$U_H = \frac{1}{\sqrt{2}} \begin{pmatrix} 1 & 1 \\ 1 & -1 \end{pmatrix} \equiv \frac{1}{\sqrt{2}} (\sigma_3 + \sigma_1)$$

with $U_H = U_H^{-1}$ and the *phase gate*

$$U_P = \begin{pmatrix} 1 & 0 \\ 0 & i \end{pmatrix}$$

are in the one-qubit Clifford group $\mathcal{C}_1$. For $n = 2$ examples are

$$U_Z = \begin{pmatrix} 1 & 0 & 0 & 0 \\ 0 & 1 & 0 & 0 \\ 0 & 0 & 1 & 0 \\ 0 & 0 & 0 & -1 \end{pmatrix}, \quad U_{XOR} = \begin{pmatrix} 1 & 0 & 0 & 0 \\ 0 & 1 & 0 & 0 \\ 0 & 0 & 0 & 1 \\ 0 & 0 & 1 & 0 \end{pmatrix} \equiv I_2 \oplus \begin{pmatrix} 0 & 1 \\ 1 & 0 \end{pmatrix}$$

with $U_{XOR}^{-1} = U_{XOR}$.   ♣

The Clifford group $\mathcal{C}_n$ on n-qubits has order

$$|\mathcal{C}_n| = 2^{n^2 + 2n + 3} \prod_{j=1}^{n} (4^j - 1).$$

Thus for $\mathcal{C}_2$ we have 92160. The two-qubit Clifford group may be generated by

$$\mathcal{C}_2 = \langle U_H \otimes U_H, U_H \otimes U_P, U_Z \rangle.$$

Consider the *Bell matrix* U_B and the inverse

$$U_B = \frac{1}{\sqrt{2}} \begin{pmatrix} 1 & 0 & 0 & 1 \\ 0 & 1 & -1 & 0 \\ 0 & 1 & 1 & 0 \\ -1 & 0 & 0 & 1 \end{pmatrix} \Rightarrow U_B^{-1} = \frac{1}{\sqrt{2}} \begin{pmatrix} 1 & 0 & 0 & -1 \\ 0 & 1 & 1 & 0 \\ 0 & -1 & 1 & 0 \\ 1 & 0 & 0 & 1 \end{pmatrix}.$$

Let I_2 be the 2×2 unit matrix. Then U_B satisfies the *Yang-Baxter equation*

$$(U_B \otimes I_2)(I_2 \otimes U_B)(U_B \otimes I_2) = (I_2 \otimes U_B)(U_B \otimes I_2)(I_2 \otimes U_B).$$

The *Bell group* $\mathcal{B}_2$ is a subgroup of the Clifford group $\mathcal{C}_2$ and has order 15360. The Bell group is generated by $\mathcal{B}_2 = \langle U_H \otimes U_H, U_H \otimes U_P, U_B \rangle$.

3.5   Applications in Quantum Theory

The dynamics in quantum mechanics is governed by the Schrödinger equation which describes the time evolution of the wave function (wave vector) and the Heisenberg equation of motion which describes the time evolution of operators. The quantum mechanical system is described by a Hamilton operator $\hat{H}$. We consider the special case that the Hamilton operator $\hat{H}$ acts in the finite dimensional vector space $\mathbb{C}^n$.

The *Schrödinger equation* is given by

$$i\hbar \frac{\partial \psi}{\partial t} = \hat{H}\psi.$$

The solution of the Schrödinger equation takes the form

$$\psi(t) = \exp(-i\hat{H}t/\hbar)\psi(0).$$

Example 3.4. Let σ_1, σ_2, σ_3 be the Pauli spin matrices. Consider the Hamilton operator

$$\hat{H} = \hbar\omega \begin{pmatrix} 0 & 1 \\ 1 & 0 \end{pmatrix} = \hbar\omega\sigma_1.$$

Let

$$\psi(0) = \begin{pmatrix} 1 \\ 0 \end{pmatrix}$$

be the initial state. Then from the solution of the Schrödinger equation we find

$$\psi(t) = \begin{pmatrix} \cos(\omega t) & -i\sin(\omega t) \\ -i\sin(\omega t) & \cos(\omega t) \end{pmatrix} \begin{pmatrix} 1 \\ 0 \end{pmatrix} = \begin{pmatrix} \cos(\omega t) \\ -i\sin(\omega t) \end{pmatrix}.$$

Thus the *probability* to find the system in the initial state after a time t is given by

$$|(\psi(t), \psi(0))|^2 = \cos^2(\omega t).$$

♣

Let A, $\hat{H}$ be $n \times n$ hermitian matrices, where $\hat{H}$ plays the role of the Hamilton operator. The *Heisenberg equation of motion* is given by

$$\frac{dA(t)}{dt} = \frac{i}{\hbar}[\hat{H}, A(t)] \equiv \frac{i}{\hbar}[\hat{H}, A](t)$$

with $A = A(t = 0) = A(0)$ and the solution of the initial value problem

$$A(t) = e^{i\hat{H}t/\hbar} A e^{-i\hat{H}t/\hbar}.$$

Let E_j $(j = 1, 2, \ldots, n^2)$ be an orthonormal basis in the *Hilbert space* $\mathcal{H}$ of the $n \times n$ matrices with scalar product

$$\langle X, Y \rangle := \text{tr}(XY^*), \qquad X, Y \in \mathcal{H}.$$

Now $A(t)$ can be expanded using this orthonormal basis as

$$A(t) = \sum_{j=1}^{n^2} c_j(t) E_j$$

and $\hat{H}$ can be expanded as

$$\hat{H} = \sum_{j=1}^{n^2} h_j E_j.$$

We find the time evolution for the coefficients $c_j(t)$, i.e. dc_j/dt, where $j = 1, 2, \ldots, n^2$. We have

$$\frac{dA(t)}{dt} = \sum_{j=1}^{n^2} \frac{dc_j}{dt} E_j.$$

Inserting this equation and the expansion for H into the Heisenberg equation of motion we arrive at

$$\sum_{j=1}^{n^2} \frac{dc_j}{dt} E_j = \frac{i}{\hbar} \sum_{k=1}^{n^2} \sum_{j=1}^{n^2} h_k c_j(t)[E_k, E_j].$$

Taking the scalar product of the left and right-hand side of this equation with E_ℓ $(\ell = 1, \ldots, n^2)$ gives

$$\sum_{j=1}^{n^2} \frac{dc_j(t)}{dt} \text{tr}(E_j E_\ell^*) = \frac{i}{\hbar} \sum_{k=1}^{n^2} \sum_{j=1}^{n^2} h_k c_j(t) \text{tr}(([E_k, E_j]) E_\ell^*)$$

where $\ell = 1, 2, \ldots, n^2$. Since $\text{tr}(E_j E_\ell^*) = \delta_{j\ell}$ we obtain

$$\frac{dc_\ell}{dt} = \frac{i}{\hbar} \sum_{k=1}^{n^2} \sum_{j=1}^{n^2} h_k c_j(t) \text{tr}(E_k E_j E_\ell^* - E_j E_k E_\ell^*)$$

where $\ell = 1, 2, \ldots, n^2$.

Example 3.5. Let

$$\hat{H} = \hbar \omega \sigma_3$$

be a Hamilton operator acting in the two dimensional Hilbert space $\mathbb{C}^2$ and ω is the frequency. We calculate the time evolution of

$$\sigma_1 := \begin{pmatrix} 0 & 1 \\ 1 & 0 \end{pmatrix}.$$

The *Heisenberg equation of motion* is given by

$$i\hbar \frac{d\sigma_1}{dt} = [\sigma_1, \hat{H}](t).$$

Since $[\sigma_1, \hat{H}] = \hbar \omega [\sigma_1, \sigma_3] = -2i\hbar\omega\sigma_2$ we obtain

$$\frac{d\sigma_1}{dt} = -2\omega\sigma_2(t).$$

Now we have to calculate the time-evolution of σ_2, i.e.

$$i\hbar \frac{d\sigma_2}{dt} = [\sigma_2, \hat{H}](t).$$

Since $[\sigma_2, \hat{H}] = \hbar \omega [\sigma_2, \sigma_3] = 2i\hbar\omega\sigma_1$ we find

$$\frac{d\sigma_2}{dt} = 2\omega\sigma_1(t).$$

To summarize: we have to solve the following system of linear matrix differential equations with constant coefficients

$$\frac{d\sigma_1}{dt} = -2\omega\sigma_2(t), \qquad \frac{d\sigma_2}{dt} = 2\omega\sigma_1(t).$$

The initial conditions of this system are $\sigma_1(t = 0) = \sigma_1$, $\sigma_2(t = 0) = \sigma_2$. Then the solution of the initial value problem is given by

$$\sigma_1(t) = \sigma_1 \cos(2\omega t) - \sigma_2 \sin(2\omega t), \quad \sigma_2(t) = \sigma_2 \cos(2\omega t) + \sigma_1 \sin(2\omega t). \quad \clubsuit$$

The solution of the Heisenberg equation of motion can also be given as

$$\sigma_1(t) = \exp(i\hat{H}t/\hbar)\sigma_1\exp(-i\hat{H}t/\hbar)$$

$$\sigma_2(t) = \exp(i\hat{H}t/\hbar)\sigma_2\exp(-i\hat{H}t/\hbar).$$

In a number of applications the Hamilton operator $\hat{H}$ of a quantum mechanical system can be written as the tensor product of two hermitian operators or the Kronecker sum of two hermitian operators. We investigate such operators and their connection with constants of motion for the Heisenberg equation of motion.

Let A, B be $n \times n$ hermitian matrices over $\mathbb{C}$. Then $A \otimes B$ is also hermitian. Note that $A \otimes B \neq B \otimes A$ in general. Let

$$\hat{H} = \hbar\omega(A \otimes B)$$

be a Hamilton operator, where $\otimes$ denotes the Kronecker product, $\hbar$ is the Planck constant and ω the frequency. The Heisenberg equation of motion for the operator $B \otimes A$ is given by

$$i\hbar\frac{d(B \otimes A)}{dt} = [B \otimes A, \hat{H}](t).$$

The solution is given by

$$(B \otimes A)(t) = e^{i\hat{H}t/\hbar}(B \otimes A)e^{-i\hat{H}t/\hbar}$$

where $(B \otimes A)(0) = B \otimes A$. Now we have

$$[B \otimes A, \hat{H}](t) = \hbar\omega[B \otimes A, A \otimes B](t)$$
$$= \hbar\omega((B \otimes A)(A \otimes B) - (A \otimes B)(B \otimes A))(t)$$
$$= \hbar\omega((BA) \otimes (AB) - (AB) \otimes (BA))(t).$$

In general, $(AB) \otimes (BA) \neq (BA) \otimes (AB)$.

Assume that $[A, B] = 0_n$, where $[,\ ]$ denotes the commutator and 0_n is the $n \times n$ zero matrix. We can simplify the Heisenberg equation of motion using this condition. We find

$$i\hbar\frac{d(B \otimes A)}{dt} = 0_{n^2}.$$

Thus $B \otimes A$ is a constant of motion.

Now assume that $[A, B]_+ = 0_n$, where $[\,,\,]_+$ denotes the anticommutator. We can simplify the Heisenberg equation of motion using this condition. From $[A, B]_+ = 0_n$ we have $AB = -BA$. Thus we also have

$$i\hbar\frac{d(B \otimes A)}{dt} = 0_n$$

in this case.

Examples are the Pauli spin matrices. Let $A = \sigma_1$ and $B = \sigma_2$. Then $[A, B]_+ = 0_2$. Another example is given by Fermi operators. They have the matrix representation $(j = 1, 2, \ldots, N)$

$$c_j^\dagger = \sigma_3 \otimes \cdots \otimes \sigma_3 \otimes \left(\frac{1}{2}\sigma_+\right) \otimes I_2 \otimes \cdots \otimes I_2$$

$$c_j = \sigma_3 \otimes \cdots \otimes \sigma_3 \otimes \left(\frac{1}{2}\sigma_-\right) \otimes I_2 \otimes \cdots \otimes I_2$$

where σ_3 appears $j - 1$ times, σ_+ and σ_- at the j-th place and

$$\sigma_+ := \sigma_1 + i\sigma_2, \qquad \sigma_- := \sigma_1 - i\sigma_2.$$

We have

$$[c_k^\dagger, c_j]_+ = \delta_{jk}I, \qquad [c_k^\dagger, c_j^\dagger]_+ = [c_k, c_j]_+ = 0.$$

Assume that $[A, B]_+ = I_n$. We can simplify the Heisenberg equation of motion using this condition. Since $BA = -AB + I_n$ we find

$$i\hbar\frac{d(B \otimes A)}{dt} = ((AB) \otimes I_n - I_n \otimes (AB))(t).$$

We have a constant of motion if

$$(AB) \otimes I_n = I_n \otimes (AB).$$

Let A, B be $n \times n$ hermitian matrices over $\mathbb{C}$. Let K be an $n \times n$ hermitian matrix over $\mathbb{C}$ and $\hat{H} = \hbar\omega K$ be a Hamilton operator, where $\hbar$ is the Planck constant and ω the frequency. The Heisenberg equation of motion for the operators A and B are given by

$$i\hbar\frac{dA}{dt} = [A, \hat{H}](t), \qquad i\hbar\frac{dB}{dt} = [B, \hat{H}](t).$$

We find the time evolution of $A \otimes B$, $B \otimes A$, $A \otimes A$ and $B \otimes B$. We obtain

$$\frac{d}{dt}(A \otimes B) = \frac{dA}{dt} \otimes B + A \otimes \frac{dB}{dt} = \frac{1}{i\hbar}([A, \hat{H}] \otimes B + A \otimes [B, \hat{H}])$$

$$= \frac{1}{i\hbar}((A\hat{H}) \otimes B - (\hat{H}A) \otimes B + A \otimes (B\hat{H}) - A \otimes (\hat{H}B))$$

$$= \frac{1}{i\hbar}((A \otimes B)(\hat{H} \otimes I_n) - (\hat{H} \otimes I_n)(A \otimes B)$$

$$+ (A \otimes B)(I_n \otimes \hat{H}) - (I_n \otimes \hat{H})(A \otimes B))$$

$$= \frac{1}{i\hbar}\left([A \otimes B, \hat{H} \otimes I_n + I_n \otimes \hat{H}]\right).$$

Analogously we obtain

$$\frac{d}{dt}(B \otimes A) = \frac{1}{i\hbar}\left([B \otimes A, \hat{H} \otimes I_n + I_n \otimes \hat{H}]\right),$$

$$\frac{d}{dt}(A \otimes A) = \frac{1}{i\hbar}\left([A \otimes A, \hat{H} \otimes I_n + I_n \otimes \hat{H}]\right),$$

$$\frac{d}{dt}(B \otimes B) = \frac{1}{i\hbar}\left([B \otimes B, \hat{H} \otimes I_n] + I_n \otimes \hat{H}]\right).$$

Assume that $[A, \hat{H}] = 2i\hbar\omega B$, $[B, \hat{H}] = -2i\hbar\omega A$. We can simplify the Heisenberg equation of motion with these conditions. With the given conditions and the *Jacobi identity* we find

$$[\hat{H}, [A, B]] = 0_n.$$

The matrices $A, B, \hat{H}$ form a basis of a Lie algebra. The equations of motion simplify to

$$\frac{d}{dt}(A \otimes B) = 2\omega(B \otimes B - A \otimes A)$$

$$\frac{d}{dt}(B \otimes A) = 2\omega(B \otimes B - A \otimes A)$$

$$\frac{d}{dt}(A \otimes A) = 2\omega(A \otimes B + B \otimes A)$$

$$\frac{d}{dt}(B \otimes B) = -2\omega(A \otimes B + B \otimes A).$$

Thus we find the constants of motion

$$\frac{d}{dt}(A \otimes B - B \otimes A) = 0_{n^2}, \qquad \frac{d}{dt}(A \otimes A + B \otimes B) = 0_{n^2}.$$

Obviously these calculations can be extended to other finite dimensional Lie algebras, where one of the generators is selected as the Hamilton operator $\hat{H}$.

Another example is when the Hamilton operator is of the form (Kronecker sum)

$$\hat{H} = \hbar\omega(A \otimes I_n + I_n \otimes B)$$

where A, B are again hermitian matrices. Then we have

$$e^{-i\hat{H}t/\hbar} = e^{-i\omega t(A \otimes I_n + I_n \otimes B)} = e^{-i\omega tA} \otimes e^{-i\omega tB}$$

which can be used to solve the Schrödinger equation

$$|\psi(t)\rangle = e^{-i\hat{H}t/\hbar}|\psi(0)\rangle.$$

Thus if the state $|\psi(0)\rangle$ is entangled it will stay entangled under the time evolution. With respect to the constants of motion we have

$$[A \otimes I_n, I_n \otimes B] = 0_{n^2}$$

and therefore $[A \otimes I_n, \hat{H}] = 0_{n^2}$ and $[I_n \otimes B, \hat{H}] = 0_{n^2}$.

Another Hamilton operator $\hat{H}$ for which the Heisenberg equation can be solved is (Steeb [58])

$$\hat{H} = \hbar\omega(\sigma_3 \otimes \cdots \otimes \sigma_3) + \Delta_1(\sigma_1 \otimes \cdots \otimes \sigma_1) + \Delta_2(\sigma_2 \otimes \cdots \otimes \sigma_2).$$

This Hamilton operator plays a role for entanglement and diabolic points.

Exercises. (1) Consider the spin-1 matrices

$$S_1 = \frac{1}{\sqrt{2}}\begin{pmatrix} 0 & 1 & 0 \\ 1 & 0 & 1 \\ 0 & 1 & 0 \end{pmatrix}, \quad S_2 = \frac{1}{\sqrt{2}}\begin{pmatrix} 0 & -i & 0 \\ i & 0 & -i \\ 0 & i & 0 \end{pmatrix}, \quad S_3 = \begin{pmatrix} 1 & 0 & 0 \\ 0 & 0 & 0 \\ 0 & 0 & -1 \end{pmatrix}.$$

(i) Show that S_1, S_2, S_3 are hermitian matrices.
(ii) Find the eigenvalues and eigenvectors of S_1, S_2, S_3.
(iii) Let $\hat{H} = \hbar\omega S_1$. Calculate $\exp(-i\hat{H}t/\hbar)$. Consider the initial state

$$\psi(t = 0) = \frac{1}{\sqrt{2}}\begin{pmatrix} 1 \\ 0 \\ 1 \end{pmatrix}.$$

Calculate the state at time t

$$\psi(t) = \exp(-i\hat{H}t/\hbar)\psi(t = 0)$$

and the probability $|\psi(t)^*\psi(t = 0)|^2$.
(iv) Solve the Heisenberg equation of motion

$$i\hbar\frac{dS_2}{dt} = [S_2, \hat{H}](t).$$

(2) Consider the spin-1 matrices given above and the Hamilton operator

$$\hat{H} = \hbar\omega(S_1 \otimes S_1 + S_2 \otimes S_2 + S_3 \otimes S_3).$$

Find the eigenvalues and eigenvectors of $\hat{H}$. Find the unitary matrix $\exp(-i\hat{H}t/\hbar)$.

(3) Consider the Hamilton operator $\hat{H} = \hbar\omega(\sigma_2 \otimes \sigma_2 \otimes \sigma_2)$. Find the time-evolution of $A = \sigma_1 \otimes \sigma_1 \otimes \sigma_1$.

(4) Show that the unitary and hermitian 4×4 matrix $\sigma_2 \otimes \sigma_2$ admits entangled states as eigenvectors and unentangled states (product states) as eigenvectors (Steeb and Hardy [62]).

(5) Let $p_j \geq 0$, $(j = 1, \dots, n)$ and $\sum_{j=1}^{n} p_j = 1$. Show that

$$\rho(\phi_1, \dots, \phi_n) = \sum_{j=1}^{n} p_j \begin{pmatrix} \cos^2(\phi_j) & \cos(\phi_j)\sin(\phi_j) \\ \cos(\phi_j)\sin(\phi_j) & \sin^2(\phi_j) \end{pmatrix}$$

is a *density matrix*. Note that $\mathrm{tr}(\rho(\phi_1, \dots, \phi_n)) = 1$. So we only have to show that the matrix ρ is positive semi-definite.

3.6 Partition Functions and Thermodynamics

In a number of applications the Hamilton operator $\hat{H}$ of a quantum mechanical system can be written as the Kronecker product of two hermitian operators or the Kronecker sum of two hermitian operators. We calculate the partition function for such operators.

Let A be an $m \times m$ hermitian matrix and let B be an $n \times n$ hermitian matrix over $\mathbb{C}$. Then $A \otimes B$, $A \otimes I_n$, $I_m \otimes B$ are also hermitian matrices, where I_m is the $m \times m$ identity matrix and $\otimes$ denotes the Kronecker product. We have

$$[A \otimes B, A \otimes I_n] = 0_{n \cdot m}, \quad [A \otimes B, I_m \otimes B] = 0_{n \cdot m}, \quad [A \otimes I_n, I_m \otimes B] = 0_{n \cdot m}.$$

Let ϵ_1, ϵ_2 and ϵ_3 be real parameters. Consider the Hamilton operator

$$\hat{H} = \hbar\omega(\epsilon_1 A \otimes B + \epsilon_2 A \otimes I_n + \epsilon_3 I_m \otimes B)$$

where $\hbar$ is the Planck constant and ω the frequency. The *partition function* $Z(\beta)$ is given by

$$Z(\beta) = \mathrm{tr}(\exp(-\beta\hat{H}))$$

where $\hat{H}$ is the (hermitian) Hamilton operator and tr denotes the trace. From the partition function we obtain the Helmholtz free energy, entropy and specific heat. We derive $Z(\beta)$ for the Hamilton operator given above and provide two applications.

Since A is an $m \times m$ matrix we can find an $m \times m$ unitary matrix U_A such that $\widetilde{A} = U_A^* A U_A$ is a diagonal matrix. We set $\mathrm{diag}(\widetilde{A}) = (\lambda_1, \lambda_2, \dots, \lambda_m)$.

Analogously for the $n \times n$ hermitian matrix B we find an $n \times n$ hermitian matrix U_B such that

$$\widetilde{B} = U_B^* B U_B$$

is a diagonal matrix. We set $\operatorname{diag}(\widetilde{B}) = (\mu_1, \mu_2, \ldots, \mu_n)$. Since A and B are hermitian the diagonal elements of $\widetilde{A}$ and $\widetilde{B}$ are real. Since U_A and U_B are unitary matrices we find that $U_A \otimes U_B$ is also a unitary matrix and

$$(U_A \otimes U_B)^* = U_A^* \otimes U_B^*.$$

Now we find

$$
\begin{aligned}
\operatorname{tr}\left(e^{-\beta \hat{H}}\right) &= \operatorname{tr}((U_A^* \otimes U_B^*) e^{-\beta H}(U_A \otimes U_B)) \\
&= \operatorname{tr}\left(e^{-\beta(U_A^* \otimes U_B^*)H(U_A \otimes U_B)}\right) \\
&= \operatorname{tr}\left(e^{-\beta \hbar \omega(\epsilon_1(U_A^* A U_A) \otimes (U_B^* B U_B) + \epsilon_2(U_A^* A U_A) \otimes I_n + \epsilon_3(I_m \otimes (U_B^* B U_B)))}\right) \\
&= \operatorname{tr}\left(e^{-\beta \hbar \omega(\epsilon_1 \widetilde{A} \otimes \widetilde{B} + \epsilon_2 \widetilde{A} \otimes I_n + \epsilon_3 I_m \otimes \widetilde{B})}\right) \\
&= \sum_{j=1}^{m} \sum_{k=1}^{n} e^{-\beta \hbar \omega(\epsilon_1 \lambda_j \mu_k + \epsilon_2 \lambda_j + \epsilon_3 \mu_k)}.
\end{aligned}
$$

This calculation can be extended straightforward to the matrix

$$A_1 \otimes A_2 \otimes A_3 + A_1 \otimes I_{n_2} \otimes I_{n_3} + I_{n_1} \otimes A_2 \otimes I_{n_3} + I_{n_1} \otimes I_{n_2} \otimes A_3$$

and so on, where A_1, A_2, A_3 are $n_1 \times n_1$, $n_2 \times n_2$, $n_3 \times n_3$ matrices, respectively.

Example 3.6. Consider the special case that $n = m = 2$ and A, B are any of the Pauli spin matrices σ_1, σ_2, σ_3. The eigenvalues of any Pauli spin matrix are $+1$ and -1. Thus for any combination $\sigma_j \otimes \sigma_k$ for $A \otimes B$ we find

$$Z(\beta) = e^{-\beta \hbar \omega \epsilon_1} 2 \cosh(\beta \hbar \omega(\epsilon_2 + \epsilon_3)) + e^{\beta \hbar \omega \epsilon_1} 2 \cosh(\beta \hbar \omega(\epsilon_2 - \epsilon_3)). \quad \clubsuit$$

Example 3.7. Consider Fermi creation and annihilation operators $c_j^\dagger$, c_j. They have the matrix representation $(j = 1, 2, \ldots, N)$

$$c_j^\dagger = \sigma_3 \otimes \cdots \otimes \sigma_3 \otimes \left(\frac{1}{2}\sigma_+\right) \otimes I_2 \otimes \cdots \otimes I_2$$

$$c_j = \sigma_3 \otimes \cdots \otimes \sigma_3 \otimes \left(\frac{1}{2}\sigma_-\right) \otimes I_2 \otimes \cdots \otimes I_2$$

where σ_3 appears $j-1$ times, σ_+ and σ_- at the j-th place and

$$\sigma_+ := \sigma_1 + i\sigma_2, \qquad \sigma_- := \sigma_1 - i\sigma_2.$$

We have

$$[c_k^\dagger, c_j]_+ = \delta_{jk} I, \qquad [c_k^\dagger, c_j^\dagger]_+ = [c_k, c_j]_+ = 0.$$

Thus

$$c_j^\dagger c_j = I_2 \otimes \cdots \otimes I_2 \otimes \begin{pmatrix} 1 & 0 \\ 0 & 0 \end{pmatrix} \otimes I_2 \otimes \cdots \otimes I_2.$$

The eigenvalues of the operator $c_j^\dagger c_j$ are given by $+1$ (2^{N-1} times) and 0 (2^{N-1} times). ♣

3.7 One Dimensional Ising Model

The one dimensional *Ising model* with cyclic boundary conditions is given by

$$\hat{H} = -J \sum_{j=1}^{N} \sigma_{3,j} \sigma_{3,j+1}$$

where $\sigma_{3,N+1} \equiv \sigma_{3,1}$. This means we impose cyclic boundary conditions. Obviously $\hat{H}$ is a $2^N \times 2^N$ diagonal matrix. Here N is the number of lattice sites and J is a nonzero real constant, the so-called exchange constant. The diagonal of the matrix only contains $+1$'s and -1's and $\operatorname{tr}(\hat{H}) = 0$. We calculate the *partition function* $Z(\beta J)$ which is defined by

$$Z_N(\beta J) := \operatorname{tr}\left(e^{-\beta \hat{H}}\right) \equiv \operatorname{tr}\left(\exp\left(\beta J \sum_{j=1}^{N} \sigma_{3,j} \sigma_{3,j+1}\right)\right)$$

where $\beta = 1/(kT)$. Here k denotes the Boltzmann constant and T is the absolute temperature. From the partition function Z_N we obtain the *Helmholtz free energy* per lattice site

$$\frac{F(\beta J)}{N} := -\frac{1}{N\beta} \ln(Z_N(\beta J)).$$

We first consider the case with four lattice points, i.e. $N = 4$. Then we extend to arbitrary N. Since

$$\sigma_{3,j} = \overbrace{I_2 \otimes \cdots \otimes I_2 \otimes \sigma_3 \otimes I_2 \otimes \cdots \otimes I_2}^{N\text{-factors}}$$

$$\underset{j\text{-th place}}{\uparrow}$$

we have for $N = 4$

$$\sigma_{3,1} = \sigma_3 \otimes I_2 \otimes I_2 \otimes I_2, \quad \sigma_{3,2} = I_2 \otimes \sigma_3 \otimes I_2 \otimes I_2$$

$$\sigma_{3,3} = I_2 \otimes I_2 \otimes \sigma_3 \otimes I_2, \quad \sigma_{3,4} = I_2 \otimes I_2 \otimes I_2 \otimes \sigma_3.$$

Obviously these matrices are diagonal matrices. We set

$$X_{12} := \sigma_{3,1}\sigma_{3,2}, \quad X_{23} := \sigma_{3,2}\sigma_{3,3}, \quad X_{34} := \sigma_{3,3}\sigma_{3,4}, \quad X_{41} := \sigma_{3,4}\sigma_{3,1}.$$

Consequently

$$X_{12} \equiv \sigma_3 \otimes \sigma_3 \otimes I_2 \otimes I_2, \quad X_{23} \equiv I_2 \otimes \sigma_3 \otimes \sigma_3 \otimes I_2$$

$$X_{34} \equiv I_2 \otimes I_2 \otimes \sigma_3 \otimes \sigma_3, \quad X_{41} \equiv \sigma_3 \otimes I_2 \otimes I_2 \otimes \sigma_3$$

which are also diagonal matrices. Let $M := \{ X_{12}, X_{23}, X_{34}, X_{41} \}$. Thus the elements of M are $2^4 \times 2^4$ matrices. We recall the identities

$$\mathrm{tr}(A + B) \equiv \mathrm{tr}(A) + \mathrm{tr}(B), \qquad \mathrm{tr}(A \otimes B) \equiv (\mathrm{tr}(A))(\mathrm{tr}(B))$$

where A and B are $n \times n$ matrices. We need the following lemmata.

Lemma 3.1. *Let X be a real $n \times n$ matrix such that $X^2 = I_n$. Let $\epsilon \in \mathbb{R}$. Then $\exp(\epsilon X) \equiv I_n \cosh(\epsilon) + X \sinh(\epsilon)$. The matrices given above are examples of such matrices.*

Lemma 3.2. *Let X and Y be arbitrary elements of M. Then $[X, Y] = 0_4$.*

Lemma 3.3. *Let X, Y, Z be arbitrary elements of M. Then*

$$\mathrm{tr}(X) = 0$$
$$\mathrm{tr}(XY) = 0 \quad if \quad X \neq Y$$
$$\mathrm{tr}(XYZ) = 0 \quad if \quad X \neq Y.$$

Lemma 3.4. *Let $X_{12}, X_{23}, X_{34}, X_{41}$ be the matrices given above. Then*

$$X_{12}X_{23}X_{34}X_{41} = I_{16} \Rightarrow \mathrm{tr}(X_{12}X_{23}X_{34}X_{41}) = 2^4 = 16.$$

Theorem 3.1. *Let $K \equiv \epsilon(X_{12} + X_{23} + X_{34} + X_{41})$. Then*

$$\mathrm{tr}(\exp(K)) \equiv \mathrm{tr}(I_{16} \cosh^4(\epsilon) + I_{16} \sinh^4(\epsilon)).$$

Proof. Using the lemmata given above we find

$$\mathrm{tr}(\exp(K)) \equiv \mathrm{tr}[\exp(\epsilon X_{12}) \exp(\epsilon X_{23}) \exp(\epsilon X_{34}) \exp(\epsilon X_{41})]$$

$$\equiv \mathrm{tr}[(I_{16} \cosh(\epsilon) + X_{12} \sinh(\epsilon))(I_{16} \cosh(\epsilon) + X_{23} \sinh(\epsilon))$$

$$\times (I_{16} \cosh(\epsilon) + X_{34} \sinh(\epsilon))(I_{16} \cosh(\epsilon) + X_{41} \sinh(\epsilon))]$$

$$\equiv \mathrm{tr}\left[I_{16} \cosh^4(\epsilon) + I_{16} \sinh^4(\epsilon)\right]. \qquad \square$$

As a consequence we obtain

$$\text{tr}(I_{16}\cosh^4(\epsilon) + I_{16}\sinh^4(\epsilon)) \equiv 2^4(\cosh^4(\epsilon) + \sinh^4(\epsilon)).$$

Thus the partition function is given by

$$Z_4(\beta J) = 2^4(\cosh^4(\beta J) + \sinh^4(\beta J)).$$

From the partition function we find the Helmholtz free energy.

Next we consider the case with arbitrary N.

Theorem 3.2. *Let*

$$\overbrace{X_{12} \equiv \sigma_3 \otimes \sigma_3 \otimes I_2 \otimes \cdots \otimes I_2}^{N\text{-factors}}$$
$$X_{23} \equiv I_2 \otimes \sigma_3 \otimes \sigma_3 \otimes \cdots \otimes I_2$$
$$\vdots$$
$$X_{N-1,N} \equiv I_2 \otimes I_2 \otimes \cdots \otimes I_2 \otimes \sigma_3 \otimes \sigma_3$$
$$X_{N1} \equiv \sigma_3 \otimes I_2 \otimes \cdots \otimes I_2 \otimes \sigma_3$$

where I_2 is the unit 2×2 matrix. Let

$$K := \epsilon(X_{12} + X_{23} + \cdots + X_{N-1,N} + X_{N1})$$

where $\epsilon \in \mathbb{R}$. Then

$$\text{tr}(\exp(K)) \equiv \text{tr}(I_{2^N}\cosh^N(\epsilon) + I_{2^N}\sinh^N(\epsilon))$$

where I_{2^N} is the unit $2^N \times 2^N$ matrix. As a consequence we obtain

$$\text{tr}(\exp(K)) \equiv 2^N\cosh^N(\epsilon) + 2^N\sinh^N(\epsilon) \equiv 2^N(\cosh^N(\epsilon) + \sinh^N(\epsilon)).$$

Remark 1. The lemmata described above can be extended to arbitrary N. Consequently, the proof of this theorem can be performed in the same way as the proof of the theorem for the case $N = 4$.

Remark 2. When we take the Pauli matrices σ_1 or σ_2 instead of σ_3, we obtain the same result for the partition function, because $\sigma_1^2 = \sigma_2^2 = I_2$.

Remark 3. The $2^N \times 2^N$ matrix K describes the one dimensional Ising model with cyclic boundary conditions. The traditional approach is as follows: in one dimension the Ising model is given by

$$H_N = -J\sum_{j=1}^{N}\sigma_j\sigma_{j+1}$$

where $\sigma_{N+1} = \sigma_1$ (cyclic boundary conditions) and $\sigma_j \in \{1, -1\}$. Here N is the number of lattice sites, and J ($J > 0$) is the real constant. The partition function is

$$Z_N(\beta J) = \sum_{\sigma_1, \ldots, \sigma_N \in \Sigma} \exp \left(\beta J \sum_{j=1}^{N} \sigma_j \sigma_{j+1} \right)$$

where the first sum runs over all possible configurations Σ. There are 2^N possible configurations. Hence, we must sum over all possible configurations, namely 2^N, instead of calculating the trace of a $2^N \times 2^N$ matrix.

Exercises. (1) Let

$$\hat{H} = -J \sum_{j=1}^{4} \sigma_{3,j} \sigma_{3,j+1}$$

and $\sigma_{3,5} = 0$. This is the open end boundary condition. Find the partition function $Z(\beta)$. Find the Helmholtz free energy.

(2) Find the eigenvalues and eigenvectors of $\hat{H} = -J(\sigma_{3,1}\sigma_{3,2} + \sigma_{3,2}\sigma_{3,3})$.

(3) Find the eigenvalues of
$$\hat{H} = -J(\sigma_{3,1}\sigma_{3,2} + \sigma_{3,2}\sigma_{3,3}) + B(\sigma_{3,1} + \sigma_{3,2} + \sigma_{3,3})$$
where B is a positive constant (the magnetic field).

(4) Given n lattice sites. Each lattice site can either be occupied by a spin up or by a spin down. Show that the number of possible configurations is 2^n. Discuss the case where a lattice site can be occupied in three different ways, for example spin up, spin down and spin horizontal.

(5) Find the eigenvalues of the Hamilton operator
$$\hat{H} = J(\sigma_{3,00}\sigma_{3,10} + \sigma_{3,10}\sigma_{3,20} + \sigma_{3,01}\sigma_{3,11} + \sigma_{3,11}\sigma_{3,21}$$
$$+ \sigma_{3,02}\sigma_{3,12} + \sigma_{3,12}\sigma_{3,22} + \sigma_{3,00}\sigma_{3,01} + \sigma_{3,01}\sigma_{3,02}$$
$$+ \sigma_{3,10}\sigma_{3,11} + \sigma_{3,11}\sigma_{3,12} + \sigma_{3,20}\sigma_{3,21} + \sigma_{3,21}\sigma_{3,22}).$$
Thus we have nine lattice sites on a 3×3 rectangular lattice. How could this Hamilton operator be written using the Kronecker product?

(6) Find the eigenvalues of the Hamilton operator
$$\hat{H} = J(\sigma_{3,0}\sigma_{3,1} + \sigma_{3,1}\sigma_{3,3} + \sigma_{3,1}\sigma_{3,4} + \sigma_{3,0}\sigma_{3,2} + \sigma_{3,2}\sigma_{3,5} + \sigma_{3,2}\sigma_{3,6}).$$
Thus the underlying lattice has a tree structure.

3.8 Fermi Systems

From statistical mechanics it is known that all equilibrium thermodynamic quantities of interest for a many body system of interacting Fermi particles can be determined from the *grand partition function* (Fetter and Walecka [21], Steeb [52])

$$Z(\beta) := \text{tr}(\exp(-\beta(\hat{H} - \mu\hat{N}_e))).$$

Here $\hat{H}$ is the Hamilton operator describing the system. We consider Fermi systems in the occupation number formalism. Here $\hat{N}_e$ is the total number operator, μ the chemical potential, and β the inverse temperature. In general, the trace cannot be evaluated. However, often, the Hamilton operator $\hat{H}$ consists of two terms, namely

$$\hat{H} = \hat{H}_0 + \hat{H}_1$$

where $\hat{H}_0$ is so chosen that the properties described by $\hat{H}_0$ alone are well-known. In most cases stated in the literature, the operator $\hat{H}_0 - \mu\hat{N}_e$ has the form

$$\hat{H}_0 - \mu\hat{N}_e = \sum_k \sum_{\sigma\in\{\uparrow,\downarrow\}} (\varepsilon(k) - \mu)c_{k,\sigma}^\dagger c_{k,\sigma}$$

where σ denotes the spin with $\sigma \in \{\uparrow,\downarrow\}$ and $\varepsilon(k)$ is the one-particle energy. We investigate a lattice system and therefore k runs over the first Brioullin zone. Here we are able to calculate the trace exactly.

The requirements of the *Pauli principle* are satisfied if the Fermion operators

$$\{ c_{k,\sigma}^\dagger, c_{j,\sigma} \ : \ k,j = 1,2,\dots,N : \sigma \in \{\uparrow,\downarrow\} \}$$

satisfy the anticommutation relations

$$[c_{k,\sigma}^\dagger, c_{j,\sigma'}]_+ = \delta_{k,j}\delta_{\sigma,\sigma'}I, \qquad [c_{k,\sigma}^\dagger, c_{j,\sigma'}^\dagger]_+ = [c_{k,\sigma}, c_{j,\sigma'}]_+ = 0$$

for all $k,j = 1,2,\dots,N$ and $\sigma,\sigma' \in \{\uparrow,\downarrow\}$. Here I denotes the unit operator.

We show that the trace calculation of $\hat{H}_0 - \mu\hat{N}_e$ can easily be performed with the aid of matrix calculation and the Kronecker product of matrices (Steeb [52], Steeb [53], Steeb [54], Villet [72]).

First we recall that

$$\text{tr}(\exp[(A_1 \otimes I \otimes \cdots \otimes I) + (I \otimes A_2 \otimes I \cdots \otimes I) + (I \otimes I \otimes \cdots \otimes A_N)])$$

$$= (\text{tr}(\exp(A_1)))(\text{tr}(\exp(A_2))) \cdots (\text{tr}(\exp(A_N))) = \prod_{k=1}^{N} \text{tr}(\exp(A_k)).$$

This formula is the main result. It remains to show that the Hamilton operator $\hat{H}_0 - \mu \hat{N}_e$ can be brought into the form

$$(A_1 \otimes I \otimes \cdots \otimes I) + (I \otimes A_2 \otimes I \cdots \otimes I) + \cdots + (I \otimes I \otimes \cdots \otimes A_N)$$

where I is the unit matrix.

Therefore the matrix representation for the Fermi operators will be given. For the sake of simplicity we first consider spinless Fermions. Then the case including the spin is described.

The requirements of the *Pauli principle* are satisfied if the spinless Fermion operators

$$\{ c_k^\dagger, c_j : k, j = 1, 2, \ldots, N \}$$

satisfy the anticommutation relations

$$[c_k^\dagger, c_j]_+ = \delta_{kj} I, \qquad [c_k^\dagger, c_j^\dagger]_+ = [c_k, c_j]_+ = 0$$

for all $k, j = 1, 2, \ldots, N$.

Let us now give a faithful matrix representation of the operators. First we discuss the case $N = 1$. A basis is given by

$$\{ c^\dagger |0\rangle, \ |0\rangle \}$$

and the corresponding dual one by

$$\{ \langle 0|c, \ \langle 0| \}$$

with $\langle 0|0 \rangle = 1$, $\langle 0|cc^\dagger|0 \rangle = 1$, $\langle 0|c|0 \rangle = 0$, $\langle 0|c^\dagger|0 \rangle = 0$, where the vector space (Hilbert space) under consideration is two dimensional, i.e. $\mathbb{C}^2$. Consequently, $c^\dagger$ and c have the faithful matrix representation

$$c^\dagger = \begin{pmatrix} 0 & 1 \\ 0 & 0 \end{pmatrix} = \frac{1}{2}(\sigma_1 + i\sigma_2) = \frac{1}{2}\sigma_+, \quad c = \begin{pmatrix} 0 & 0 \\ 1 & 0 \end{pmatrix} = \frac{1}{2}(\sigma_1 - i\sigma_2) = \frac{1}{2}\sigma_-.$$

The two *state vectors* are

$$c^\dagger |0\rangle = \begin{pmatrix} 1 \\ 0 \end{pmatrix}, \qquad |0\rangle = \begin{pmatrix} 0 \\ 1 \end{pmatrix}.$$

Here σ_1, σ_2, σ_3, σ_+, σ_- are the Pauli spin matrices. The *number operator* $\hat{n}$ defined by

$$\hat{N} := c^\dagger c$$

becomes

$$\hat{N} := c^\dagger c = \begin{pmatrix} 1 & 0 \\ 0 & 0 \end{pmatrix}$$

with the eigenvalues 0 and 1. The extension to the case $N > 1$ leads to

$$c_k^\dagger = \overbrace{\sigma_3 \otimes \sigma_3 \otimes \cdots \otimes \sigma_3 \otimes \left(\frac{1}{2}\sigma_+\right)}^{N\text{-times}} \otimes I_2 \otimes I_2 \otimes \cdots \otimes I_2$$

$$c_k = \sigma_3 \otimes \sigma_3 \otimes \cdots \otimes \sigma_3 \otimes \left(\frac{1}{2}\sigma_-\right) \otimes I_2 \otimes I_2 \otimes \cdots \otimes I_2$$

$$k\text{-th place}$$

where I_2 is the 2×2 unit matrix. One can easily calculate that the anti-commutation relations are fulfilled. The *number operator*

$$\hat{N}_k := c_k^\dagger c_k$$

with quantum number k is found to be

$$\hat{N}_k = c_k^\dagger c_k = I_2 \otimes \cdots \otimes I_2 \otimes \begin{pmatrix} 1 & 0 \\ 0 & 0 \end{pmatrix} \otimes I_2 \otimes \cdots \otimes I_2.$$

Finally, one has

$$\sum_{k=1}^{N} \lambda_k \hat{N}_k = \begin{pmatrix} \lambda_1 & 0 \\ 0 & 0 \end{pmatrix} \otimes I_2 \otimes \cdots \otimes I_2 + I_2 \otimes \begin{pmatrix} \lambda_2 & 0 \\ 0 & 0 \end{pmatrix} \otimes I_2 \otimes \cdots \otimes I_2$$

$$+ \cdots + I_2 \otimes \cdots \otimes I_2 \otimes \begin{pmatrix} \lambda_N & 0 \\ 0 & 0 \end{pmatrix}$$

where $\lambda_k \in \mathbb{R}$. The total number operator is defined as

$$\hat{N}_e := \sum_{k=1}^{N} \hat{N}_k \equiv \sum_{k=1}^{N} c_k^\dagger c_k.$$

The underlying vector space is given by $\mathbb{C}^{2^N}$. Thus finding the trace

$$\text{tr}\left(\exp\left(\sum_{k=1}^{N} \lambda_k \hat{n}_k\right)\right)$$

reduces to the trace calculation in a subspace. In occupation number formalism this subspace is given by the basis $\left\{ c^\dagger|0\rangle, |0\rangle \right\}$ and in matrix calculation it is given by the basis

$$\left\{ \begin{pmatrix} 1 \\ 0 \end{pmatrix}, \ \begin{pmatrix} 0 \\ 1 \end{pmatrix} \right\}.$$

Consequently, we find that

$$\mathrm{tr}\left(\exp\left(\sum_{k=1}^{N} \lambda_k \hat{n}_k \right) \right) = \prod_{k=1}^{N} \mathrm{tr}\begin{pmatrix} e^{\lambda_k} & 0 \\ 0 & 1 \end{pmatrix} = \prod_{k=1}^{N}(e^{\lambda_k} + 1).$$

Example 3.8. For an approximative solution we consider the *Hubbard model*. The Hamilton operator can be written in *Bloch representation* as

$$\hat{H} = \sum_{k\sigma} \varepsilon(k) c_{k\sigma}^\dagger c_{k\sigma} + U \sum_{k_1 k_2 k_3 k_4} \delta(k_1 - k_2 + k_3 - k_4) c_{k_1\uparrow}^\dagger c_{k_2\uparrow} c_{k_3\downarrow}^\dagger c_{k_4\downarrow}$$

where k runs over the first Brioullin zone. As an approximative method we use the following inequality, which is sometimes called *Bogolyubov inequality*,

$$\Omega \le \mathrm{tr}((\hat{H} - \mu\hat{N}_e)W_t) + \frac{1}{\beta}\mathrm{tr}(W_t \ln(W_t))$$

for the grand potential Ω. W_t is the so-called trial density matrix. Consider the case where

$$W_t = \frac{\exp(-\beta \sum_k (E_\uparrow(k) c_{k\uparrow}^\dagger c_{k\uparrow} + E_\downarrow(k) c_{k\downarrow}^\dagger c_{k\downarrow}))}{\mathrm{tr}(\exp(-\beta \sum_k (E_\uparrow(k) c_{k\uparrow}^\dagger c_{k\uparrow} + E_\downarrow(k) c_{k\downarrow}^\dagger c_{k\downarrow})))}.$$

Both $E_\uparrow(k)$ and $E_\downarrow(k)$ play the role of real variation parameters.

For example, we have to calculate traces such as

$$\mathrm{tr}(c_{k\uparrow}^\dagger c_{k\uparrow} W_t)$$

and

$$\sum_{k_1 k_2 k_3 k_4} \delta(k_1 - k_2 + k_3 - k_4)\mathrm{tr}(c_{k_1\uparrow}^\dagger c_{k_2\uparrow} c_{k_3\downarrow}^\dagger c_{k_4\downarrow} W_t).$$

Now we show how, within the framework of the developed calculation, the trace can be evaluated. Since we have included the spin, the matrix representation of the Fermi operators

$$\left\{ c_{k\sigma}^\dagger, \ c_{k\sigma} : k = 1, \dots, N; \ \sigma = \uparrow, \downarrow \right\}$$

becomes

$$c_{k\uparrow}^\dagger = \underbrace{\sigma_3 \otimes \cdots \otimes \sigma_3 \otimes \overset{\text{k-th place}}{\left(\frac{1}{2}\sigma_+\right)} \otimes I_2 \otimes \cdots \otimes I_2}_{2N \text{ times}}$$

$$c_{k\downarrow}^\dagger = \sigma_3 \otimes \cdots \otimes \sigma_3 \otimes \left(\frac{1}{2}\sigma_+\right) \otimes I_2 \otimes \cdots \otimes I_2.$$
$$(k+N)\text{-th place}$$

Using the identity

$$\exp\begin{pmatrix} \lambda & 0 \\ 0 & 0 \end{pmatrix} \equiv \begin{pmatrix} \exp(\lambda) & 0 \\ 0 & 1 \end{pmatrix}$$

it follows that

$$\exp\left(\sum_k E_\uparrow(k)c_{k\uparrow}^\dagger c_{k\uparrow} + \sum_k E_\downarrow(k)c_{k\downarrow}^\dagger c_{k\downarrow}\right)$$

$$\equiv \begin{pmatrix} e^{E_\uparrow(1)} & 0 \\ 0 & 1 \end{pmatrix} \otimes \cdots \otimes \begin{pmatrix} e^{E_\uparrow(N)} & 0 \\ 0 & 1 \end{pmatrix} \otimes \begin{pmatrix} e^{E_\downarrow(N+1)} & 0 \\ 0 & 1 \end{pmatrix} \otimes \cdots \otimes \begin{pmatrix} e^{E_\downarrow(2N)} & 0 \\ 0 & 1 \end{pmatrix}.$$

As an abbreviation we set $-\beta E_\sigma(k) \to E_\sigma(k)$. Since

$$c_{k\uparrow}^\dagger c_{k\uparrow} = I_2 \otimes I_2 \otimes \cdots \otimes \overset{\text{k-th place}}{\begin{pmatrix} 1 & 0 \\ 0 & 0 \end{pmatrix}} \otimes I_2 \otimes \cdots \otimes I_2$$

we have

$$\text{tr}(c_{k\uparrow}^\dagger c_{k\uparrow} W_t) = \frac{\exp(E_\uparrow(k))}{\exp(E_\uparrow(k)) + 1}.$$

Now we cast the interaction term in a form more convenient form

$$\sum_{k_1 k_2 k_3 k_4} \delta(k_1 - k_2 + k_3 - k_4)c_{k_1\uparrow}^\dagger c_{k_2\uparrow} c_{k_3\downarrow}^\dagger c_{k_4\downarrow}$$

$$\equiv \sum_{k_1 k_3} c_{k_1\uparrow}^\dagger c_{k_1\uparrow} c_{k_3\downarrow}^\dagger c_{k_3\downarrow} + \sum_{\substack{k_1 k_2 k_3 k_4 \\ k_1 \neq k_2}} \delta(k_1 - k_2 + k_3 - k_4)c_{k_1\uparrow}^\dagger c_{k_2\uparrow} c_{k_3\downarrow}^\dagger c_{k_4\downarrow}.$$

For the first term on the right-hand side we find that $c_{k_1\uparrow}^\dagger c_{k_1\uparrow} c_{k_3\downarrow}^\dagger c_{k_3\downarrow}$ takes the form

$$I_2 \otimes I_2 \otimes \cdots \otimes \begin{pmatrix} 1 & 0 \\ 0 & 0 \end{pmatrix} \otimes I_2 \otimes \cdots \otimes I_2 \otimes \begin{pmatrix} 1 & 0 \\ 0 & 0 \end{pmatrix} \otimes I_2 \otimes \cdots \otimes I_2$$

where the first diagonal matrix $\text{diag}(1,0)$ is at the k_1-th place and the second at the $k_3 + N$-th place. Calculating the trace provides

$$\frac{1}{N}\sum_{k_1 k_3}\text{tr}(c^\dagger_{k_1\uparrow}c_{k_1\uparrow}c^\dagger_{k_3\downarrow}c_{k_3\downarrow}W_t) = \frac{1}{N}\sum_{k_1 k_3}\frac{e^{E_\uparrow(k_1)}}{1+e^{E_\uparrow(k_1)}}\cdot\frac{e^{E_\downarrow(k_3)}}{1+e^{E_\downarrow(k_3)}}.$$

For the second term on the right-hand side there is no contribution to the trace since

$$\text{tr}\left(\begin{pmatrix}0 & -1\\0 & 0\end{pmatrix}\begin{pmatrix}e^\lambda & 0\\0 & 1\end{pmatrix}\right) = 0, \qquad \frac{1}{2}\sigma_+\sigma_3 = \begin{pmatrix}0 & -1\\0 & 0\end{pmatrix}.$$

The described method can be extended in order to obtain better approximations. To this end we make a unitary transformation given by

$$U := \exp(iS)$$

with $S = S^\dagger = S^*$, and followed then by the trace determination. In this case too the trace calculation can easily be performed within the framework of usual matrix calculation. ♣

We now calculate the grand canonical partition function Z and grand thermodynamic potential for the Hamilton operator

$$\hat{H} = U\sum_{j=1}^{N}\hat{n}_{j\uparrow}\hat{n}_{j\downarrow}$$

where U is a positive constant (the repulsion of the electrons at the same lattice site j), N is the number of lattice sites and $\hat{n}_{j\sigma} := c^\dagger_{j\sigma}c_{j\sigma}$. For these calculations we need the following theorems

Theorem 3.3. *Let A_1, A_2, B_1, B_2 be $n \times n$ matrices. Let*

$$X := A_1 \otimes I_n \otimes B_1 \otimes I_n + I_n \otimes A_2 \otimes I_n \otimes B_2.$$

Then $\text{tr}(\exp(X)) = \text{tr}(\exp(A_1 \otimes B_1))\text{tr}(\exp(A_2 \otimes B_2))$.

An extension of this formula is

Theorem 3.4. *Let A_1, $A_2,\ldots,A_N$, B_1, $B_2,\ldots,B_N$ be $n \times n$ matrices. Then*

$$\text{tr}(\exp(A_1 \otimes I_n \otimes \cdots \otimes I_n \otimes B_1 \otimes I_n \otimes \cdots \otimes I_n$$
$$+ I_n \otimes A_1 \otimes \cdots \otimes I_n \otimes I_n \otimes B_2 \otimes \cdots \otimes I_n$$
$$\vdots$$
$$+ I_n \otimes I_n \otimes \cdots \otimes A_N \otimes I_n \otimes I_n \otimes \cdots \otimes B_N))$$
$$= (\text{tr}(\exp(A_1 \otimes B_1)))(\text{tr}(\exp A_2 \otimes B_2)))\cdots(\text{tr}(\exp(A_N \otimes B_N))).$$

For our purpose we need a further extension.

Theorem 3.5. *Let A_1, A_2, B_1, B_2, be $n \times n$ matrices. Assume that*

$$[A_1, C] = [A_2, C] = [B_1, C] = [B_2, C] = 0_n.$$

Then

$$\mathrm{tr}(\exp(A_1 \otimes I_n \otimes B_1 \otimes I_n + I_n \otimes A_2 \otimes I_n \otimes B_2 + C \otimes I_n \otimes I_n \otimes I_n$$
$$+ I_n \otimes C \otimes I_n \otimes I_n + I_n \otimes I_n \otimes C \otimes I_n + I_n \otimes I_n \otimes I_n \otimes C))$$
$$= \prod_{j=1}^{2} \mathrm{tr}(\exp(A_j \otimes B_j + C \otimes I_n + I_n \otimes C)).$$

An extension is as follows

Theorem 3.6. *Let $A_1, A_2, \ldots, A_N, B_1, B_2, \ldots, B_N, C$ be $n \times n$ matrices.*
Assume that $[A_j, C] = [B_j, C] = 0_n$ for $j = 1, 2, \ldots, N$. Then

$$\mathrm{tr}(\exp(A_1 \otimes I_n \otimes \cdots \otimes I_n \otimes B_1 \otimes I_n \otimes \cdots \otimes I_n$$
$$+ I_n \otimes A_2 \otimes \cdots \otimes I_n \otimes I_n \otimes B_2 \otimes \cdots \otimes I_n$$
$$\vdots$$
$$+ I_n \otimes I_n \otimes \cdots \otimes A_N \otimes I_n \otimes I_n \otimes \cdots \otimes B_N$$
$$+ C \otimes I \otimes \cdots \otimes I_n \otimes I_n \otimes I_n \otimes \cdots \otimes I_n$$
$$+ I_n \otimes C \otimes \cdots \otimes I_n \otimes I_n \otimes I_n \otimes \cdots \otimes I_n$$
$$\vdots$$
$$+ I_n \otimes I_n \otimes \cdots \otimes I_n \otimes I_n \otimes I_n \otimes \cdots \otimes C))$$
$$= \prod_{j=1}^{N} \mathrm{tr}(\exp(A_j \otimes B_j + C \otimes I_n + I_n \otimes C)).$$

Before we describe the connection with Fermi systems consider a special case of the identity. Assume that

$$A_1 = \cdots = A_N = B_1 = \cdots = B_N.$$

We put $A = A_1 = \cdots = B_N$. Then the right-hand side of the identity takes the form

$$(\mathrm{tr}(\exp(A \otimes A + C \otimes I_n + I_n \otimes C)))^N.$$

Since $[A, C] = 0_n$ it follows that

$$(\mathrm{tr}(\exp(A \otimes A + C \otimes I_n + I_n \otimes C)))^N \equiv (\mathrm{tr}(\exp(A \otimes A)\exp(C \otimes I_n)\exp(I_n \otimes C)))^N.$$

Assume that the matrices A and C can be written as $A = \sqrt{a}X$, $C = bY$, where $a \in \mathbb{R}^+$, $b \in \mathbb{R}$ and $X^2 = X$, $Y^2 = Y$. This means that X and Y are idempotent. Then we obtain

$$[\text{tr}(\exp(A \otimes A + C \otimes I_n + I_n \otimes C))]^N =$$

$$[\text{tr}(I \otimes I_n + (e^a - 1)(X \otimes X))(I_n \otimes I_n + (e^b - 1)(Y \otimes I_n))(I_n \otimes I_n + (e^b - 1)(I_n \otimes Y))]^N.$$

Using the identity $(R \otimes S)(U \otimes V) \equiv (RU) \otimes (SV)$ it follows that

$$\begin{aligned}
[\text{tr}(\exp&(A \otimes A + C \otimes I_n + I_n \otimes C))]^N \\
&= [\text{tr}(I_n \otimes I_n + (e^b - 1)(Y \otimes I_n + I_n \otimes Y) + (e^a - 1)(X \otimes X) \\
&\quad + (e^b - 1)^2(Y \otimes Y) + (e^a - 1)(e^b - 1)(XY \otimes X + X \otimes XY) \\
&\quad + (e^a - 1)(e^b - 1)^2(XY \otimes XY))]^N.
\end{aligned}$$

If we assume that I_n is the 2×2 unit matrix and

$$X = Y = \begin{pmatrix} 1 & 0 \\ 0 & 0 \end{pmatrix}$$

then

$$(\text{tr}(\exp(A \otimes A + C \otimes I_n + I_n \otimes C)))^N = (1 + 2e^b + e^{a+2b})^N.$$

Now we have

$$c_{j\uparrow}^\dagger = \underbrace{\sigma_3 \otimes \cdots \otimes \sigma_3 \otimes \overset{\text{j-th place}}{\left(\frac{1}{2}\sigma_+\right)} \otimes I_2 \otimes \cdots \otimes I_2}_{2N \text{ times}}$$

$$c_{j\downarrow}^\dagger = \underbrace{\sigma_3 \otimes \cdots \otimes \sigma_3 \otimes \overset{\text{$(j + N)$-th place}}{\left(\frac{1}{2}\sigma_+\right)} \otimes I_2 \otimes \cdots \otimes I_2}_{2N \text{ times}}$$

where $j = 1, 2, \ldots, 2N$ and σ_+. For the Fermi annihilation operators with spin up and down, respectively, we have to replace σ_+ by σ_-. Consider now the Hamilton operator

$$\hat{K} = a \sum_{j=1}^N c_{j\uparrow}^\dagger c_{j\uparrow} c_{j\downarrow}^\dagger c_{j\downarrow} + b \sum_{j=1}^N (c_{j\uparrow}^\dagger c_{j\uparrow} + c_{j\downarrow}^\dagger c_{j\downarrow})$$

where $a, b \in \mathbb{R}$. Since $\sigma_3^2 = I_2$ and

$$\left(\frac{1}{2}\sigma_+\right)\left(\frac{1}{2}\sigma_-\right) = \begin{pmatrix} 1 & 0 \\ 0 & 0 \end{pmatrix}$$

we obtain

$$j\text{-th place}$$

$$c_{j\uparrow}^{\dagger} c_{j\uparrow} = I_2 \otimes \cdots \otimes I_2 \otimes \begin{pmatrix} 1 & 0 \\ 0 & 0 \end{pmatrix} \otimes I_2 \otimes \cdots \otimes I_2$$

$$(j+N)\text{-th place}$$

$$c_{j\downarrow}^{\dagger} c_{j\downarrow} = I_2 \otimes \cdots \otimes I_2 \otimes \begin{pmatrix} 1 & 0 \\ 0 & 0 \end{pmatrix} \otimes I_2 \otimes \cdots \otimes I_2$$

and

$$j\text{-th place} \qquad\qquad (j+N)\text{-th place}$$

$$c_{j\uparrow}^{\dagger} c_{j\uparrow} c_{j\downarrow}^{\dagger} c_{j\downarrow} = I_2 \otimes \cdots \otimes I_2 \otimes \begin{pmatrix} 1 & 0 \\ 0 & 0 \end{pmatrix} \otimes I_2 \otimes \cdots \otimes I_2 \otimes \begin{pmatrix} 1 & 0 \\ 0 & 0 \end{pmatrix} \otimes I_2 \otimes \cdots \otimes I_2.$$

Consequently, the problem to calculate $\operatorname{tr}(\exp(\hat{K}))$ has been solved. Using the equations given above we find that the grand thermodynamic potential per lattice site is given by

$$\frac{\Omega}{N} = -\frac{1}{\beta} \ln(1 + 2\exp(\beta\mu) + \exp(\beta(2\mu - U)))$$

where μ is the *chemical potential*. The number of electrons per lattice site $n_e \equiv N_e/N$ is defined by

$$n_e := -\frac{1}{N} \frac{\partial \Omega}{\partial \mu}.$$

We find

$$n_e = \frac{2(\exp(\beta\mu) + \exp(\beta(2\mu - U)))}{1 + 2\exp(\beta\mu) + \exp(\beta(2\mu - U))}$$

where $0 \leq n_e \leq 2$. Because of the Pauli principle we have $\mu \to \infty$ as $n_e \to 2$.

Exercises. (1) Show that $[\sigma_3, c^{\dagger}]_+ = 0_2$, $[\sigma_3, c]_+ = 0_2$.

(2) Find the eigenvalues of the operators $\hat{n}_k$ and $\hat{N}_e$.

(3) Let $k = 1, 2, \ldots, N$. Show that $c_k^{\dagger}$ and c_k can be written as

$$c_k^{\dagger} = \frac{1}{2} \left(\prod_{j=1}^{k-1} \sigma_{3,j} \right) \sigma_{+,k}, \qquad c_k = \frac{1}{2} \left(\prod_{j=1}^{k-1} \sigma_{3,j} \right) \sigma_{-,k}$$

where $\sigma_{3,j} := I_2 \otimes \cdots \otimes I_2 \otimes \sigma_3 \otimes I_2 \otimes \cdots \otimes I_2$ with σ_3 at the j-th position.

(4) Show that

$$\sum_{j=1}^{N} c_j^\dagger c_j = \frac{1}{4} \sum_{j=1}^{N} \sigma_{+,j} \sigma_{-,j}.$$

(5) Let $k = 1, 2, \ldots, N$. Show that

$$c_k^\dagger = \frac{1}{2}(-1)^{k-1} \left(\prod_{j=1}^{k-1} \sigma_{3,j} \right) \sigma_{+,k}, \qquad c_k = \frac{1}{2}(-1)^{k-1} \left(\prod_{j=1}^{k-1} \sigma_{3,j} \right) \sigma_{-,k}$$

is also a faithful representation of the Fermi operators.

(6) Consider the Hamilton operator

$$\hat{H} = W \sum_{j=1}^{N} \hat{n}_j - J \sum_{j=1}^{N} (c_j^\dagger c_{j-1} + c_{j-1}^\dagger c_j) + V \sum_{j=1}^{N} \hat{n}_j \hat{n}_{j-1}$$

with cyclic boundary conditions and $\hat{n}_j := c_j^\dagger c_j$. Here W, J and V are constants. The *Jordan-Wigner transformation* for c_j is defined as

$$c_j = \exp \left(i\pi \sum_{l=1}^{j-1} S_{+,l} S_{-,l} \right) S_{-,j}$$

where

$$S_{-,j} := \frac{1}{2}(\sigma_{1,j} - i\sigma_{2,j}) = \frac{1}{2}\sigma_{-,j}, \qquad S_{+,j} := \frac{1}{2}(\sigma_{1,j} + i\sigma_{2,j}) = \frac{1}{2}\sigma_{+,j}.$$

Find $c_j^\dagger$ and $\hat{n}_j$. Show that the Hamilton operator takes the form

$$\hat{H} = -\sum_{j=1}^{N} \left(J(S_{+,j} S_{-,j-1} + S_{-,j} S_{+,j-1}) - \frac{V}{4} \sigma_{3,j} \sigma_{3,j-1} \right)$$

$$+ \frac{W+V}{2} \sum_{j=1}^{N} \sigma_{3,j} + \frac{N}{2} \left(W + \frac{V}{2} \right) I.$$

3.9   Dimer Problem

Let M, N be natural numbers. Consider a $2M \times 2N$ square lattice. In how many ways $\mathcal{N}$ can we fill a $2M$ by $2N$ square lattice with nonoverlapping dimers? All sites are occupied and no two dimers overlap. A dimer is a horizontal or vertical bond which occupies two vertices. In other words a dimer is a rigid rot just long enough to cover two neighbouring vertices either horizontally or vertically.

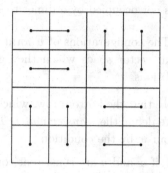

We consider the case of periodic boundary conditions which means that the first column is to be identified with the $2M + 1$-th column and the first row with the $2N + 1$-th row and $N, M \to \infty$.

There are several methods to solve the dimer problem. We follow the *transfer matrix method* (Lieb [42], Percus [47]). We cast the problem in such a form that we can apply the Kronecker product. One specifies a dimer arrangement by listing the type of dimer at each vertex i.e. up, down, to the left or the right. This introduces the possibility of two dimensional topological problems. To avoid these, one constructs composite vertices which can be ordered in a one dimensional space. These composite vertices are simply the rows of the square lattice. The configurations of a given row u form the base vectors of the space in which the computations will be performed with the help of the Kronecker product. It is only necessary to determine whether or not there is a vertical dimer from any given vertex of a row. There are then 2^{2M} base vectors of configurations of a given row. A vertex j with an upward vertical dimer will be specified by the unit vector

$$\begin{pmatrix} 1 \\ 0 \end{pmatrix}$$

or if not occupied by an upward vertical dimer, by the unit vector

$$\begin{pmatrix} 0 \\ 1 \end{pmatrix}.$$

The configuration of u is then given by

$$u = u(1) \otimes u(2) \otimes \cdots \otimes u(2M)$$

where

$$
u(j) = \begin{cases} \begin{pmatrix} 1 \\ 0 \end{pmatrix} & \text{if } j \text{ is occupied by a dimer to its upper neighbour} \\[2ex] \begin{pmatrix} 0 \\ 1 \end{pmatrix} & \text{if } j \text{ is not occupied by a dimer to its upper neighbour} \end{cases}
$$

and $j = 1, 2, \ldots, 2M$. The configurations of u form an orthonormal basis for the 2^{2M} dimensional vector space which they generate. We impose periodic boundary conditions.

Definition 3.2. Suppose that the bottom row, which we call the first, has the configuration u_1. We define the *transfer matrix* $T(w, u)$ between a row u and its upper neighbour w by the condition

$$
T(w, u) := \begin{cases} 1 & \text{if } w \text{ and } u \text{ are consistent configurations} \\ 0 & \text{otherwise} \end{cases}
$$

By consistent we mean that the designated pair of rows can be continued to a legitimate configuration of the lattice. The quantity

$$
T u_1 = \sum_{u_2} T(u_2, u_1) u_2
$$

is then the sum of all possible configurations of the second row given that the first row is u_1. Analogously

$$
T^2 u_1 = T \left(\sum_{u_2} T(u_2, u_1) u_2 \right) = \sum_{u_3} \sum_{u_2} T(u_2, u_1) T(u_3, u_2) u_3
$$

is the sum of all possible configurations of the third row given that the first is u_1. Iterating this process we see that the sum of all possible configurations of the $2N + 1$-th row is given by $T^{2N} u_1$. However owing to periodic boundary conditions the configuration of the $2N + 1$-th row must be u_1.

The number of possible configurations $\mathcal{N}$ corresponding to u_1 is therefore the coefficient of u_1 in $T^{2N} u_1$ or by orthonormality

$$
\mathcal{N}(u_1) = (u_1, T^{2N} u_1).
$$

Summing over all u_1 we find that $\mathcal{N} = \text{tr}(T^{2N})$, where tr denotes the trace.

Now we have to evaluate the transfer matrix T. Suppose that we are given a row configuration u and ask for those upper neighbour configurations w which are consistent with u. Now u will interfere with w only by virtue of its vertical dimers which terminate at w. This is why the configuration is sufficient to determine consistency. The row w is characterized first of all by its horizontal dimers. Suppose $S = \{\alpha\}$ is the set of adjacent vertex pairs on which the horizontal dimers of w are placed. A dimer can be placed on w at $(i, i+1)$ only if there are no vertical dimers from u at $(i, i+1)$. This means the partial configuration of u for this pair is given by

$$\begin{pmatrix} 0 \\ 1 \end{pmatrix} \otimes \begin{pmatrix} 0 \\ 1 \end{pmatrix}.$$

Thus the transfer matrix T must yield 0 for each configuration except

$$\begin{pmatrix} 0 \\ 1 \end{pmatrix} \otimes \begin{pmatrix} 0 \\ 1 \end{pmatrix}.$$

Since the $(i, i+1)$ pair on w cannot then be occupied by vertical dimers, T must map the vector

$$\begin{pmatrix} 0 \\ 1 \end{pmatrix} \otimes \begin{pmatrix} 0 \\ 1 \end{pmatrix}$$

into itself. The projection matrix $\bar{H}_\alpha$ which has these properties is given by

$$\bar{H}_\alpha := I_2 \otimes I_2 \otimes \cdots \otimes I_2 \otimes \underset{i}{\begin{pmatrix} 0 & 0 \\ 0 & 1 \end{pmatrix}} \otimes \underset{i+1}{\begin{pmatrix} 0 & 0 \\ 0 & 1 \end{pmatrix}} \otimes I_2 \otimes \cdots \otimes I_2$$
$$\phantom{\bar{H}_\alpha :=} \underset{1}{} \quad \underset{2}{} \quad \cdots \qquad\qquad\qquad\qquad\qquad\qquad \cdots \quad \underset{2M}{}$$

owing to the eigenvalue equations

$$\begin{pmatrix} 0 & 0 \\ 0 & 1 \end{pmatrix} \begin{pmatrix} 0 \\ 1 \end{pmatrix} = 1 \begin{pmatrix} 0 \\ 1 \end{pmatrix}, \qquad \begin{pmatrix} 0 & 0 \\ 0 & 1 \end{pmatrix} \begin{pmatrix} 1 \\ 0 \end{pmatrix} = 0 \begin{pmatrix} 1 \\ 0 \end{pmatrix}.$$

Obviously $\bar{H}_\alpha$ is a $2^{2M} \times 2^{2M}$ symmetric matrix over $\mathbb{R}$.

Consider a vertex j on w which is not occupied by a horizontal dimer. Its configuration is completely determined by that of the corresponding vertex j on u: a vertical dimer on u means that there cannot be one on w, an absent vertical dimer on u requires a vertical dimer at j on w (since it is not occupied by a horizontal dimer). The transfer matrix must then reverse the configuration of j. The permutation matrix P_j which carries this out takes the form

$$P_j := I_2 \otimes I_2 \otimes \cdots \otimes I_2 \otimes \underset{j}{\begin{pmatrix} 0 & 1 \\ 1 & 0 \end{pmatrix}} \otimes I_2 \otimes \cdots \otimes I_2.$$
$$ \underset{1}{} \quad \underset{2}{} \quad \cdots \qquad\qquad\qquad \cdots \quad \underset{2M}{}$$

It follows that the transfer matrix is given by

$$T = \sum_S \prod_{\alpha \in S} \bar{H}_\alpha \prod_{j \in \bar{S}} P_j.$$

Using the property (see exercise (2)) $(A \otimes A)(B \otimes I_2)(I_2 \otimes B) = C \otimes C$ the transfer matrix T can be simplified as $\bar{H}_{i,i+1} = H_{i,i+1} P_i P_{i+1}$, where

$$H_{i,i+1} = I_2 \otimes \cdots \otimes I_2 \otimes \begin{pmatrix} 0 & 0 \\ 1 & 0 \end{pmatrix} \otimes \begin{pmatrix} 0 & 0 \\ 1 & 0 \end{pmatrix} \otimes I_2 \otimes \cdots \otimes I_2.$$

Consequently the transfer matrix T is

$$T = \sum_S \prod_{\alpha \in S} H_\alpha \prod_{j=1}^{2M} P_j.$$

Owing to the method of construction, the product over α includes only nonoverlapping adjacent-element pairs. However, since

$$\begin{pmatrix} 0 & 0 \\ 1 & 0 \end{pmatrix} \begin{pmatrix} 0 & 0 \\ 1 & 0 \end{pmatrix} = \begin{pmatrix} 0 & 0 \\ 0 & 0 \end{pmatrix}$$

we find that

$$H_\alpha H_\beta = 0 \quad \text{if } \alpha \text{ and } \beta \text{ overlap}.$$

Thus the product over α may be extended to include any set of adjacent pairs whose union is the set S. Consequently

$$T = \prod_{\alpha=(1,2)}^{(2M,1)} (I + H_\alpha) \prod_{j=1}^{2M} P_j = \left(\exp \left(\sum_{i=1}^{2M} H_{i,i+1} \right) \right) \prod_{j=1}^{2M} P_j.$$

It is convenient to consider T^2 instead of T. Using the results from exercise (3) we find that

$$H_{j,j+1} = S_{-,j} S_{-,j+1}, \quad P_j P_{j+1} H_{j,j+1} P_j P_{j+1} = S_{+,j} S_{+,j+1}$$

where

$$S_{-,j} := I_2 \otimes \cdots \otimes I_2 \otimes \begin{pmatrix} 0 & 0 \\ 1 & 0 \end{pmatrix} \otimes I_2 \otimes \cdots \otimes I_2$$

$$S_{+,j} := I_2 \otimes \cdots \otimes I_2 \otimes \begin{pmatrix} 0 & 1 \\ 0 & 0 \end{pmatrix} \otimes I_2 \otimes \cdots \otimes I_2.$$

Then we find that

$$T^2 = \exp \left(\sum_{j=1}^{2M} (S_{-,j} S_{-,j+1}) \right) \exp \left(\sum_{j=1}^{2M} (S_{+,j} S_{+,j+1}) \right).$$

Let $\lambda_{\max}$ be the maximum eigenvalue of T^2. In the asymptotic limit $N \to \infty$,

$$\mathcal{N} = (\operatorname{tr}(T^2))^N \simeq (\lambda_{\max}(T^2))^N$$

so that we only have to find the maximum eigenvalue $\lambda_{\max}$ of T^2. To find the maximum eigenvalue of T^2 we apply the *Paulion to Fermion transformations*. This is done by appending a sign which depends multiplicatively on the states of the preceding vertices. We define

$$c_k := \prod_{j=1}^{k-1} (-\sigma_{3,j}) S_{-,k}, \qquad c_k^\dagger := \prod_{j=1}^{k-1} (-\sigma_{3,j}) S_{+,k}$$

where

$$\sigma_{3,k} = I_2 \otimes \cdots \otimes I_2 \otimes \begin{pmatrix} 1 & 0 \\ 0 & -1 \end{pmatrix} \otimes I_2 \otimes \cdots \otimes I_2.$$

We find that

$$S_{-,j} S_{-,j+1} = -c_j c_{j+1}, \qquad j = 1, \ldots, 2M - 1$$

$$S_{+,j} S_{+,j+1} = c_j^\dagger c_{j+1}^\dagger, \qquad j = 1, \ldots, 2M - 1.$$

On the other hand for $j = 2M$ it is necessary to introduce the hermitian matrix

$$A := \sum_{j=1}^{2M} c_j^\dagger c_j = \sum_{j=1}^{2M} S_{+,j} S_{-,j}.$$

Using the result from exercises (5) and (6) we find

$$T^2 = \exp\left(-\sum_{j=1}^{2M-1} c_j c_{j+1} + (-1)^A c_{2M} c_1 \right) \exp\left(\sum_{j=1}^{2M-1} c_j^\dagger c_{j+1}^\dagger - (-1)^A c_{2M}^* c_1^* \right)$$

where

$$(-1)^A \equiv \exp(i\pi A).$$

Now the eigenvector belonging to the maximum eigenvalue of T^2 is also an eigenvector of A. The corresponding eigenvalue of the matrix A is odd. Thus the matrix T^2 takes the form

$$T^2 = \exp\left(-\sum_{j=1}^{2M} c_j c_{j+1} \right) \exp\left(\sum_{j=1}^{2M} c_j^* c_{j+1}^* \right).$$

To diagonalize T^2, we introduce the one dimensional *discrete Fourier transformation*

$$c_j = \frac{1}{(2Mi)^{1/2}} \sum_{k=k_0}^{2Mk_0} \exp(ijk)S_k, \qquad c_j^\dagger = \frac{1}{(-2Mi)^{1/2}} \sum_{k=k_0}^{2Mk_0} \exp(-ijk)S_k^*$$

where $k_0 = (2\pi)/(2M)$. It follows that

$$S_k S_\ell^* + S_\ell^* S_k = \frac{1}{2M} \sum_{j,j'} \exp(-ijk + ij'\ell)(c_j c_{j'}^\dagger + c_{j'}^\dagger c_j)$$

$$= \frac{1}{2M} \sum_{j,j'} \exp(-ijk + ij'\ell)\delta_{j,j'} I = \delta_{k\ell} I.$$

All other anticommutators vanish. The exponents of T^2 again become diagonal in the S_k's, resulting in

$$T^2 = \prod_{k=k_0}^{Mk_0} \Lambda_k$$

where

$$\Lambda_k = \exp\left(2S_k S_{-k} \sin(k)\right) \exp\left(2S_{-k}^* S_k^* \sin(k)\right).$$

To find the maximum eigenvalue of T^2, we can restrict our attention to a special subspace which is invariant under T^2. It is defined by first choosing any column vector $\mathbf{x}_0$ which satisfies $S_k \mathbf{x}_0 = \mathbf{0}$ for all k, and then constructing the unit vectors

$$\mathbf{e}(\delta_1, \ldots, \delta_m) = \prod_{k=k_0}^{Mk_0} (S_{-k}^* S_k^*)^{\delta_k} \mathbf{x}_0$$

where $\delta_k = 0$ or 1. We find that

$$S_{-k}^* S_k^* = I_2 \otimes \cdots \otimes I_2 \otimes \begin{pmatrix} 0 & 0 \\ 1 & 0 \end{pmatrix} \otimes I_2 \otimes \cdots \otimes I_2$$

$$S_k S_{-k} = I_2 \otimes \cdots \otimes I_2 \otimes \begin{pmatrix} 0 & 1 \\ 0 & 0 \end{pmatrix} \otimes I_2 \otimes \cdots \otimes I_2.$$

Let

$$\sigma_+ = \begin{pmatrix} 0 & 2 \\ 0 & 0 \end{pmatrix}, \qquad \sigma_- = \begin{pmatrix} 0 & 0 \\ 2 & 0 \end{pmatrix}.$$

Hence we find

$$\Lambda_k = I_2 \otimes \cdots \otimes I_2 \otimes \exp\left(\sigma_- \sin(k)\right) \exp\left(\sigma_+ \sin(k)\right) \otimes I_2 \otimes \cdots \otimes I_2$$

$$= I_2 \otimes \cdots \otimes I_2 \otimes \left[I_2 + \sigma_- \sin(k)\right]\left[I_2 + \sigma_+ \sin(k)\right] \otimes I_2 \otimes \cdots \otimes I_2$$

$$= I_2 \otimes \cdots \otimes I_2 \otimes \begin{pmatrix} 1 & 2\sin(k) \\ 2\sin(k) & 1 + 4\sin^2(k) \end{pmatrix} \otimes I_2 \otimes \cdots \otimes I_2.$$

The maximum eigenvalue for Λ_k is found to be

$$\left(\sin(k) + (1 + \sin^2(k))^{1/2}\right)^2.$$

Thus the number of configurations $\mathcal{N}(2M, 2N)$ is given by

$$\mathcal{N}(2M, 2N) \simeq \left[\lambda_{\max}(T^2)\right]^N = \prod_{k=k_0}^{Mk_0} \left[\sin(k) + (1 + \sin^2(k))^{1/2}\right]^{2N}$$

$$= \exp\left(2N \sum_{k=k_0}^{Mk_0} \ln(\sin(k) + (1 + \sin^2(k))^{1/2})\right).$$

As $M \to \infty$ we obtain

$$\mathcal{N} = \exp\left(\frac{2NM}{\pi} \int_0^\pi \ln(\sin(k) + (1 + \sin^2(k))^{1/2}) dk\right).$$

The integration over k yields

$$\mathcal{N}(2M, 2N) \simeq \exp\left(\frac{G}{\pi}(2M)(2N)\right)$$

where $G = 0.915965594\ldots$ (Catalan's constant).

Exercises. (1) Let $N = M = 1$. Find $\mathcal{N}$. Let $N = M = 2$. Find $\mathcal{N}$.

(2) Let

$$A = \begin{pmatrix} 0 & 0 \\ 1 & 0 \end{pmatrix}, \qquad B = \begin{pmatrix} 0 & 1 \\ 1 & 0 \end{pmatrix}, \qquad C = \begin{pmatrix} 0 & 0 \\ 0 & 1 \end{pmatrix}.$$

Show that $(A \otimes A)(B \otimes I_2)(I_2 \otimes B) = C \otimes C$.

(3) Show that $P_j P_j = I_{2^{2M} \times 2^{2M}}$ and

$$P_j P_{j+1} H_{j,j+1} P_j P_{j+1} = I_2 \otimes \cdots \otimes I_2 \otimes \begin{pmatrix} 0 & 1 \\ 0 & 0 \end{pmatrix} \otimes \begin{pmatrix} 0 & 1 \\ 0 & 0 \end{pmatrix} \otimes I_2 \otimes \cdots \otimes I_2.$$

(4) Show that T^2 is a hermitian matrix.

(5) Show that $c_j^\dagger c_k^\dagger + c_k^\dagger c_j^\dagger = 0$, $c_j c_k + c_k c_j = 0$, $c_j^\dagger c_k + c_k c_j^\dagger = \delta_{jk} I$, where $j, k = 1, 2, \ldots, 2M$.

(6) Show that the eigenvalues of the hermitian matrix A are integers, that A commutes with T^2, that the matrix $(-1)^A \equiv \exp(i\pi A)$ commutes with any monomial of even degree in the c_j's and $c_j^\dagger$'s and that

$$S_{-,2M} S_{-,1} = (-1)^A c_{2M} c_1, \qquad S_{+,2M} S_{+,1} = -(-1)^A c_{2M}^\dagger c_1^\dagger.$$

(7) Show that $S_k S_{-k}$ and $S_{-k}^* S_k^*$ do commute with $S_\ell S_{-\ell}$ and $S_{-\ell}^* S_\ell^*$ when $\ell \neq \pm k$.

(8) Show that the number of ways to cover a chessboard with dominoes, each domino filling two squares is given by 12 988 816.

3.10   Two Dimensional Ising Model

We have studied the one dimensional Ising model. Here we discuss the two dimensional Ising model. The model considered is a plane square lattice having N points, at each of which is a spin with its axis perpendicular to the lattice plane. The spin can have two opposite orientations, so that the total number of possible configurations of the spins in the lattice is 2^N. To each lattice point (with integral coordinates k, l) we assign a variable $\sigma_{k,l}$ which takes two values ± 1, corresponding to the two possible orientations of the spin. If we take into account only the interaction between adjoining spins, the energy of the configuration is given by

$$E(\sigma) := -J \sum_{k,l=1}^{n} \left(\sigma_{k,l} \sigma_{k,l+1} + \sigma_{k,l} \sigma_{k+1,l} \right)$$

where n is the number of points in a lattice line, the lattice being considered as a large square, and $N = n^2$.

In our approach we follow Onsager [46] (see also Kaufman [37], Huang [33]) since the Kronecker product is used. Consider a square lattice of

$$N = n^2$$

spins consisting of n rows and n columns and periodic boundary conditions and the rows and the columns, i.e. we enlarge the lattice by one row and one column with the requirement that the configuration of the $(n+1)$th row and column be identical with that of the first row and column respectively. This boundary condition endows the lattice with the topology of a torus. Let μ_j $(j = 1, \ldots, n)$ denote the collection of all the spin coordinates of the j-th row

$$\mu_j \equiv \{\sigma_1, \sigma_2, \ldots, \sigma_n\} \quad j\text{-th row.}$$

The toroidal (periodic) boundary condition implies the definition

$$\mu_{n+1} \equiv \mu_1.$$

A configuration of the entire lattice is then specified by $\{\mu_1, \ldots, \mu_n\}$. There are 2^N configurations. The j-th row interacts only with the $(j-1)$-th and the $(j+1)$-th row. Let $E(\mu_{j+1}, \mu_j)$ be the interaction energy between the j-th and the $(j+1)$-th row. Let $E(\mu_j)$ be the interaction energy of the spins within the j-th row. We can write

$$E(\mu, \mu') = -J \sum_{k=1}^{n} \sigma_k \sigma_k', \qquad E(\mu) = -J \sum_{k=1}^{n} \sigma_k \sigma_{k+1}$$

where μ and μ' respectively denote the collection of spin coordinates in two neighbouring rows $\mu \equiv \{\sigma_1, \ldots, \sigma_n\}$, $\mu' \equiv \{\sigma_1', \ldots, \sigma_n'\}$. The toroidal boundary condition implies that in each row $\sigma_{n+1} \equiv \sigma_1$. The total energy of the lattice for the configuration $\{\mu_1, \ldots, \mu_n\}$ is then given by

$$\sum_{j=1}^{n} (E(\mu_j, \mu_{j+1}) + E(\mu_j)).$$

The *partition function* $Z(\beta)$ is defined as

$$Z(\beta) := \sum_{\mu_1} \cdots \sum_{\mu_n} \exp\left(-\beta \sum_{j=1}^{n} (E(\mu_j, \mu_{j+1}) + E(\mu_j))\right).$$

Let a $2^n \times 2^n$ symmetric matrix P be so defined that its matrix elements are

$$\langle \mu | P | \mu' \rangle = \exp\left(-\beta(E(\mu, \mu') + E(\mu))\right).$$

Using this definition we find

$$Z(\beta) = \sum_{\mu_1} \cdots \sum_{\mu_n} \langle \mu_1 | P | \mu_2 \rangle \langle \mu_2 | P | \mu_3 \rangle \cdots \langle \mu_n | P | \mu_1 \rangle = \sum_{\mu_1} \langle \mu_1 | P^n | \mu_1 \rangle$$

$$= \operatorname{tr}(P^n)$$

where tr denotes the trace. The trace of a matrix is independent of the representation of the matrix. Thus the trace may be evaluated by bringing the symmetric matrix P into its diagonal form, i.e.

$$P = \operatorname{diag}(\lambda_1, \lambda_2, \ldots, \lambda_{2^n})$$

where λ_1, λ_2, ..., λ_{2^n} are the 2^n real eigenvalues of P. The eigenvalues depend on β. The matrix P^n is then also diagonal, with the diagonal matrix elements $(\lambda_1)^n$, $(\lambda_2)^n$, ..., $(\lambda_{2^n})^n$. It follows that

$$Z(\beta) = \sum_{j=1}^{2^n} (\lambda_j)^n.$$

We expect that the eigenvalues of the matrix P are in general of the order of $\exp(n)$ when n is large, since $E(\mu, \mu')$ and $E(\mu)$ are of the order of n. If $\lambda_{\max}$ is the largest eigenvalue of P, it follows that

$$\lim_{n\to\infty} \frac{1}{n} \ln(\lambda_{\max}) = \text{finite number.}$$

If all the eigenvalues λ_j are positive, we obtain

$$(\lambda_{\max})^n \leq Z(\beta) \leq 2^n (\lambda_{\max})^n$$

or

$$\frac{1}{n} \ln(\lambda_{\max}) \leq \frac{1}{n^2} \ln(Z(\beta)) \leq \frac{1}{n} \ln(\lambda_{\max}) + \frac{1}{n} \ln(2).$$

It follows that

$$\lim_{N\to\infty} \frac{1}{N} \ln(Z(\beta)) = \lim_{n\to\infty} \frac{1}{n} \ln(\lambda_{\max})$$

where $N = n^2$. We find that this is true and that all the eigenvalues λ_j are positive. Thus it is sufficient to find the largest eigenvalue of the matrix P.

We obtain the matrix elements of P in the form

$$\langle \sigma_1, \ldots, \sigma_n | P | \sigma_1', \ldots, \sigma_n' \rangle = \prod_{k=1}^{n} \exp(\beta J \sigma_k \sigma_{k+1}) \exp(\beta J \sigma_k \sigma_k').$$

We define two $2^n \times 2^n$ matrices V_1' and V_2 whose matrix elements are respectively given by

$$\langle \sigma_1, \ldots, \sigma_n | V_1' | \sigma_1', \ldots, \sigma_n' \rangle = \prod_{k=1}^{n} \exp(\beta J \sigma_k \sigma_k')$$

$$\langle \sigma_1, \ldots, \sigma_n | V_2 | \sigma_1', \ldots, \sigma_n' \rangle = \delta_{\sigma_1, \sigma_1'} \cdots \delta_{\sigma_n, \sigma_n'} \prod_{k=1}^{n} \exp(\beta J \sigma_k \sigma_{k+1})$$

where $\delta_{\sigma, \sigma'}$ is the Kronecker symbol. Thus V_2 is a diagonal matrix in the present representation. It follows that $P = V_2 V_1'$. Let σ_1, σ_2 and σ_3 be the three Pauli spin matrices. The $2^n \times 2^n$ matrices $\sigma_{1,j}$, $\sigma_{2,j}$, $\sigma_{3,j}$ are defined as

$$\sigma_{1,j} := I_2 \otimes \cdots \otimes I_2 \otimes \sigma_1 \otimes I_2 \otimes \cdots \otimes I_2 \qquad n \text{ factors}$$
$$\sigma_{2,j} := I_2 \otimes \cdots \otimes I_2 \otimes \sigma_2 \otimes I_2 \otimes \cdots \otimes I_2 \qquad n \text{ factors}$$
$$\sigma_{3,j} := I_2 \otimes \cdots \otimes I_2 \otimes \sigma_3 \otimes I_2 \otimes \cdots \otimes I_2 \qquad n \text{ factors}$$

where $j = 1, \ldots, n$ and σ_1, σ_2, σ_3 are at the j-th position. The $2^n \times 2^n$ matrix V_1' is the Kronecker product of n 2×2 identical matrices

$$V_1' := A \otimes A \otimes \cdots \otimes A$$

where $\langle \sigma | A | \sigma' \rangle = \exp(\beta J \sigma \sigma')$. Since $\sigma, \sigma' \in \{1, -1\}$ we have the symmetric 2×2 matrix

$$A = \begin{pmatrix} \exp(\beta J) & \exp(-\beta J) \\ \exp(-\beta J) & \exp(\beta J) \end{pmatrix} \equiv \exp(\beta J) I_2 + \exp(-\beta J) \sigma_1.$$

Obviously, A can be written as

$$A = \sqrt{2 \sinh(2\beta J)} \exp(\theta \sigma_1)$$

where $\tanh(\theta) \equiv \exp(-2\beta J)$. It follows that

$$V_1' = (2 \sinh(2\beta J))^{n/2} \exp(\theta \sigma_1) \otimes \exp(\theta \sigma_1) \otimes \cdots \otimes \exp(\theta \sigma_1).$$

Applying the identity

$$\exp(\theta \sigma_1) \otimes \exp(\theta \sigma_1) \otimes \cdots \otimes \exp(\theta \sigma_1) \equiv \exp(\theta \sigma_{1,1}) \exp(\theta \sigma_{1,2}) \cdots \exp(\theta \sigma_{1,n})$$

we obtain

$$V_1' = (2 \sinh(2\beta J))^{n/2} V_1, \qquad V_1 = \prod_{j=1}^{n} \exp(\theta \sigma_{1,j}).$$

For the matrix V_2 we find

$$V_2 = \prod_{j=1}^{n} \exp(\beta J \sigma_{3,j} \sigma_{3,j+1})$$

where $\sigma_{3,n+1} = \sigma_{3,1}$ (boundary condition). Therefore

$$P = (2 \sinh(2\beta J))^{n/2} V_2 V_1.$$

This completes the formulation of the two dimensional Ising model.

We introduce a class of matrices to solve the two dimensional Ising model (Huang [33]). Let $2n$ matrices Γ_μ, $\mu = 1, \ldots, 2n$ be a set of matrices with the following anticommutation rule

$$\Gamma_\mu \Gamma_\nu + \Gamma_\nu \Gamma_\mu = 2\delta_{\mu\nu} I$$

where $\nu = 1, \ldots, 2n$. A representation of the matrices $\{\Gamma_\mu\}$ by $2^n \times 2^n$ matrices is given by

$$\begin{aligned}
\Gamma_1 &= \sigma_{3,1} & \Gamma_2 &= \sigma_{2,1} \\
\Gamma_3 &= \sigma_{1,1}\sigma_{3,2} & \Gamma_4 &= \sigma_{1,1}\sigma_{2,2} \\
\Gamma_5 &= \sigma_{1,1}\sigma_{1,2}\sigma_{3,3} & \Gamma_6 &= \sigma_{1,1}\sigma_{1,2}\sigma_{2,3} \\
&\ \ \vdots & &\ \ \vdots
\end{aligned}$$

Therefore

$$\begin{aligned}
\Gamma_{2j-1} &= \sigma_{1,1}\sigma_{1,2}\cdots\sigma_{1,j-1}\sigma_{3,j}, & j &= 1, \ldots, n \\
\Gamma_{2j} &= \sigma_{1,1}\sigma_{1,2}\ldots\sigma_{1,j-1}\sigma_{2,j}, & j &= 1, \ldots, n.
\end{aligned}$$

Another representation is obtained by interchanging the roles of $\sigma_{1,j}$ and $\sigma_{2,j}$ $(j = 1, \ldots, n)$. The $2^n \times 2^n$ matrices V_1 and V_2 are matrices that transform one set of $\{\Gamma_\mu\}$ into another equivalent set. Let a definite set $\{\Gamma_\mu\}$ be given and let ω be the $2n \times 2n$ matrix describing a linear orthogonal transformation among the members of $\{\Gamma_\mu\}$, i.e.

$$\Gamma_\mu' = \sum_{\nu=1}^{2n} \omega_{\mu,\nu} \Gamma_\nu$$

where $\omega_{\mu\nu} \in \mathbb{C}$ are complex numbers satisfying

$$\sum_{\mu=1}^{2n} \omega_{\mu,\nu}\omega_{\mu,\lambda} = \delta_{\nu,\lambda}.$$

This relation can be written in matrix form as $\omega^T \omega = I$, where ω^T is the transpose of ω and I is the $2n \times 2n$ unit matrix. If Γ_μ is regarded as a component of a vector in a $2n$ dimensional space, then ω induces a rotation in that space

$$\begin{pmatrix} \Gamma_1' \\ \Gamma_2' \\ \vdots \\ \Gamma_{2n}' \end{pmatrix} = \begin{pmatrix} \omega_{11} & \omega_{12} & \cdots & \omega_{1,2n} \\ \omega_{21} & \omega_{22} & \cdots & \omega_{2,2n} \\ \vdots & & & \\ \omega_{2n,1} & \omega_{2n,2} & \cdots & \omega_{2n,2n} \end{pmatrix} \begin{pmatrix} \Gamma_1 \\ \Gamma_2 \\ \vdots \\ \Gamma_{2n} \end{pmatrix}.$$

Therefore $\Gamma'_\mu = S(\omega)\Gamma_\mu S^{-1}(\omega)$, where $S(\omega)$ is a nonsingular $2^n \times 2^n$ matrix. There is a mapping

$$\omega \leftrightarrow S(\omega)$$

which provides $S(\omega)$ as a $2^n \times 2^n$ matrix representation of a rotation in a $2n$ dimensional space. It follows that

$$S(\omega)\Gamma_\mu S^{-1}(\omega) = \sum_{\nu=1}^{2n} \omega_{\mu,\nu}\Gamma_\nu.$$

We call ω a *rotation* and $S(\omega)$ the spin representative of the rotation ω. If ω_1 and ω_2 are two rotations then $\omega_1\omega_2$ is also a rotation. Furthermore

$$S(\omega_1\omega_2) = S(\omega_1)S(\omega_2).$$

Consider a rotation in a two dimensional plane of the $2n$ dimensional space. A rotation in the plane μ, ν through the angle θ is defined by the transformation

$$\begin{aligned}
\Gamma'_\lambda &= \Gamma_\lambda & (\lambda \neq \mu, \lambda \neq \nu) \\
\Gamma'_\mu &= \Gamma_\mu \cos(\theta) - \Gamma_\nu \sin(\theta) & (\mu \neq \nu) \\
\Gamma'_\nu &= \Gamma_\mu \sin(\theta) + \Gamma_\nu \cos(\theta) & (\mu \neq \nu)
\end{aligned}$$

where θ is a complex number. The rotation matrix, denoted by $\omega(\mu\nu|\theta)$, is defined by

$$\omega(\mu\nu|\theta) := \begin{pmatrix} & \vdots & & \vdots & \\ \cdots & \cos(\theta) & \cdots & \sin(\theta) & \cdots \\ & \vdots & & \vdots & \\ \cdots & -\sin(\theta) & \cdots & \cos(\theta) & \cdots \\ & \vdots & & \vdots & \end{pmatrix}$$

where the matrix elements $\cos(\theta)$ are in the μ-th column and μ-th row and ν-th column and ν-th row, respectively. The matrix element $\sin(\theta)$ is in the ν-th column and μ-th row and the matrix element $-\sin(\theta)$ is in the μ-th column and ν-th row. The matrix elements not displayed are unity along the diagonal and zero everywhere else. Thus the matrix $\omega(\mu, \nu|\theta)$ is a *Givens matrix*. The matrix $\omega(\mu\nu|\theta)$ is called the plane rotation in the plane $\mu\nu$.

The properties of ω and $S(\omega)$ that are relevant to the solution of the two dimensional Ising model are summarized in the following lemmas (Huang [33]).

Lemma 3.5. *If* $\omega(\mu\nu|\theta) \leftrightarrow S_{\mu\nu}(\theta)$ *then*
$$S_{\mu\nu}(\theta) = \exp\left(-\theta\Gamma_\mu\Gamma_\nu/2\right).$$

Proof. Since $\Gamma_\mu\Gamma_\nu = -\Gamma_\nu\Gamma_\mu$ for $\mu \neq \nu$ we have $\Gamma_\mu\Gamma_\nu\Gamma_\mu\Gamma_\nu = -I$ and

$$\exp\left(-\theta\Gamma_\mu\Gamma_\nu/2\right) = I\cos\left(\theta/2\right) - \Gamma_\mu\Gamma_\nu\sin\left(\theta/2\right).$$

Since $(\Gamma_\mu\Gamma_\nu)(\Gamma_\nu\Gamma_\mu) = (\Gamma_\nu\Gamma_\mu)(\Gamma_\mu\Gamma_\nu) = I$ we find

$$\exp\left(\theta\Gamma_\mu\Gamma_\nu/2\right)\exp\left(-\theta\Gamma_\mu\Gamma_\nu/2\right) = I.$$

Thus

$$S_{\mu\nu}^{-1}(\theta) = \exp\left(\theta\Gamma_\mu\Gamma_\nu/2\right)$$

and

$$S_{\mu\nu}(\theta)\Gamma_\lambda S_{\mu\nu}^{-1}(\theta) = \Gamma_\lambda, \qquad \lambda \neq \mu, \lambda \neq \nu$$

$$S_{\mu\nu}(\theta)\Gamma_\mu S_{\mu\nu}^{-1}(\theta) = \Gamma_\mu\cos(\theta) + \Gamma_\nu\sin(\theta),$$

$$S_{\mu\nu}(\theta)\Gamma_\nu S_{\mu\nu}^{-1}(\theta) = \Gamma_\mu\sin(\theta) - \Gamma_\nu\cos(\theta). \qquad \square$$

Lemma 3.6. *The eigenvalues of the $2n \times 2n$ matrix $\omega(\mu\nu|\theta)$ are 1 $(2n-2$-fold degenerate), and $\exp(\pm i\theta)$ (nondegenerate). The eigenvalues of $S_{\mu\nu}(\theta)$ are $\exp(\pm i\theta/2)$ (each 2^{n-1}-fold degenerate).*

Proof. The first part is trivial. The second part can be proved by choosing a special representation for Γ_μ and Γ_ν, since the eigenvalues of $S_{\mu\nu}(\theta)$ are independent of the representation. We use the representation given above with σ_1 and σ_3 interchanged, i.e. $\Gamma_\mu = \sigma_{3,1}\sigma_{1,2}$, $\Gamma_\nu = \sigma_{3,1}\sigma_{2,2}$. Then

$$\Gamma_\mu\Gamma_\nu = \sigma_{1,2}\sigma_{2,2} = i\sigma_{3,2} = I_2 \otimes \begin{pmatrix} i & 0 \\ 0 & -i \end{pmatrix} \otimes I_2 \otimes \cdots \otimes I_2$$

since $\sigma_{3,1}\sigma_{3,1} = I$. Therefore

$$S_{\mu\nu}(\theta) = I\cos(\theta/2) - \Gamma_\mu\Gamma_\nu\sin(\theta/2) = I_2 \otimes \begin{pmatrix} e^{-i\theta/2} & 0 \\ 0 & e^{i\theta/2} \end{pmatrix} \otimes I_2 \otimes \cdots \otimes I_2.$$

Consequently, $S_{\mu\nu}(\theta)$ is diagonal. The diagonal elements are either $\exp(i\theta/2)$ or $\exp(-i\theta/2)$, each appearing 2^{n-1} times each. $\qquad \square$

Lemma 3.7. *Let ω be a matrix product of n commuting plane rotations*

$$\omega = \omega(\alpha\beta|\theta_1)\omega(\gamma\delta|\theta_2)\cdots\omega(\mu\nu|\theta_n)$$

where $\{\alpha, \beta, \ldots, \mu, \nu\}$ is a permutation of the set of integers $\{1, 2, \ldots, 2n\}$, and $\theta_1, \ldots, \theta_n$ are complex numbers. Then

(a) $\omega \leftrightarrow S(\omega)$, with

$$S(\omega) = \exp\left(-\theta_1\Gamma_\alpha\Gamma_\beta/2\right)\exp\left(-\theta_2\Gamma_\gamma\Gamma_\delta/2\right)\cdots\exp\left(-\theta_n\Gamma_\mu\Gamma_\nu/2\right).$$

(b) The $2n$ eigenvalues of ω are given by

$$\exp(\pm i\theta_1), \quad \exp(\pm i\theta_2), \quad \ldots, \exp(\pm i\theta_n).$$

(c) The 2^n eigenvalues of $S(\omega)$ are the values

$$\exp\left(i(\pm\theta_1 \pm \theta_2 \pm \cdots \pm \theta_n)/2\right)$$

with the signs $\pm$ are chosen independently.

By this lemma, the eigenvalues of $S(\omega)$ can be obtained from those of ω. With the help of these lemmas we can express $V_2 V_1$ in terms of $S(\omega)$.

Now we can find the solution. The partition function is given by

$$\lim_{N \to \infty} \frac{1}{N} \ln(Z(\beta)) = \frac{1}{2} \ln[2 \sinh(2\beta J)] + \lim_{n \to \infty} \frac{1}{n} \ln(\Lambda)$$

where $\Lambda =$ largest eigenvalue of V and $V = V_1 V_2$. These formulas are valid if all eigenvalues of V are positive and if

$$\lim_{n \to \infty} \frac{1}{n} \ln(\Lambda)$$

exists. Thus we have to diagonalize the matrix V. Using the representation given above, we find that

$$\Gamma_{2j} \Gamma_{2j-1} = \sigma_{2,j} \sigma_{3,j} = i\sigma_{1,j}, \qquad j = 1, \ldots, n$$

and

$$V_1 = \prod_{j=1}^{n} \exp(\theta \sigma_{1,j}) = \prod_{j=1}^{n} \exp(-i\theta \Gamma_{2j} \Gamma_{2j-1}).$$

Thus V_1 is a spin representative of a product of commuting plane rotations. Furthermore

$$\Gamma_{2j+1} \Gamma_{2j} = \sigma_{1,j} \sigma_{3,j+1} \sigma_{2,j} = i\sigma_{3,j} \sigma_{3,j+1} \qquad j = 1, \ldots, n-1$$

$$\Gamma_1 \Gamma_{2n} = \sigma_{3,1}(\sigma_{1,1} \cdots \sigma_{1,n-1}) \sigma_{2,n} = -i\sigma_{3,1} \sigma_{3,n}(\sigma_{1,1} \cdots \sigma_{1,n}).$$

We have

$$V_2 = \left(\prod_{j=1}^{n-1} \exp(\beta J \sigma_{3,j} \sigma_{3,j+1}) \right) \exp(\beta J \sigma_{3,n} \sigma_{3,1}).$$

Therefore we can write

$$V_2 = \exp(i\beta J U \Gamma_1 \Gamma_{2n}) \prod_{j=1}^{n-1} \exp(-i\beta J \Gamma_{2j+1} \Gamma_{2j})$$

where the $2^n \times 2^n$ unitary matrix U is defined by

$$U := \sigma_{1,1} \sigma_{1,2} \cdots \sigma_{1,n} \equiv \sigma_1 \otimes \sigma_1 \otimes \cdots \otimes \sigma_1.$$

Owing to the first factor in $\exp(i\beta J U \Gamma_1 \Gamma_2)$, the matrix V_2 is not a spin representative of a product of commuting plane rotations. This factor comes from the toroidal boundary condition imposed on the problem (i.e. the condition that $\sigma_{n+1} \equiv \sigma_1$ in every row of the lattice). Thus for the symmetric matrix $V = V_2 V_1$ we obtain

$$V = \exp(i\phi U \Gamma_1 \Gamma_{2n}) \left(\prod_{j=1}^{n-1} \exp(-i\phi \Gamma_{2j+1} \Gamma_{2j}) \right) \left(\prod_{k=1}^{n} \exp(-i\phi \Gamma_{2k} \Gamma_{2k-1}) \right)$$

where $\phi = \beta J$, $J > 0$, $\theta = \tanh^{-1}(\exp(-2\phi))$. Now

$$\exp(i\phi \Gamma_1 \Gamma_{2n} U) = \frac{1}{2}(I + U) \exp(i\phi \Gamma_1 \Gamma_{2n}) + \frac{1}{2}(I - U) \exp(-i\phi \Gamma_1 \Gamma_{2n}).$$

Thus V takes the form

$$V = \frac{1}{2}(I + U)V^+ + \frac{1}{2}(I - U)V^-$$

where

$$V^{\pm} = \exp(\pm i\phi \Gamma_1 \Gamma_{2n}) \left(\prod_{j=1}^{n-1} \exp(-i\phi \Gamma_{2j+1} \Gamma_{2j}) \right) \left(\prod_{k=1}^{n} \exp(-i\theta \Gamma_{2k} \Gamma_{2k-1}) \right).$$

Thus both matrices V^+ and V^- are spin representatives of rotations.

We find a representation in which U is diagonal. Now

$$[U, V^+] = 0_{2^n}, \quad [U, V^-] = 0_{2^n}, \quad [V^+, V^-] = 0_{2^n}.$$

Hence the three matrices U, V^+, V^- can be simultaneously diagonalized. We first transform V into the representation in which U is diagonal (but in which $V^{\pm}$ are not necessarily diagonal)

$$R V R^{-1} \equiv \tilde{V} = \frac{1}{2}(I + \tilde{U})\tilde{V}^+ + \frac{1}{2}(I - \tilde{U})\tilde{V}^-,$$

$$\tilde{U} \equiv R U R^{-1}, \qquad \tilde{V}^{\pm} \equiv R V^{\pm} R^{-1}.$$

Since $U^2 = I$ the eigenvalues of U are either $+1$ or -1. Since

$$U = \sigma_1 \otimes \sigma_1 \otimes \cdots \otimes \sigma_1$$

a diagonal form of U is $\sigma_3 \otimes \sigma_3 \otimes \cdots \otimes \sigma_3$. The eigenvalues are $+1$ (2^{n-1} times) and -1 (2^{n-1} times). Other diagonal forms of U may be obtained by permuting the relative positions of the eigenvalues along the diagonal. We choose R in such a way that all the eigenvalues 1 are in one submatrix, and -1 in the other, so that the matrix $\tilde{U}$ can be represented in the form

$$\tilde{U} = \begin{pmatrix} I & 0 \\ 0 & -I \end{pmatrix} \equiv I \oplus (-I)$$

where I is the $2^{n-1} \times 2^{n-1}$ unit matrix and $\oplus$ is the direct sum. Since $\widetilde{V}^{\pm}$ commute with $\widetilde{U}$, they must have the forms

$$\widetilde{V}^{\pm} = \begin{pmatrix} \mathcal{A}^{\pm} & 0 \\ 0 & \mathcal{B}^{\pm} \end{pmatrix} \equiv \mathcal{A}^{\pm} \oplus \mathcal{B}^{\pm}$$

where $\mathcal{A}^{\pm}$ and $\mathcal{B}^{\pm}$ are $2^{n-1} \times 2^{n-1}$ matrices. They are not necessarily diagonal. The $2^n \times 2^n$ matrix $\frac{1}{2}(I + \widetilde{U})$ annihilates the lower submatrix and $\frac{1}{2}(I - \widetilde{U})$ annihilates the upper submatrix, i.e.

$$\frac{1}{2}(I + \widetilde{U})\widetilde{V}^{+} = \begin{pmatrix} \mathcal{A}^{+} & 0 \\ 0 & 0 \end{pmatrix}, \qquad \frac{1}{2}(I - \widetilde{U})\widetilde{V}^{-} = \begin{pmatrix} 0 & 0 \\ 0 & \mathcal{B}^{-} \end{pmatrix}.$$

Therefore we have the direct sum

$$\widetilde{V} = \begin{pmatrix} \mathcal{A}^{+} & 0 \\ 0 & \mathcal{B}^{-} \end{pmatrix} \equiv \mathcal{A}^{+} \oplus \mathcal{B}^{-}.$$

To diagonalize V, it is sufficient to diagonalize $\widetilde{V}$, which has the same eigenvalues as V. Moreover, to diagonalize V it is sufficient to diagonalize $\mathcal{A}^{+}$ and $\mathcal{B}^{-}$ separately. The combined set of their eigenvalues constitutes the set of eigenvalues of V. We first diagonalize $\widetilde{V}^{+}$ and $\widetilde{V}^{-}$ separately and independently, thereby obtaining twice too many eigenvalues for each. To find the eigenvalues of $\frac{1}{2}(I + \widetilde{U})\widetilde{V}^{+}$ and $\frac{1}{2}(I - \widetilde{U})\widetilde{V}^{-}$, we then decide which eigenvalues so obtained are to be discarded. This last step will not be necessary. We show that as $n \to \infty$ a knowledge of the eigenvalues of $\widetilde{V}^{+}$ and $\widetilde{V}^{-}$ suffices to determine the largest eigenvalue of V. The set of eigenvalues of $\widetilde{V}^{\pm}$, however, is respectively equal to the set of eigenvalues of $V^{\pm}$.

To find the eigenvalues of the matrices V^{+} and V^{-} we first find the eigenvalues of the rotations, of which V^{+} and V^{-} are spin representatives. These rotations are denoted by Ω^{+} and Ω^{-}, respectively which are both $2n \times 2n$ matrices. Thus we have the one-to-one mapping $V^{\pm} \leftrightarrow \Omega^{\pm}$. We find that

$$\Omega^{\pm} = \omega(1, 2n|\pm 2i\phi) \left(\prod_{j=1}^{n-1} \omega(2j+1, 2j| -2i\phi) \right) \left(\prod_{k=1}^{n} \omega(2k, 2k-1| -2i\theta) \right).$$

The eigenvalues of $\Omega^{\pm}$ are the same as that of $\omega^{\pm} = \Delta\Omega^{\pm}\Delta^{-1}$, where matrix Δ is the square root

$$\Delta := \sqrt{\prod_{k=1}^{n} \omega(2k, 2k-1| -2i\theta)} = \prod_{k=1}^{n} \omega(2k, 2k-1| -i\theta).$$

Thus

$$\omega^\pm = \Delta\chi^\pm\Delta$$

$$\Delta = \omega(12|i\theta)\omega(34|i\theta)\cdots\omega(2n-1,2n|i\theta)$$

$$\chi^\pm = \omega(1,2n|\pm 2i\phi)\left(\omega(23|2i\phi)\omega(45|2i\phi)\cdots\omega(2n-2,2n-1|2i\phi)\right).$$

We have $\Delta = J \oplus J \oplus \cdots \oplus J$ and J is the 2×2 matrix

$$J := \begin{pmatrix} \cosh(\theta) & i\sinh(\theta) \\ -i\sinh(\theta) & \cosh(\theta) \end{pmatrix}$$

and

$$\chi^\pm = \begin{pmatrix} a & 0 & 0 & \cdots & & & \pm b \\ 0 & K & 0 & & & & \\ 0 & 0 & K & & & & \\ \vdots & & & \ddots & & & \vdots \\ & & & & & & 0 \\ & & & & & K & 0 \\ \mp b & & & \cdots & & 0 & a \end{pmatrix}.$$

Here K is the 2×2 hermitian matrix

$$K := \begin{pmatrix} \cosh(2\phi) & i\sinh(2\phi) \\ -i\sinh(2\phi) & \cosh(2\phi) \end{pmatrix}$$

and $a := \cosh(2\phi)$, $b := i\sinh(2\phi)$. Consequently, we obtain

$$\omega^\pm = \Delta\chi^\pm\Delta = \begin{pmatrix} A & B & 0 & 0 & \cdots & 0 & \mp B^* \\ B^* & A & B & 0 & & 0 & 0 \\ 0 & B^* & A & B & & & \vdots \\ \vdots & & & & & & \\ & 0 & 0 & & & & B \\ \mp B & 0 & & \cdots & & B^* & A \end{pmatrix}$$

where A and B are 2×2 matrices given by

$$A := \begin{pmatrix} \cosh(2\phi)\cosh(2\theta) & -i\cosh(2\phi)\sinh(2\theta) \\ i\cosh(2\phi)\sinh(2\theta) & \cosh(2\phi)\cosh(2\theta) \end{pmatrix}$$

$$B := \begin{pmatrix} -\frac{1}{2}\sinh(2\phi)\sinh(2\theta) & i\sinh(2\phi)\sinh^2(\theta) \\ -i\sinh(2\phi)\cosh^2(\theta) & -\frac{1}{2}\sinh(2\phi)\sinh(2\theta) \end{pmatrix}.$$

To find the eigenvalues of the matrix $\omega^\pm$, we make the following ansatz for an eigenvector of $\omega^\pm$

$$\mathbf{v} = \begin{pmatrix} z\mathbf{u} \\ z^2\mathbf{u} \\ \vdots \\ z^n\mathbf{u} \end{pmatrix}$$

where z is a nonzero complex number and $\mathbf{u}$ is a two-component vector

$$\mathbf{u} = \begin{pmatrix} u_1 \\ u_2 \end{pmatrix}.$$

From the eigenvalue equation $\omega^{\pm}\mathbf{v} = \lambda\mathbf{v}$ we obtain the following n eigenvalue equations

$$(zA + z^2B \mp z^nB^*)\mathbf{u} = z\lambda\mathbf{u}$$
$$(z^2A + z^3B + zB^*)\mathbf{u} = z^2\lambda\mathbf{u}$$
$$(z^3A + z^4B + z^2B^*)\mathbf{u} = z^4\lambda\mathbf{u}$$

$$\vdots$$

$$(z^{n-1}A + z^nB + z^{n-2}B^*)\mathbf{u} = z^{n-1}\lambda\mathbf{u}$$
$$(z^nA \mp zB + z^{n-1}B^*)\mathbf{u} = z^n\lambda\mathbf{u}.$$

There are only three independent eigenvalue equations

$$(A + zB \mp z^{n-1}B^*)\mathbf{u} = \lambda\mathbf{u}$$
$$(A + zB + z^{-1}B^*)\mathbf{u} = \lambda\mathbf{u}$$
$$(A \mp z^{1-n}B + z^{-1}B^*)\mathbf{u} = \lambda\mathbf{u}.$$

These eigenvalue equations are solved by setting $z^n = \mp 1$. These three equations then become the same one, namely

$$(A + zB + z^{-1}B^*)\mathbf{u} = \lambda\mathbf{u}$$

where the sign $\mp$ is associated with $\omega^{\pm}$. Thus, for ω^+ and for ω^-, there are n values of z, namely

$$z_k = \exp(2i\pi k/n), \qquad k = 0, 1, \ldots, 2n-1$$

where $k = 1, 3, 5, \ldots, 2n-1$, for ω^+ and $k = 0, 2, 4, \ldots, 2n-2$, for ω^-. For each k, two eigenvalues λ_k are determined by the equation

$$(A + z_kB + z_k^{-1}B^*)\mathbf{u} = \lambda_k\mathbf{u}$$

and λ_k is to be associated with $\omega^{\pm}$. This determines $2n$ eigenvalues each for $\omega^{\pm}$. To find λ_k, we note that $\det(A) = 0$, $\det(B) = \det(B^*) = 0$ and

$$\det(A + z_kB + z_k^{-1}B^*) = 1.$$

Since the determinant of a 2×2 matrix is the product of the two eigenvalues we find that the two values of λ_k are given by

$$\lambda_k = \exp(\pm\gamma_k), \qquad k = 0, 1, \ldots, 2n-1.$$

The two values of γ_k may be found from the equation

$$\frac{1}{2}\mathrm{tr}(A + z_k B + z_k^{-1} B^*) = \frac{1}{2}(\exp(\gamma_k) + \exp(-\gamma_k)) \equiv \cosh(\gamma_k)$$

where we have used the fact that the trace of a 2×2 matrix is the sum of the eigenvalues. Evaluating the trace we obtain

$$\cosh(\gamma_k) = \cosh(2\phi)\cosh(2\theta) - \cos(\pi k/n)\sinh(2\phi)\sinh(2\theta)$$

where $k = 0, 1, \ldots, 2n - 1$. If γ_k is a solution, then $-\gamma_k$ is also a solution. This possibility has already been taken into account. Therefore we define γ_k to be the positive solution. It can be shown that

$$\gamma_k = \gamma_{2n-k}, \qquad 0 < \gamma_0 < \gamma_1 < \cdots < \gamma_n.$$

The inequality $0 < \gamma_0 < \gamma_1 < \cdots < \gamma_n$ can be seen by noting that

$$\frac{\partial \gamma_k}{\partial k} = \frac{\pi}{n} \frac{\sin(\pi k/n)}{\sin(\gamma_k)}$$

which is positive for $k \leq n$.

The eigenvalues of $\Omega^\pm$ are the same as those of $\omega^\pm$, respectively. Therefore $\Omega^\pm$ are products of commuting plane rotations. The 2^n eigenvalues of V^+ and V^- are

eigenvalues of V^+ are $\exp\left((\pm\gamma_0 \pm \gamma_2 \pm \gamma_4 \pm \cdots \pm \gamma_{2n-2})/2\right)$

eigenvalues of V^- are $\exp\left((\pm\gamma_1 \pm \gamma_3 \pm \gamma_5 \pm \cdots \pm \gamma_{2n-1})/2\right)$

where all possible choices of the signs $\pm$ are to be made independently. Next we derive the eigenvalues of V. The set of eigenvalues of V consists of one-half the set of eigenvalues of V^+ and one-half that of V^-. The eigenvalues of $V^\pm$ are all positive and of order $\exp(n)$. Therefore all eigenvalues of V are positive and of order $\exp(n)$. We only need the largest eigenvalue of V. Let F and G be invertible $2^n \times 2^n$ matrices such that

$$F\left(\frac{1}{2}(I + \tilde{U})\tilde{V}^+\right)F^{-1} = V_D^+, \qquad G\left(\frac{1}{2}(I - \tilde{U})\tilde{V}^-\right)G^{-1} = V_D^-$$

where $V_D^\pm$ are diagonal matrices with half the eigenvalues of

$$\exp((\pm\gamma_0 \pm \gamma_2 \pm \cdots \pm \gamma_{2n-2})/2)$$

and

$$\exp((\pm\gamma_1 \pm \gamma_3 \pm \cdots \pm \gamma_{2n-1})/2)$$

respectively, appearing along the diagonal. It is possible to choose F and G in such a way that $F\widetilde{U}F^{-1}$ and $G\widetilde{U}\,G^{-1}$ remain diagonal matrices. Then F and G permute the eigenvalues of $\widetilde{U}$ along the diagonal. The convention has been adopted that $\widetilde{U}$ is given by $I \oplus (-I)$. Hence F and G either leave $\widetilde{U}$ unchanged or simply interchange the two sub-matrices I and $-I$. That is, F and G either commute or anticommute with $\widetilde{U}$. It follows that

$$V_D^+ = \frac{1}{2}(I \pm \widetilde{U})F\widetilde{V}^+F^{-1}, \qquad V_D^- = \frac{1}{2}(I \pm \widetilde{U})G\widetilde{V}^-G^{-1}$$

where the signs $\pm$ can be definitely determined by an explicit calculation. Now, we may write

$$\frac{1}{2}(I \pm \widetilde{U}) = \frac{1}{2}(I \pm \sigma_{3,1}\sigma_{3,2}\cdots\sigma_{3,n})$$

$$F\widetilde{V}^+F^{-1} = \prod_{k=1}^{n} \exp\left(\frac{1}{2}\gamma_{2k-1}\sigma_{3,P_k}\right)$$

$$G\widetilde{V}^-G^{-1} = \prod_{k=1}^{n} \exp\left(\frac{1}{2}\gamma_{2k-2}\sigma_{3,Q_k}\right)$$

where P and Q are two definite permutations of the integers $1, 2, \ldots, n$. The permutation P maps k into P_k and Q maps k into Q_k. The matrices $F\widetilde{V}^+F^{-1}$ and $G\widetilde{V}^-G^{-1}$ are arrived at by noting that they must respectively have the same eigenvalues as V^+ and V^-, except for possible different orderings of the eigenvalues. Since the eigenvalues of $\sigma_{3,k}$ are ± 1, we have

$$\frac{1}{2}(I \pm \widetilde{U}) = \begin{cases} 1 \text{ if an even number of } \sigma_{3,k} \text{ are } \pm 1 \\ 0 \text{ if an odd number of } \sigma_{3,k} \text{ are } \pm 1 \end{cases}.$$

This condition is invariant under any permutation that maps $\sigma_{3,k}$ into σ_{3,P_k}. Therefore the eigenvalues of V_D^+ consists of those eigenvalues

$$\exp((\pm\gamma_0 \pm \gamma_2 \pm \cdots \pm \gamma_{2n-2})/2)$$

for which an even (odd) number of $-$ signs appears in the exponents, if the $+(-)$ sign is chosen in $\frac{1}{2}(I \pm \widetilde{V})F\widetilde{V}^+F^{-1}$. We conclude that the largest eigenvalue of V_D^+ is

$$\exp\left(\frac{1}{2}(\pm\gamma_0 + \gamma_2 + \gamma_4 + \cdots + \gamma_{2n-2})\right)$$

where the $\pm$ sign corresponds to the $\pm$ sign in $\frac{1}{2}(I \pm \widetilde{V})F\widetilde{V}^+F^{-1}$. As $n \to \infty$, these two possibilities give the same result, since γ_0 is negligible compared to the entire exponent. A similar conclusion can be found for V_D^-. Therefore we conclude that as $n \to \infty$ the largest eigenvalue of V_D^+ is

$$\exp\left(\frac{1}{2}(\gamma_0 + \gamma_2 + \gamma_4 + \cdots + \gamma_{2n-2})\right)$$

and the largest eigenvalue of V_D^- is

$$\exp\left(\frac{1}{2}(\gamma_1 + \gamma_3 + \gamma_5 + \cdots + \gamma_{2n-1})\right).$$

Then the largest eigenvalue of V is given by

$$\Lambda = \exp\left(\frac{1}{2}(\gamma_1 + \gamma_3 + \gamma_5 + \cdots + \gamma_{2n-1})\right)$$

since $\gamma_k = \gamma_{2n-k}$ and $0 < \gamma_0 < \gamma_1 < \cdots < \gamma_n$. We define

$$\mathcal{L} := \lim_{n \to \infty} \frac{1}{n} \ln(\Lambda) = \lim_{n \to \infty} \frac{1}{2n}(\gamma_1 + \gamma_2 + \gamma_5 + \cdots + \gamma_{2n-1}).$$

Let $\gamma(\nu) \equiv \gamma_{2k-1}$, $\nu := \frac{\pi}{n}(2k-1)$. As $n \to \infty$, ν becomes a continuous variable, and we have

$$\sum_{k=1}^{n} \gamma_{2k-1} \to \frac{n}{2\pi} \int_0^{2\pi} \gamma(\nu)d\nu.$$

Therefore

$$\mathcal{L} = \frac{1}{4\pi} \int_0^{2\pi} \gamma(\nu)d\nu = \frac{1}{2\pi} \int_0^{\pi} \gamma(\nu)d\nu$$

where we have used that $\gamma_k = \gamma_{n-k}$. Thus $\gamma(\nu) = \gamma(2\pi - \nu)$. We note that $\gamma(\nu)$ is the positive solution of the equation

$$\cosh(\gamma(\nu)) = \cosh(2\phi)\cosh(2\theta) - \cos(\nu)\sinh(2\phi)\sinh(2\theta)$$

with $\phi = \beta J$, $J > 0$, $\theta = \tanh^{-1}(\exp(-2\phi))$. Since

$$\sinh(2\theta) = \frac{1}{\sinh(2\phi)}, \qquad \cosh(2\theta) = \coth(2\phi)$$

we find that $\cosh(\gamma(\nu)) = \cosh(2\phi)\coth(2\phi) - \cos(\nu)$. In the following we use the identity

$$|z| = \frac{1}{\pi} \int_0^{\pi} \ln(2\cosh(z) - 2\cos(t))dt.$$

Then we find that $\gamma(\nu)$ has the integral representation

$$\gamma(\nu) = \frac{1}{\pi} \int_0^{\pi} \ln(2\cosh(2\phi)\coth(2\phi) - 2\cos(\nu) - 2\cos(\nu'))d\nu'.$$

Therefore

$$\mathcal{L} = \frac{1}{2\pi^2} \int_0^\pi d\nu \int_0^\pi d\nu' \ln(2\cosh(2\phi)\coth(2\phi) - 2(\cos(\nu) + \cos(\nu'))).$$

The integral remains unchanged if we let the region of integration be the rectangle

$$0 \leq \frac{\nu + \nu'}{2} \leq \pi, \qquad 0 \leq \nu - \nu' \leq \pi.$$

Let $\delta_1 \equiv (\nu + \nu')/2$, $\delta_2 \equiv \nu - \nu'$. Then

$$\mathcal{L} = \frac{1}{2\pi^2} \int_0^\pi d\delta_1 \int_0^\pi d\delta_2 \ln(2\cosh(2\phi)\coth(2\phi) - 4\cos(\delta_1)\cos(2\delta_2))$$

$$= \frac{1}{2\pi} \int_0^{\pi/2} d\delta_1 \ln(2\cos(\delta_1)) + \frac{1}{2\pi} \int_0^{\pi/2} d\delta_2 \cosh^{-1}\left(\frac{D}{2\cos(\delta_2)}\right)$$

where $D := \cosh(2\phi)\coth(2\phi)$. Since $\cosh^{-1}(x) \equiv \ln(x + \sqrt{x^2 - 1})$ we can write

$$\mathcal{L} = \frac{1}{2\pi} \int_0^\pi \ln(D(1 + \sqrt{1 - \kappa^2 \cos^2 \delta}))d\delta$$

where

$$\kappa = \frac{2}{D} = \frac{\exp(2\beta J) - \exp(-2\beta J)}{(\exp(2\beta J) + \exp(-2\beta J))^2}.$$

In the last integral $\cos^2(\delta)$ may be replaced by $\sin^2(\delta)$ without altering the value of the integral. Therefore

$$\mathcal{L} = \frac{1}{2} \ln\left(\frac{2\cosh^2(2\beta J)}{\sinh(2\beta J)}\right) + \frac{1}{2\pi} \int_0^\pi d\phi \ln\left(\frac{1}{2}(1 + \sqrt{1 - \kappa^2 \sin^2(\phi)})\right).$$

Since $\mathcal{L} := \lim_{n\to\infty} \frac{1}{n} \ln(\Lambda)$ and

$$\lim_{N\to\infty} \frac{1}{N} \ln(Z(\beta)) = \frac{1}{2} \ln(2\sinh(2\beta J)) + \lim_{n\to\infty} \frac{1}{n} \ln(\Lambda)$$

we find the partition function per lattice site. The Helmholtz free energy is given by

$$F(\beta) = -\frac{1}{\beta} \ln(Z(\beta)).$$

Exercises. (1) Show that the matrix P is symmetric over $\mathbb{R}$.

(2) Show that for $j \neq k$

$$[\sigma_{1,j}, \sigma_{1,k}] = [\sigma_{2,j}, \sigma_{2,k}] = [\sigma_{3,j}, \sigma_{3,k}] = 0$$

$$[\sigma_{1,j}, \sigma_{2,k}] = [\sigma_{1,j}, \sigma_{3,k}] = [\sigma_{2,j}, \sigma_{3,k}] = 0.$$

(3) Show that

$$\exp(\theta\sigma_1) \otimes \exp(\theta\sigma_1) \otimes \cdots \otimes \exp(\theta\sigma_1) \equiv \exp(\theta(\sigma_{1,1} + \sigma_{1,2} + \cdots + \sigma_{1,n})).$$

(4) Show that the matrices $\{\Gamma_\mu\}$ have the following properties:

(a) The dimensionality of Γ_μ cannot be smaller than $2^n \times 2^n$.

(b) If $\{\Gamma_\mu\}$ and $\{\Gamma'_\mu\}$ are two sets of matrices with $\Gamma_\mu\Gamma_\nu + \Gamma_\nu\Gamma_\mu = 2\delta_{\nu\mu}I$, there exists a nonsingular matrix S such that

$$\Gamma_\mu = S\Gamma'_\mu S^{-1}.$$

Show that the converse is also true.

(c) Any $2^n \times 2^n$ matrix is a linear combination of the unit matrix, the matrices Γ_μ (chosen to be $2^n \times 2^n$), and all the independent products $\Gamma_\mu\Gamma_\nu, \Gamma_\mu\Gamma_\nu\Gamma_\rho, \ldots$.

(5) Show that the set $\{\Gamma'_\mu\}$ satisfies $\Gamma'_\mu\Gamma'_\nu + \Gamma'_\nu\Gamma'_\mu = 2\delta_{\mu,\nu}I$.

(6) Show that $\omega(\mu\nu|\theta) = \omega(\nu\mu| - \theta)$, $\omega^T(\mu\nu|\theta)\omega(\mu\nu|\theta) = I$.

(7) Show that

$$U^2 = I, \quad U(I+U) = I+U, \quad U(I-U) = -(I-U), \quad U = i^n\Gamma_1\Gamma_2\cdots\Gamma_{2n}.$$

(8) Show that U commutes with a product of an even number of Γ_μ and anticommutes with a product of an odd number of Γ_μ.

(9) Show that the three matrices U, V^+ and V^- commute with one another.

(10) Discuss γ_k as a function of ϕ for $n \to \infty$.

3.11   One Dimensional Heisenberg Model

In this section we study the one dimensional isotropic Heisenberg model with cyclic boundary conditions and the Yang-Baxter relation. The Yang-Baxter relation plays a central role in the investigation of exactly solvable models in statistical physics. The Yang-Baxter relation includes the Kronecker product of matrix valued square matrices. We follow Takhtadzhyan and Faddeev [67], Kulish and Sklyanin [39], Baxter [8], Sogo and Wadati [51], Barouch [5].

The one dimensional isotropic Heisenberg model describes a system of N interacting particles with spin $1/2$ on a one dimensional lattice. Let σ_j $j = 1, 2, 3$ be the Pauli spin matrices and let I_2 be the 2×2 unit matrix. We set

$$\sigma_{j,n} := I_2 \otimes \cdots \otimes I_2 \otimes \sigma_j \otimes I_2 \otimes \cdots \otimes I_2, \qquad j = 1, 2, 3$$

where σ_j stands at the n-th place and $n = 1, 2, \ldots, N$. Then the Hamilton operator (*isotropic Heisenberg model*) is defined as

$$\hat{H}_N := \frac{J}{4} \sum_{n=1}^{N} (\sigma_{1,n}\sigma_{1,n+1} + \sigma_{2,n}\sigma_{2,n+1} + \sigma_{3,n}\sigma_{3,n+1} - I)$$

where I is the $2^N \times 2^N$ unit matrix. We assume periodic boundary conditions $\sigma_{j,N+1} = \sigma_{j,1}$, where $j = 1, 2, 3$. In the following we omit the index N. The operator $\hat{H}$ is a $2^N \times 2^N$ hermitian matrix. Thus the underlying vector space is

$$\mathcal{H} = \prod_{n=1}^{N} \otimes \eta_n, \qquad \eta_n \equiv \mathbb{C}^2$$

with $\dim(\mathcal{H}) = 2^N$.

Depending on the sign of the exchange constant J we distinguish the ferromagnetic case $J < 0$ and the antiferromagnetic case $J > 0$. We find the eigenvectors and eigenvalues of the Hamilton operator $\hat{H}$ and investigate their asymptotic behaviour as $N \to \infty$.

Let $\hat{A}$ be an operator ($2^N \times 2^N$ matrix). The equations of motion are given by the *Heisenberg equation of motion*, i.e.

$$i\hbar \frac{d\hat{A}(t)}{dt} = [\hat{A}, \hat{H}](t).$$

Definition 3.3. A $2^N \times 2^N$ matrix $\hat{A}$ is called a *constant of motion* (for the isotropic Heisenberg model $\hat{H}$) if

$$[\hat{A}, \hat{H}] = 0_{2^N}.$$

For the one dimensional isotropic Heisenberg model there exists a so-called *local transition matrix*

$$L_n(\lambda) := \begin{pmatrix} \lambda I + \frac{i}{2}\sigma_{3,n} & \frac{i}{2}\sigma_{-,n} \\ \frac{i}{2}\sigma_{+,n} & \lambda I - \frac{i}{2}\sigma_{3,n} \end{pmatrix}$$

where $\sigma_{+,n} := \sigma_{1,n} + i\sigma_{2,n}$, $\sigma_{-,n} := \sigma_{1,n} - i\sigma_{2,n}$ and $n = 1, 2, \ldots, N$. The entries of the 2×2 matrix L_n are $2^N \times 2^N$ matrices, i.e. I is the $2^N \times 2^N$ unit matrix. Here λ is a so-called *spectral parameter*. Thus L_n is a matrix valued 2×2 matrix.

The constants of motion can be found with the help of the L_n. We describe this later.

Let

$$S_{1,N} := \frac{1}{2}\sum_{j=1}^{N}\sigma_{1,j}, \qquad S_{2,N} := \frac{1}{2}\sum_{j=1}^{N}\sigma_{2,j}, \qquad S_{3,N} := \frac{1}{2}\sum_{j=1}^{N}\sigma_{3,j}.$$

The Heisenberg equation of motion for $S_{1,N}$, $S_{2,N}$, $S_{3,N}$ can be expressed with the help of L_n as follows. Consider the linear problem

$$L_n\psi_n = \psi_{n+1}, \qquad M_n\psi_n = \frac{d}{dt}\psi_n$$

where $n = 1, 2, \ldots, N$. Then the *compatability condition* (Steeb [60]) leads to the equation of motion

$$\left(\frac{d}{dt}L_n\right)\psi_n = (M_{n+1}L_n - L_nM_n)\,\psi_n$$

or as an operator equation

$$\frac{d}{dt}L_n = M_{n+1}L_n - L_nM_n.$$

The construction of the matrix-valued 2×2 matrix M_n is rather cumbersome. Here we refer to Barouch [5].

In order to introduce the Yang-Baxter relation we need the following two definitions. First we define the multiplication of an $n \times n$ matrix over $\mathbb{C}$ with a matrix valued $n \times n$ matrix.

Definition 3.4. Let A be an $n \times n$ matrix, where $a_{jk} \in \mathbb{C}$. Let B be an $n \times n$ matrix, where the entries B_{jk} are $m \times m$ matrices. Then we define

$$AB = C$$

where

$$C_{jk} = \sum_{l=1}^{n} a_{jl}B_{lk}.$$

Thus C is a matrix valued $n \times n$ matrix.

Furthermore we have to define the Kronecker product of two matrix valued 2×2 matrices.

Definition 3.5. Let A and B be 2×2 matrices where the entries A_{jk} and B_{jk}, respectively are $m \times m$ matrices. We define

$$A \otimes B := \begin{pmatrix} A_{11}B & A_{12}B \\ A_{21}B & A_{22}B \end{pmatrix}, \qquad A_{jk}B = \begin{pmatrix} A_{jk}B_{11} & A_{jk}B_{12} \\ A_{jk}B_{21} & A_{jk}B_{22} \end{pmatrix}.$$

Thus $A \otimes B$ is a matrix valued 4×4 matrix. We say that the Kronecker product is taken in the *auxiliary space*.

The matrix $L_n(\lambda)$ can also be written in the form (exercise (2))

$$L_n(\lambda) = \lambda I_2 \otimes I + \frac{i}{2} \sum_{j=1}^{3} \sigma_j \otimes \sigma_{j,n}$$

where I is the $2^N \times 2^N$ unit matrix.

Theorem 3.7. *Let*

$$R(\lambda) = \frac{1}{\lambda + i} \left(\left(\frac{\lambda}{2} + i \right) I \otimes I + \frac{\lambda}{2} \sum_{j=1}^{3} \sigma_j \otimes \sigma_j \right).$$

Then

$$R(\lambda - \mu) \left(L_n(\lambda) \otimes L_n(\mu) \right) = \left(L_n(\mu) \otimes L_n(\lambda) \right) R(\lambda - \mu)$$

holds, where the Kronecker product is taken in the auxiliary space.

This is the famous *Yang-Baxter relation* (also known as the factorization equation or star-triangle equation). Thus $R(\lambda)$ is a c-number matrix, i. e. a 4×4 matrix over $\mathbb{C}$. The proof of this theorem is by straightforward calculation.

The matrix R can be written in the form

$$R(\lambda) = \begin{pmatrix} 1 & 0 & 0 & 0 \\ 0 & b(\lambda) & c(\lambda) & 0 \\ 0 & c(\lambda) & b(\lambda) & 0 \\ 0 & 0 & 0 & 1 \end{pmatrix}$$

where $b(\lambda) := i/(\lambda + i)$, $c(\lambda) := \lambda/(\lambda + i)$. The Yang-Baxter relation is a sufficient condition for a system to be integrable (Baxter [8]).

Definition 3.6. The *monodromy matrix* $T_N(\lambda)$ is defined by

$$T_N(\lambda) := L_N(\lambda) \cdots L_1(\lambda) \equiv \prod_{n=1}^{\widehat{N}} L_n(\lambda).$$

Similar to the local transition matrix $L_n(\lambda)$ it satisfies the relation

$$R(\lambda - \mu)(T_N(\lambda) \otimes T_N(\mu)) = (T_N(\mu) \otimes T_N(\lambda))R(\lambda - \mu)$$

since $L_n(\lambda)$'s with different n commute.

The matrix valued 2×2 matrix $\mathcal{T}(\lambda)$ can be written as

$$\mathcal{T}_N(\lambda) = \begin{pmatrix} A_N(\lambda) & B_N(\lambda) \\ C_N(\lambda) & D_N(\lambda) \end{pmatrix}.$$

The $2^N \times 2^N$ matrices $A_N(\lambda)$, $B_N(\lambda)$, $C_N(\lambda)$, and $D_N(\lambda)$ act in the space $\mathcal{H}$. We set

$$T_N(\lambda) := A_N(\lambda) + D_N(\lambda) = \mathrm{tr}(\mathcal{T}_N(\lambda)).$$

The following commutation relations hold (exercise (4))

$$[T_N(\lambda), T_N(\mu)] = 0, \qquad [B_N(\lambda), B_N(\mu)] = 0$$

and

$$A_N(\lambda)B_N(\mu) = \frac{1}{c(\mu - \lambda)} B_N(\mu)A_N(\lambda) - \frac{b(\mu - \lambda)}{c(\lambda - \mu)} B_N(\lambda)A_N(\mu)$$

$$D_N(\lambda)B_N(\mu) = \frac{1}{c(\lambda - \mu)} B_N(\mu)D_N(\lambda) - \frac{b(\lambda - \mu)}{c(\lambda - \mu)} B_N(\lambda)D_N(\mu).$$

The family of commuting matrices $T_N(\lambda)$ contains the momentum operator $\hat{P}_N$ and the Hamilton operator $\hat{H}$. Putting $\lambda = i/2$ in $L_n(\lambda)$ and using the definition of the monodromy matrix we find that the matrix $i^{-N}T_N(i/2)$ is unitary and coincides with the *cyclic shift matrix*, i.e.

$$\exp(-i\hat{P}_N)\sigma_{j,n} \exp(i\hat{P}_N) = \sigma_{j,n+1}$$

where $j = 1, 2, 3$ and $n = 1, 2, \ldots, N$ with $N + 1 = 1$. Its eigenvalues are given by $\exp(i\pi j)$, $0 \le P_j < 2\pi$, $(j = 1, \ldots, 2^N)$. Therefore

$$\hat{P}_N = \frac{1}{i} \ln(i^{-N}T_N(i/2)).$$

The Hamilton operator $\hat{H}$ is obtained by expanding the $T_N(\lambda)$ in the neighbourhood of the point $\lambda = i/2$ and taking into account that

$$\exp(-i\hat{P}_N)\sigma_{j,n} \exp(i\hat{P}_N) = \sigma_{j,n+1}.$$

We find

$$\hat{H} = \frac{iJ}{2} \frac{d}{d\lambda} \ln(T_N(\lambda))|_{\lambda = \frac{i}{2}} - \frac{NJ}{2}I.$$

To find higher order constants of motion we have to expand $T_N(\lambda)$ with respect to λ.

Next we study states for the one dimensional isotropic Heisenberg model $\hat{H}$. Consider the vectors

$$\boldsymbol{\omega}_n := \begin{pmatrix} 1 \\ 0 \end{pmatrix}, \quad n = 1, \ldots, N, \quad \boldsymbol{\Omega}_N = \prod_{n=1}^{N} \otimes \boldsymbol{\omega}_n, \quad \boldsymbol{\Omega}_N \in \mathcal{H}.$$

In the following we omit the index N. We find the following relations hold

$$A_N(\lambda)\boldsymbol{\Omega} = \left(\lambda + \frac{i}{2}\right)^N \boldsymbol{\Omega}, \qquad D_N(\lambda)\boldsymbol{\Omega} = \left(\lambda - \frac{i}{2}\right)^N \boldsymbol{\Omega}, \qquad C_N(\lambda)\boldsymbol{\Omega} = 0.$$

It follows that the vector

$$\boldsymbol{\Psi}_N(\lambda_1, \lambda_2, \ldots, \lambda_\ell) := B_N(\lambda_1)B_N(\lambda_2) \cdots B_N(\lambda_\ell)\boldsymbol{\Omega}$$

is an eigenvector of the family of operators $T_N(\lambda)$ if the complex numbers $\lambda_1, \ldots, \lambda_\ell$ satisfy the system of equations

$$\left(\frac{\lambda_j - \frac{i}{2}}{\lambda_j + \frac{i}{2}}\right) = \prod_{\substack{k=1 \\ k \neq j}}^{\ell} \frac{\lambda_j - \lambda_k - i}{\lambda_j - \lambda_k + i} \qquad j = 1, \ldots, \ell.$$

We call these eigenvectors *Bethe's vectors*. The corresponding eigenvalue $\Lambda(\lambda; \lambda_1, \ldots, \lambda_\ell)$ has the form

$$\Lambda(\lambda; \lambda_1, \ldots, \lambda_\ell) = \left(\lambda + \frac{i}{2}\right)^N \prod_{j=1}^{\ell} \frac{\lambda - \lambda_j - i}{\lambda - \lambda_j} + \left(\lambda - \frac{i}{2}\right)^N \prod_{j=1}^{\ell} \frac{\lambda - \lambda_j + i}{\lambda - \lambda_j}.$$

Since the matrices $B_N(\lambda)$ commute, i.e. $[B(\lambda), B(\mu)] = 0$ both the vector $\boldsymbol{\Psi}_N(\lambda_1, \ldots, \lambda_\ell)$ and the eigenvalue $\Lambda(\lambda; \lambda_1, \ldots, \lambda_\ell)$ are symmetric functions of $\lambda_1, \ldots, \lambda_\ell$.

The eigenvalues of the operators $\hat{P}_N$ and $\hat{H}_N$ have the form

$$p(\lambda_1, \ldots, \lambda_\ell) = \frac{1}{i} \sum_{j=1}^{\ell} \ln\left(\frac{\lambda_j + \frac{i}{2}}{\lambda_j - \frac{i}{2}}\right) \mod 2\pi$$

$$h(\lambda_1, \ldots, \lambda_\ell) = -\frac{J}{2} \sum_{j=1}^{\ell} \frac{1}{\lambda_j^2 + \frac{1}{4}}.$$

In addition to the spectral parameter λ, it is convenient to introduce the variable $p(\lambda)$ by

$$\exp(ip(\lambda)) = \frac{\lambda + \frac{i}{2}}{\lambda - \frac{i}{2}}.$$

Thus

$$p(\lambda) = -2\text{arctg}(2\lambda) + \pi \quad \text{modulo } 2\pi.$$

In the variables P_j the momentum p and the energy h have the form

$$p(\lambda_1, \ldots, \lambda_\ell) = \sum_{j=1}^{\ell} P_j \quad \text{mod } 2\pi$$

$$h(\lambda_1, \ldots, \lambda_\ell) = -J \sum_{j=1}^{\ell} (1 - \cos(P_j)).$$

The vector $\boldsymbol{\Omega}$ plays the role of a vacuum. The matrix $B_N(\lambda)$ has the meaning of the creation operator with the momentum $p(\lambda)$ and the energy

$$h(\lambda) = \frac{J}{2} \frac{dp(\lambda)}{d\lambda} = -J(1 - \cos(p(\lambda))).$$

A necessary condition for Bethe's vector not to vanish is $\ell \leq N/2$. We can calculate the normalization of Bethe's vector using the commutation relation

$$[C_N(\lambda), B_N(\mu)] = \frac{b(\lambda - \mu)}{c(\lambda - \mu)} (A_N(\mu)D_N(\lambda) - A_N(\lambda)D_N(\mu)).$$

We can also prove the simpler assertion that Bethe's vectors with different collections $(\lambda_1, \ldots, \lambda_\ell)$, $(\lambda'_1, \ldots, \lambda'_{\ell'})$ are orthogonal.

In addition to $\hat{P}$ and $\hat{H}$, among the observable values of the system, we have the total spin operators

$$\hat{S}_j := \frac{1}{2} \sum_{n=1}^{N} \sigma_{j,n}, \qquad j = 1, 2, 3.$$

We omit the index N in the cases when this cannot cause misunderstanding. The hermitian matrices $\hat{P}$ and $\hat{H}$ commute with $\hat{S}_j$. We find the eigenvalue equation $\hat{S}_+ \boldsymbol{\Psi} = 0\boldsymbol{\Psi}$, where

$$\hat{S}_+ := \frac{1}{2} \sum_{n=1}^{N} \sigma_{+,n}, \qquad \hat{S}_- := \frac{1}{2} \sum_{n=1}^{N} \sigma_{-,n}.$$

Thus $\boldsymbol{\Psi}$ is an eigenstate of S_+ with eigenvalue 0.

For the isotropic Heisenberg model the Bethe's ansatz does not determine all eigenvectors of the Hamilton operator. Together with Bethe's vectors, vectors of the form $\hat{S}_-^m \boldsymbol{\Psi}$ are also eigenvectors, where $1 \leq m \leq 2L$ and L is the spin of the representation to which $\mathbf{\Psi}$ belongs. Moreover, the eigenvalue equation

$$\hat{S}_3 \mathbf{\Psi} = \left(\frac{N}{2} - \ell\right) \mathbf{\Psi}$$

holds. From the formula for the square of the spin S^2 with eigenvalues $L(L+1)$, $L \geq 0$

$$\hat{S}^2 = \sum_{j=1}^{3} \hat{S}_j^2 = \hat{S}_- \hat{S}_+ + \hat{S}_3(\hat{S}_3 + I)$$

it follows that for Bethe's vectors $L = N/2 - \ell$. Therefore the important inequality follows $\ell \leq \frac{N}{2}$. We find that

$$[\hat{S}_j, T(\lambda)] = 0, \quad j = 1, 2, 3$$

and

$$[\hat{S}_3, B(\lambda)] = -B(\lambda), \qquad [\hat{S}_+, B(\lambda)] = A(\lambda) - D(\lambda).$$

By construction the vector $\mathbf{\Omega}$ is also an eigenvector for the operators $\hat{S}_+$, $\hat{S}_3$. We find the eigenvalue equations

$$\hat{S}_+ \mathbf{\Omega} = 0\mathbf{\Omega}, \qquad \hat{S}_3 \mathbf{\Omega} = \frac{N}{2}\mathbf{\Omega}.$$

We have, after carrying $\hat{S}_+$ through all operators $B(\lambda_j)$ to the vector $\mathbf{\Omega}$,

$$\hat{S}_+ \mathbf{\Psi} = \sum_{j=1}^{\ell} B(\lambda_1) \ldots B(\lambda_{j-1})(A(\lambda_j) - D(\lambda_j))B(\lambda_{j+1}) \ldots B(\lambda_\ell)\mathbf{\Omega}.$$

Using the permutation relations given above we carry the matrices $A(\lambda_j)$ and $D(\lambda_j)$ through the $B(\lambda_k)$ to the vector $\mathbf{\Omega}$. We arrive at the state

$$\hat{S}_+ \mathbf{\Psi} = \sum_{j=1}^{\ell} M_j(\lambda_1, \ldots, \lambda_\ell)B(\lambda_1) \cdots B(\lambda_{j-1})B(\lambda_{j+1}) \cdots B(\lambda_\ell)\mathbf{\Omega}.$$

To obtain $M_1(\lambda_1, \ldots, \lambda_\ell)$ we have to carry $A(\lambda_1) - D(\lambda_1)$ through the chain $B(\lambda_2) \cdots B(\lambda_\ell)$ to the vector $\mathbf{\Omega}$. Therefore

$$M_1(\lambda_1, \ldots, \lambda_\ell) = \left(\lambda_1 + \frac{i}{2}\right)^N \prod_{j=2}^{\ell} \frac{\lambda_1 - \lambda_j - i}{\lambda_1 - \lambda_j} - \left(\lambda_1 - \frac{i}{2}\right)^N \prod_{j=2}^{\ell} \frac{\lambda_1 - \lambda_j + i}{\lambda_1 - \lambda_j}.$$

The remaining coefficients $M_j(\lambda_1 \ldots, \lambda_\ell)$ are obtained from $M_1(\lambda_1, \ldots, \lambda_\ell)$ by the corresponding permutation of the numbers $\lambda_1, \ldots, \lambda_\ell$. They have the following form

$$M_j(\lambda_1, \ldots, \lambda_\ell) = \left(\lambda_j + \frac{i}{2}\right)^N \prod_{\substack{k=1 \\ k \neq j}}^{\ell} \frac{\lambda_j - \lambda_k - i}{\lambda_j - \lambda_k} - \left(\lambda_j - \frac{i}{2}\right)^N \prod_{\substack{k=1 \\ k \neq j}}^{\ell} \frac{\lambda_j - \lambda_k + i}{\lambda_j - \lambda_k}.$$

where $j = 1, \ldots, \ell$. The system of equations

$$\left(\frac{\lambda_j - \frac{i}{2}}{\lambda_j + \frac{i}{2}}\right) = \prod_{\substack{k=1 \\ k \neq j}}^{\ell} \frac{\lambda_j - \lambda_k - i}{\lambda_j - \lambda_k + i}, \qquad j = 1, \ldots, \ell$$

means exactly that $M_j(\lambda_1, \ldots, \lambda_\ell) = 0$, $(j = 1, \ldots, \ell)$. For further reading we refer to Takhtadzhyan and Faddeev [67], Kulish and Sklyanin [39], Baxter [8], Sogo and Wadati [51], Barouch [5].

Exercises. (1) Show that

$$\left(I_2 \otimes I_2 - \sum_{j=1}^{3} \sigma_j \otimes \sigma_j\right)^2 \equiv 4 \left(I_2 \otimes I_2 - \sum_{j=1}^{3} \sigma_j \otimes \sigma_j\right).$$

(2) Let $J < 0$. Show that $\hat{H}$ can be represented in the form

$$\hat{H} = -J \sum_{n=1}^{N} (\sigma_{1,n} \otimes \sigma_{1,n+1} + \sigma_{2,n} \otimes \sigma_{2,n+1} + \sigma_{3,n} \otimes \sigma_{3,n+1} - I)^2.$$

(3) Let

$$S_{N,1} := \frac{1}{2} \sum_{j=1}^{N} \sigma_{1,j}, \qquad S_{N,2} := \frac{1}{2} \sum_{j=1}^{N} \sigma_{2,j}, \qquad S_{N,3} := \frac{1}{2} \sum_{j=1}^{N} \sigma_{3,j}.$$

Find the equations of motion for $S_{N,1}$, $S_{N,2}$ and $S_{N,3}$.

(4) Show that the matrix $L_n(\lambda)$ can also be written in the form

$$L_n(\lambda) = \lambda I_2 \otimes I + \frac{i}{2} \sum_{j=1}^{3} \sigma_j \otimes \sigma_{j,n}$$

where I is the $2^N \times 2^N$ unit matrix.

(5) Show that R can be written in the form

$$R(\lambda) = \begin{pmatrix} 1 & 0 & 0 & 0 \\ 0 & b(\lambda) & c(\lambda) & 0 \\ 0 & c(\lambda) & b(\lambda) & 0 \\ 0 & 0 & 0 & 1 \end{pmatrix}$$

where $b(\lambda) := i/(\lambda + i)$, $c(\lambda) := \lambda/(\lambda + i)$.

(6) Show that the following commutation relations hold

$$[T_N(\lambda), T_N(\mu)] = 0, \qquad [B_N(\lambda), B_N(\mu)] = 0$$

and

$$A_N(\lambda)B_N(\mu) = \frac{1}{c(\mu - \lambda)} B_N(\mu)A_N(\lambda) - \frac{b(\mu - \lambda)}{c(\lambda - \mu)} B_N(\lambda)A_N(\mu)$$

$$D_N(\lambda)B_N(\mu) = \frac{1}{c(\lambda - \mu)} B_N(\mu)D_N(\lambda) - \frac{b(\lambda - \mu)}{c(\lambda - \mu)} B_N(\lambda)D_N(\mu).$$

(7) Show that

$$\sigma_{j,n}\sigma_{k,n} = \delta_{jk}I + i\sum_{l=1}^{3} \epsilon_{jkl}\sigma_{l,n}, \quad [\sigma_{j,m}, \sigma_{k,n}] = 2i\sum_{l=1}^{3} \epsilon_{jkl}\sigma_{l,m}\delta_{mn}$$

where

$$\epsilon_{jkl} := \begin{cases} 0 & \text{if two indices } jkl \text{ are the same} \\ 1 & \text{if } jkl \text{ is an even permutation of 1,2,3} \\ -1 & \text{if } jkl \text{ is an odd permutation of 1,2,3.} \end{cases}$$

(8) Show that the following relations for the state Ω hold

$$A_N(\lambda)\Omega = \left(\lambda + \frac{i}{2}\right)^N \Omega, \quad D_N(\lambda)\Omega = \left(\lambda - \frac{i}{2}\right)^N \Omega, \quad C_N(\lambda)\Omega = 0\Omega.$$

(9) Show that the eigenvalues of the operators $\hat{P}_N$ and $\hat{H}_N$ have the form

$$p(\lambda_1, \ldots, \lambda_\ell) = \frac{1}{i}\sum_{j=1}^{\ell} \ln\left(\frac{\lambda_j + i/2}{\lambda_j - i/2}\right) \quad \text{mod } 2\pi$$

$$h(\lambda_1, \ldots, \lambda_\ell) = -\frac{J}{2}\sum_{j=1}^{\ell} \frac{1}{\lambda_j^2 + 1/4}.$$

(10) Show that the matrices $\hat{P}$ and $\hat{H}$ commute with $\hat{S}_j$.

(11) Show that $[\hat{S}_j, T(\lambda)] = 0$, $j = 1, 2, 3$ and

$$[\hat{S}_3, B(\lambda)] = -B(\lambda), \qquad [\hat{S}_+, B(\lambda)] = A(\lambda) - D(\lambda).$$

(12) Let $J < 0$. Show that the Hamilton operator $\hat{H}$ annihilates the vector Ω, i.e. $\hat{H}\Omega = 0\Omega$. The vector Ω is the ground state — the *ferromagnetic vacuum*.

3.12   Hopf Algebras

First we introduce the concept of a *Hopf algebra* (Abe [1], Sweedler [66], Dăscălescu et al [16]).

Let $\mathbb{F}$ be a field. In most cases we have $\mathbb{F} = \mathbb{R}$ or $\mathbb{F} = \mathbb{C}$. A Hopf algebra is an associative algebra $\mathcal{A}$ over a field $\mathbb{F}$ with five basic maps, namely four homomorphisms

$$\mu : \mathcal{A} \otimes \mathcal{A} \to \mathcal{A} \quad \text{(multiplication)}$$

$$\Delta : \mathcal{A} \to \mathcal{A} \otimes \mathcal{A} \quad \text{(coproduct, co-multiplication)}$$

$$\eta : \mathbb{F} \to \mathcal{A} \quad \text{(inclusion, unit map)}$$

$$\varepsilon : \mathcal{A} \to \mathbb{F} \quad \text{(co-unit map, counit)}$$

and one antihomomorphism

$$S : \mathcal{A} \to \mathcal{A} \quad \text{(antipode)}.$$

They satisfy the following relations for any $a \in \mathcal{A}$

$$(\Delta \otimes \mathrm{id})\Delta(a) = (\mathrm{id} \otimes \Delta)\Delta(a)$$

$$(\varepsilon \otimes \mathrm{id})\Delta(a) = (\mathrm{id} \otimes \varepsilon)\Delta(a) = \mathrm{id}(a) = a$$

$$\mu(S \otimes \mathrm{id})\Delta(a) = \mu(\mathrm{id} \otimes S)\Delta(a) = \eta \circ \varepsilon(a) = \varepsilon(a)1$$

where id is the identity map. One uses the notation

$$(\mathcal{A}, \mu, \Delta, \eta, \varepsilon, S)$$

to denote a Hopf algebra.

Now $(\mathcal{A}, \mu, \Delta', \eta, \varepsilon, S^{-1})$ is also a Hopf algebra, where Δ' denotes the opposite coproduct, which maps any $a \in \mathcal{A}$ to $\mathcal{A} \otimes \mathcal{A}$ as

$$\Delta'(a) = \sum_j c_j \otimes b_j \quad \text{if} \quad \Delta(a) = \sum_j b_j \otimes c_j$$

and S^{-1} is defined as the inverse of S

$$S(S^{-1}(a)) = S^{-1}(S(a)) = a.$$

Let G be a finite group and $\mathbb{F}$ a field. The set $\mathcal{A} = \mathrm{Map}(G, \mathbb{F})$ of all functions defined on G with values in $\mathbb{F}$ becomes a $\mathbb{F}$-algebra when we define the scalar product and the sum and product of functions by

$$(\alpha f)(x) = \alpha f(x), \qquad (f + g)(x) = f(x) + g(x),$$

$$(fg)(x) = f(x)g(x), \quad f, g \in \mathcal{A}, \quad \alpha \in \mathbb{F}, \ x \in G.$$

In general, a $\mathbb{F}$-algebra $\mathcal{A}$ can be characterized as a $\mathbb{F}$-linear space $\mathcal{A}$ together with two $\mathbb{F}$-linear maps

$$\mu : \mathcal{A} \otimes \mathcal{A} \to \mathcal{A}, \qquad \eta : \mathbb{F} \to \mathcal{A}$$

which satisfy axioms corresponding to the associative law and the unitary property respectively. If we identify $\mathcal{A} \otimes \mathcal{A}$ with $\mathrm{Map}(G \times G, \mathbb{F})$ where $\mathcal{A} = \mathrm{Map}(G, \mathbb{F})$, and if the operations of G are employed in defining the $\mathbb{F}$-linear maps

$$\Delta : \mathcal{A} \to \mathcal{A} \otimes \mathcal{A}, \qquad \epsilon : \mathcal{A} \to \mathbb{F}$$

respectively by

$$\Delta f(x \otimes y) := f(xy), \qquad \epsilon f := f(e)$$

for $x, y \in G$ and where e is the identity element of the group G, then Δ and ϵ become homomorphisms of $\mathbb{F}$-algebras having properties which are dual to μ and η, respectively.

In general, a $\mathbb{F}$-linear space A with $\mathbb{F}$-linear maps $\mu, \eta, \Delta, \epsilon$ defined as above is called a $\mathbb{F}$-bialgebra.

We can define a $\mathbb{F}$-linear endomorphism of $\mathcal{A} = \mathrm{Map}(G, \mathbb{F})$

$$S : \mathcal{A} \to \mathcal{A}, \qquad (Sf)(x) := f(x^{-1}), \quad f \in \mathcal{A}, \ x \in G$$

such that the equalities

$$\mu(1 \otimes S)\Delta = \mu(S \otimes 1)\Delta = \eta \circ \epsilon$$

hold. A $\mathbb{F}$-bialgebra on which we can define a $\mathbb{F}$-linear map S as above is called a $\mathbb{F}$-Hopf algebra. A $\mathbb{F}$-Hopf algebra is an algebraic system which simultaneously admits structures of a $\mathbb{F}$-algebra as well as its dual, where these two structures are related by a certain specific law. For a finite group G, its group ring over a field $\mathbb{F}$ is the dual space of the $\mathbb{F}$-linear space $\mathcal{A} = \mathrm{Map}(G, \mathbb{F})$ where its $\mathbb{F}$-algebra structure is given by the dual $\mathbb{F}$-linear maps of Δ and ϵ. Now, the group ring admits a $\mathbb{F}$-Hopf algebra structure when we take the dual $\mathbb{F}$-linear maps of μ, η, and S. In other words, the group ring is the dual $\mathbb{F}$-Hopf algebra of $\mathrm{Map}(G, \mathbb{F})$.

3.13   Quantum Groups

Quantum groups and quantum algebras are mathematical structures which have found applications in theoretical physics, in particular statistical physics (Kassel [36], Takhtajan [68], Zachos [75]). For example, knot theory and quantum algebras are closely related. A quantum group is a noncommutative and noncocommutative Hopf algebra. Algebraically, these structures are described as continuous deformations of the familiar Lie (super) algebras widely used in physics, a description which uses many ideas from classical q analysis.

A quasitriangular Hopf algebra is a Hopf algebra equipped with an element $R \in \mathcal{A} \otimes \mathcal{A}$ which is the solution of the algebraic version of the Yang-Baxter equation.

Definition 3.7. Let $\mathcal{H} = (\mathcal{A}, \mu, \Delta, \eta, \varepsilon, S)$ be a Hopf algebra and R (intertwiner) be an invertible element in $\mathcal{A} \otimes \mathcal{A}$. Then the pair $(\mathcal{H}, R)$ is called a quasitriangular Hopf algebra if for any $a \in \mathcal{A}$ we have

$$R\Delta(a) = \Delta'(a)R, \quad (\Delta \otimes \mathrm{id})R = R_{13}R_{23}, \quad (\mathrm{id} \otimes \Delta)R = R_{13}R_{12}.$$

Here, for example, R_{13} acts in the first and third sections in $\mathcal{A} \times \mathcal{A} \times \mathcal{A}$.

By definition, the three relations

$$R_{12}R_{13}R_{23} = R_{23}R_{13}R_{12},$$

$$(S \otimes \mathrm{id})R = (\mathrm{id} \otimes S^{-1})R = R^{-1}, \quad (\varepsilon \otimes \mathrm{id})R = (\mathrm{id} \otimes \varepsilon)R = 1$$

are satisfied. The first equation is the *Yang-Baxter equation*.

Two dimensional statistical models, field theory and S-matrix theory can be described by the quantum Yang-Baxter equation, which applies quantum algebras. Thus the quantum Yang-Baxter equation plays an important role. With solutions of quantum Yang-Baxter equation one can construct exactly solvable models and find their eigenvalues and eigenstates. On the other hand, any solution of the quantum Yang-Baxter equation can be generally used to find the new quasi-triangular Hopf algebra. Many multiparameter solutions (4×4 matrices) of the quantum Yang-Baxter equation have been obtained. Corresponding to the case of the standard one-parameter R matrix, the algebras related to the standard two-parameter R matrix have also been discussed.

The starting point in the construction of the Yang-Baxter equation is the 2×2 matrix

$$T := \begin{pmatrix} a & b \\ c & d \end{pmatrix}$$

where a, b, c and d are noncommutative linear operators. We may consider a, b, c and d as $n \times n$ matrices over the complex or real numbers. In other words, T is a matrix-valued matrix. Let I be the 2×2 identity matrix

$$I := \begin{pmatrix} 1 & 0 \\ 0 & 1 \end{pmatrix}$$

where 1 is the unit operator (identity matrix). Now we define the 4×4 matrices

$$T_1 := T \otimes I, \qquad T_2 := I \otimes T$$

where $\otimes$ denotes the Kronecker product. Thus T_1 and T_2 are matrix (operator) valued 4×4 matrices. Applying the rules for the Kronecker product, we find the matrices

$$T_1 = T \otimes I = \begin{pmatrix} a & 0 & b & 0 \\ 0 & a & 0 & b \\ c & 0 & d & 0 \\ 0 & c & 0 & d \end{pmatrix}, \qquad T_2 = I \otimes T = \begin{pmatrix} a & b & 0 & 0 \\ c & d & 0 & 0 \\ 0 & 0 & a & b \\ 0 & 0 & c & d \end{pmatrix}.$$

The algebra related to this quantum matrix is governed by the *Yang-Baxter equation*

$$R_q T_1 T_2 = T_2 T_1 R_q$$

where R_q is an R-matrix and q is a nonzero complex number.

Example 3.9. Consider the 4×4 matrix

$$R_q := \begin{pmatrix} 1 & 0 & 0 & 0 \\ 0 & -1 & 0 & 0 \\ 0 & 1+q & q & 0 \\ 0 & 0 & 0 & 1 \end{pmatrix}$$

where q is a nonzero complex number. The Yang-Baxter equation gives rise to the relations of the algebra elements a, b, c and d

$$ab = q^{-1}ba, \quad dc = qcd, \quad bc = -qcb,$$

$$bd = -db, \quad ac = -ca, \quad [a,d] = (1+q^{-1})bc$$

where all the commutative relations have been omitted. Here $[\,,\,]$ denotes the commutator. The quantum matrix T can be considered as a linear transformation of the plane $\mathcal{A}_q(2)$ with coordinates (x, ξ) satisfying

$$x\xi = -\xi x.$$

It is straightforward to prove that the coordinate transformations deduced by T

$$\begin{pmatrix} x' \\ \xi' \end{pmatrix} = \begin{pmatrix} a & b \\ c & d \end{pmatrix} \begin{pmatrix} x \\ \xi \end{pmatrix}$$

provides $x'\xi' = -\xi'x'$. As there is no nilpotent element in the quantum matrix T, there exists no constraint on the coordinates x and ξ. This quantum plane is the same as the one related to $GL_q(2)$. From the algebraic relations for a, b, c, and d we can define an element δ of the algebra,

$$\delta := ad - q^{-1}bc.$$

Then δ satisfies the following relations $[a, \delta] = 0$, $[d, \delta] = 0$,

$$\{c, \delta\}_q \equiv qc\delta + \delta c = 0, \qquad \{\delta, b\}_q \equiv q\delta b + b\delta = 0.$$

The element δ commutes only with a and b and hence is not the centre of the algebra.   ♣

If we consider the R-matrix

$$R_q := \begin{pmatrix} q & 0 & 0 & 0 \\ 0 & 1 & 0 & 0 \\ 0 & q - q^{-1} & 1 & 0 \\ 0 & 0 & 0 & q \end{pmatrix}$$

then we obtain from the Yang-Baxter equation the relations

$$ab = qba, \quad ac = qca, \quad bc = cb,$$

$$bd = qdb, \quad cd = qdc, \quad ad - da = (q - q^{-1})bc.$$

In chapter 5 we give an implementation with SymbolicC++. Two tasks have to be performed: The first would be the implementation of the relations for a, b, c and d. Then these commutation relations can be used to evaluate a new commutation relation. An example is the definition of the operator δ and evaluation of the commutation relations for δ.

3.14   Lax Representation

In this section we show how the Kronecker product can be used to find new Lax representations for first order autonomous systems of ordinary differential equations (Steeb, Hardy and Stoop [63]).

A number of interesting dynamical systems can be written in *Lax representation* (Steeb [60], Steeb and Lai [61])

$$\frac{dL}{dt} = [A, L](t)$$

where A and L are given by $n \times n$ matrices. The time-dependent matrices A and L are called a *Lax pair*. An example is the Toda lattice. Given two Lax representations we show how the Kronecker product can be used to find a new Lax representation. We then give an application. Finally, we discuss some extensions.

Let X and R be $m \times m$ matrices and Y and P be $n \times n$ matrices. Then we have $(X \otimes Y)(R \otimes P) \equiv (XR) \otimes (YP)$. This identity will be used in the following.

Theorem 3.8. *Given two Lax representations*

$$\frac{dL}{dt} = [A, L](t), \qquad \frac{dM}{dt} = [B, M](t)$$

where L and A are $m \times m$ matrices and M and B are $n \times n$ matrices. Let I_m be the $m \times m$ unit matrix and I_n be the $n \times n$ unit matrix. Then we find the Lax representation

$$\frac{d}{dt}(L \otimes M) = [A \otimes I_n + I_m \otimes B, L \otimes M](t).$$

We call this the Kronecker product Lax representation.

Proof.   This can be seen as follows: The right-hand side can be written as

$$[A \otimes I_n + I_m \otimes B, L \otimes M] = [A \otimes I_n, L \otimes M] + [I_m \otimes B, L \otimes M].$$

Thus we find that

$$[A \otimes I_n + I_n \otimes B, L \otimes M] = (AL) \otimes M - (LA) \otimes M + L \otimes (BM) - L \otimes (MB).$$

On the other hand we have

$$\frac{d}{dt}(L \otimes M) = \frac{dL}{dt} \otimes M + L \otimes \frac{dM}{dt}.$$

Inserting $dL/dt = [A, L](t)$ and $dM/dt = [B, M](t)$ into this equation completes the proof.   □

First integrals of $dL/dt = [A, L](t)$ can be found from

$$F_k = \text{tr}(L^k), \qquad k = 1, 2, \ldots \quad .$$

Since $\text{tr}(X^k \otimes Y^j) = \text{tr}(X^k)\text{tr}(Y^j)$, where X is an $m \times m$ matrix and Y is an $n \times n$ matrix, we find that first integrals of the Kronecker product Lax representation are given by

$$F_{kj} = \text{tr}(L^k)\text{tr}(M^j), \qquad k, j = 1, 2, \ldots \quad .$$

Obviously, we can extend this to more than two Lax pairs. For example, given three Lax representations

$$\frac{dL}{dt} = [A, L](t), \qquad \frac{dM}{dt} = [B, M](t), \qquad \frac{dN}{dt} = [C, N](t)$$

we find that

$$\frac{d}{dt}(L \otimes M \otimes N) = [A \otimes I_n \otimes I_p + I_m \otimes B \otimes I_p + I_m \otimes I_n \otimes C, L \otimes M \otimes N](t)$$

where C and N are $p \times p$ matrices.

Example 3.10. Consider the nonlinear system of ordinary differential equations

$$\frac{du_1}{dt} = (\lambda_3 - \lambda_2)u_2u_3, \qquad \frac{du_2}{dt} = (\lambda_1 - \lambda_3)u_3u_1, \qquad \frac{du_3}{dt} = (\lambda_2 - \lambda_1)u_1u_2$$

where $\lambda_j \in \mathbb{R}$. This system describes *Euler's rigid body motion*. The first integrals are given by $I_1(\mathbf{u}) = u_1^2 + u_2^2 + u_3^2$, $I_2(\mathbf{u}) = \lambda_1 u_1^2 + \lambda_2 u_2^2 + \lambda_3 u_3^2$. A Lax representation is given by

$$\frac{dL}{dt} = [L, \lambda L](t)$$

where

$$L := \begin{pmatrix} 0 & -u_3 & u_2 \\ u_3 & 0 & -u_1 \\ -u_2 & u_1 & 0 \end{pmatrix}, \qquad \lambda L := \begin{pmatrix} 0 & -\lambda_3 u_3 & \lambda_2 u_2 \\ \lambda_3 u_3 & 0 & -\lambda_1 u_1 \\ -\lambda_2 u_2 & \lambda_1 u_1 & 0 \end{pmatrix}.$$

Then $\text{tr}(L)^k$ $(k = 1, 2, \ldots)$ provides only one first integral. We obtain

$$\text{tr}(L) = 0, \qquad \text{tr}(L^2) = -2(u_1^2 + u_2^2 + u_3^2) = -2I_1.$$

Since L does not depend on λ we cannot find I_2. This Lax representation can be applied to the product Lax representation when we set $M = L$. ♣

To overcome the problem of finding only one first integral we consider now

$$\frac{d(L + Ay)}{dt} = [L + Ay, \lambda L + By](t)$$

where y is a dummy variable and A and B are time-independent diagonal matrices, i.e. $A = \text{diag}(A_1, A_2, A_3)$ and $B = \text{diag}(B_1, B_2, B_3)$ with $A_j, B_j \in \mathbb{R}$. The equation decomposes into various powers of y, namely

$$y^0 : \frac{dL}{dt} = [L, \lambda L]$$
$$y^1 : 0 = [L, B] + [A, \lambda L]$$
$$y^2 : [A, B] = 0.$$

The last equation is satisfied identically since A and B are diagonal matrices. The second equation leads to

$$\lambda_i = \frac{B_j - B_k}{A_j - A_k}$$

where (i, j, k) are permutations of (1,2,3). It can be satisfied by setting

$$B_j = A_j^2$$

and $\lambda_i = A_j + A_k$. Consequently the original Lax pair $L, \lambda L$ satisfies the extended Lax pair $L + Ay, \lambda L + By$. Now $\text{tr}[(L + Ay)^2]$ and $\text{tr}[(L + Ay)^3]$ provide both first integrals given above. For this extended Lax pair the concept of the Kronecker product described above can also be applied.

For some dynamical systems such as the energy level motion we find an extended Lax representation

$$\frac{dL}{dt} = [A, L](t), \qquad \frac{dK}{dt} = [A, K](t)$$

where L and K do not commute. For this extended system of Lax representations we can also apply the Kronecker product technique given above in order to find new Lax representations.

3.15 Signal Processing

The development of a fast Fourier transform algorithm has made a significant impact on the field of digital signal processing. Many researchers have since developed fast transform implementations corresponding to a wide variety of discrete unitary transforms. At present fast algorithms are known for a variety of unitary transforms such as Hadamard, Haar, Slant, discrete cosine, and Hartley transforms, etc. (see Regalia and Mitra [49] and Elliott and Rao [20] and reference therein). The discrete Fourier transform is suited to frequency domain analysis and filtering, the discrete cosine transform to data compression, the Slant transform to image coding, the Hadamard and Haar transforms to dyadic-invariant signal processing, and these and others to generalized spectral analysis. Fast transform algorithms are developed by recognizing various patterns of the elements of a discrete unitary transform matrix. The existence of such patterns implies some redundancy among the matrix elements which can be exploited in developing sparse matrix factorizations. The product of sparse matrices can result in a much simplified computational algorithm compared to the direct implementation of a matrix equation.

The Kronecker product representations lead to efficient computer implementations for numerous discrete unitary transforms and Kronecker products can be defined in terms of matrix factorizations and play a central role in generalized spectral analysis. A wide class of discrete unitary transforms can be generated using recursion formulas of generalized Kronecker products with matrix permutations (i.e. element reordering). The Kronecker product decomposition of various unitary transforms plays a central role in developing fast transform algorithms.

Regalia and Mitra [49] proposed a generalization of the Kronecker product and described its utility with some examples from the field of signal processing. A large class of discrete unitary transforms can be developed from a single recursion formula. Closed-form expressions are derived for sparse matrix factorizations in terms of this generalized matrix product. When applied to discrete unitary transform matrices, fast transform algorithms can be developed directly upon recognizing patterns in the matrices. They also derived some apparently novel properties of Hadamard transformations and polyadic permutations in the context of Kronecker products. They also showed the invariance of Hadamard matrices under a bit-permuted ordering similarity transformation, and established a simple result of the Kronecker decomposability of any polyadic permutation matrix. Closed form expressions relating this matrix product to sparse matrix factorizations are obtained, which are particularly useful in factorizing "patterned" matrices. Many unitary matrices, for example, are patterned matrices, for which the sparse matrix factorization equivalent of this matrix product directly yields fast transform algorithms. They also describe a fast transform algorithm to developing equivalent filter bank representations.

We start with the connection of the Kronecker product and the *bit-reversed order*. Let

$$a, b, c, d, e, f \in \{0, 1\}.$$

Consider the Kronecker product of the three vectors

$$\begin{pmatrix} a \\ b \end{pmatrix} \otimes \begin{pmatrix} c \\ d \end{pmatrix} \otimes \begin{pmatrix} e \\ f \end{pmatrix} = (ace \; acf \; ade \; adf \; bce \; bcf \; bde \; bdf)^T .$$

On the other hand

$$\begin{pmatrix} b \\ a \end{pmatrix} \otimes \begin{pmatrix} d \\ c \end{pmatrix} \otimes \begin{pmatrix} f \\ e \end{pmatrix} = (bdf \; bde \; bcf \; bce \; adf \; ade \; acf \; ace)^T .$$

Thus we find the *bit-reversed order*.

Regalia and Mitra [49] introduced the Kronecker product of an $m \times n$ matrix A and a $k \times l$ matrix B as the $mk \times nl$ matrix

$$A \otimes B := \begin{pmatrix} Ab_{00} & Ab_{01} & \dots & Ab_{0,l-1} \\ Ab_{10} & Ab_{11} & \dots & Ab_{1,l-1} \\ \vdots & \vdots & \ddots & \vdots \\ Ab_{k-1,0} & Ab_{k-1,1} & \dots & Ab_{k-1,l-1} \end{pmatrix}.$$

This definition of $A \otimes B$ can be transformed into the definition given in chapter 2 using permutation matrices. In the definition we count from zero, which is useful when we consider software implementation in C++.

The generalization of the definition of the Kronecker product is as follows:

Definition 3.8. Given a set of N ($m \times r$) matrices A_i, $i = 0, 1, \dots, N-1$, denoted by $\{A\}_N$, and a ($N \times l$) matrix B, we define the ($mN \times rl$) matrix ($\{A\}_N \otimes B$) as

$$\{A\}_N \otimes B := \begin{pmatrix} A_0 \otimes \mathbf{b}_0 \\ A_1 \otimes \mathbf{b}_1 \\ \vdots \\ A_{N-1} \otimes \mathbf{b}_{N-1} \end{pmatrix}$$

where $\mathbf{b}_i$ denotes the ith row vector of the matrix B. If each matrix A_i is identical, then the definition reduces to the usual Kronecker product of matrices.

Example 3.11. We have

$$\{A\}_2 = \left\{ \begin{pmatrix} 1 & 1 \\ 1 & -1 \end{pmatrix}, \begin{pmatrix} 1 & -i \\ 1 & i \end{pmatrix} \right\}, \quad B = \begin{pmatrix} 1 & 1 \\ 1 & -1 \end{pmatrix} \Rightarrow \{A\}_2 \otimes B = \begin{pmatrix} 1 & 1 & 1 & 1 \\ 1 & -1 & 1 & -1 \\ 1 & -i & -1 & i \\ 1 & i & -1 & -i \end{pmatrix}$$

which is a 4×4 discrete Fourier transform matrix with the rows arranged in bit-reversed order. ♣

We have assumed that the number of matrices in the set $\{A\}_N$ matches the number of rows in the matrix B. B need not be a single matrix, but may also be a periodic sequence of $(k \times l)$ matrices (and hence a periodic sequence of row vectors).

The matrix product $\{A\} \otimes \{B\}$ is obtained as the Kronecker product of each matrix in $\{A\}$ with each row vector in $\{B\}$, and is easily verified to yield a periodic sequence of $(mk \times rl)$ matrices, which thus admits a finite representation in one period. To affix notation, a set of matrices will always be indicated using the bracket notation (e.g. $\{A\}$), whereas if the set consists of a single matrix, the brackets will be omitted.

Example 3.12. Let

$$\{A\} = \left\{ \begin{array}{l} \begin{pmatrix} 1 & 1 \\ 1 & -1 \end{pmatrix} \\ \begin{pmatrix} 1 & -i \\ 1 & i \end{pmatrix} \\ \begin{pmatrix} 1 & e^{-i\pi/4} \\ 1 & -e^{-i\pi/4} \end{pmatrix} \\ \begin{pmatrix} 1 & e^{-i3\pi/4} \\ 1 & -e^{-i3\pi/4} \end{pmatrix} \end{array} \right\}, \quad \{B\} = \left\{ \begin{array}{l} \begin{pmatrix} 1 & 1 \\ 1 & -1 \end{pmatrix} \\ \begin{pmatrix} 1 & -i \\ 1 & i \end{pmatrix} \end{array} \right\}.$$

Then

$$\{A\} \otimes \{B\} = \left\{ \begin{array}{l} \begin{pmatrix} 1 & 1 & 1 & 1 \\ 1 & -1 & 1 & -1 \\ 1 & -i & -1 & i \\ 1 & i & -1 & -i \end{pmatrix} \\ \begin{pmatrix} 1 & e^{-i\pi/4} & -i & e^{-i3\pi/4} \\ 1 & e^{-i\pi/4} & -i & -e^{-i3\pi/4} \\ 1 & e^{-i3\pi/4} & i & e^{-i\pi/4} \\ 1 & -e^{-i3\pi/4} & i & -e^{-i\pi/4} \end{pmatrix} \end{array} \right\}.$$

Each matrix in the resulting set is a 4×4 matrix. ♣

One finds that this matrix product can result in a single matrix only if the rightmost multiplicand is a single matrix. The following algebraic property of this generalized Kronecker product is straightforward to prove

$$(\{A\} \otimes \{B\}) \otimes \{C\} = \{A\} \otimes (\{B\} \otimes \{C\}).$$

Definition 3.9. Let $\{A\}_N$ be a sequence of $(m \times n)$ matrices and E be a single $(n \times r)$ matrix. Then

$$\{A\}E := \begin{pmatrix} A_0 E \\ A_1 E \\ \vdots \\ A_{N-1} E \end{pmatrix}$$

where each matrix in the sequence in $(m \times r)$.

This leads to the identity $(\{A\}E) \otimes (\{B\}F) \equiv (\{A\} \otimes \{B\})(E \otimes F)$. The next two identities are useful in developing sparse matrix factorizations.

$$\{A\}_N \otimes I_N = \bigoplus_{j=0}^{N-1} A_j$$

where $\oplus$ denotes the direct sum. Note that both sides of the equation yield a block-diagonal matrix containing the matrices A_j.

Let p sets of matrices be denoted by $\{A^{(k)}\}_{N_k}$, $k = 0, 1, \ldots, p-1$, where each matrix is $(m \times n)$, and the kth set has $N_k = m^k$ matrices. Consider the matrix R defined by

$$R := \{A^{(p-1)}\}_{m^{p-1}} \otimes \{A^{(p-2)}\}_{m^{(p-2)}} \otimes \cdots \otimes \{A^{(1)}\}_m \otimes A^{(0)}.$$

The last matrix $A^{(0)}$ is a single matrix, as is R. The matrix R admits the sparse matrix factorization

$$R = \prod_{k=0}^{p-1} \left(\bigoplus_{i=0}^{m^{p-k-1}-1} \left(I_{n^k} \otimes A_i^{(p-k-1)} \right) \right).$$

If each matrix $A_i^{(k)}$ is (para-)unitary, then R is (para-)unitary. This result follows from the fact that the Kronecker product, direct sum, or matrix product of (para-)unitary matrices is a (para-)unitary matrix.

If $m = n$, such that each matrix $A_i^{(k)}$, $i = 0, \ldots, m^k - 1; k = 0, \ldots, p - 1$, is square, then

$$\det(R) = \prod_{k=0}^{p-1} \prod_{i=0}^{m^k-1} \left(\det \left(A_i^{(k)} \right) \right)^{m^k}.$$

This relation can be found by successive application of the algebra of determinants.

Example 3.13. We find a fast transform algorithm for an 8×8 complex *BIFORE matrix*. The matrix is given by

$$R_8 := \begin{pmatrix} 1 & 1 & 1 & 1 & 1 & 1 & 1 & 1 \\ 1 & -1 & 1 & -1 & 1 & -1 & 1 & -1 \\ 1 & -i & 1 & i & 1 & -i & -1 & i \\ 1 & i & -1 & -i & 1 & i & -1 & -i \\ 1 & 1 & -i & -i & -1 & -1 & i & i \\ 1 & -1 & -i & i & -1 & 1 & i & -i \\ 1 & 1 & i & i & -1 & -1 & -i & -i \\ 1 & -1 & i & -i & -1 & 1 & -i & i \end{pmatrix}.$$

Thus we can write

$$R_8 = \left\{ \begin{array}{c} \begin{pmatrix} 1 & 1 \\ 1 & -1 \end{pmatrix} \\ \begin{pmatrix} 1 & -i \\ 1 & i \end{pmatrix} \\ \begin{pmatrix} 1 & 1 \\ 1 & -1 \end{pmatrix} \\ \begin{pmatrix} 1 & 1 \\ 1 & -1 \end{pmatrix} \end{array} \right\} \otimes \begin{pmatrix} 1 & 1 & 1 & 1 \\ 1 & -1 & 1 & -1 \\ 1 & -i & -1 & i \\ 1 & i & -1 & -i \end{pmatrix}.$$

The 4×4 matrix on the right-hand side is a 4-point complex BIFORE transform. Denoting this matrix by R_4 and continuing the process, we obtain

$$R_4 = \begin{pmatrix} 1 & 1 & 1 & 1 \\ 1 & -1 & 1 & -1 \\ 1 & -i & -1 & i \\ 1 & i & -1 & -i \end{pmatrix} = \left\{ \begin{array}{c} \begin{pmatrix} 1 & 1 \\ 1 & -1 \end{pmatrix} \\ \begin{pmatrix} 1 & -i \\ 1 & i \end{pmatrix} \end{array} \right\} \otimes \begin{pmatrix} 1 & 1 \\ 1 & -1 \end{pmatrix}.$$

♣

We find that a recursion formula for the family of complex BIRORE matrices is expressed in the present notation. In particular, we have

$$R_N = \{B\}_{N/2} \otimes R_{N/2}, \qquad R_1 = 1$$

where N is a power of two, and $\{B\}_{N/2}$ has $N/2$ matrices in the set, with the ith matrix (counting form $i = 0$ to $N/2 - 1$) given by

$$B_i = \begin{cases} \begin{pmatrix} 1 & -i \\ 1 & i \end{pmatrix}, & i = 1 \\ \begin{pmatrix} 1 & 1 \\ 1 & -1 \end{pmatrix}, & \text{otherwise.} \end{cases}$$

Discrete Fourier transform matrices may also be expressed using the recursion relation given above provided the rows are permuted into bit-reversed order (which, in a practical implementation, requires the output samples to be sorted if they are desired in normal order). In this case the ith matrix of $\{B\}_{N/2}$ is given by

$$B_j = \begin{pmatrix} 1 & W_N^{\ll j \gg} \\ 1 & -W_N^{\ll j \gg} \end{pmatrix}, \qquad j = 0, 1, \ldots, N/2 - 1$$

where $W_N := \exp(-i2\pi/N)$ and $\ll j \gg$ is the decimal number obtained from a bit reversal of a $\log_2(N/2)$-bit binary representation of j. The matrix sets in the example given above are recognized as $\{B\}_4$ and $\{B\}_2$, for example. Although the recursion relation corresponds to radix-2 Fast Fourier Transform algorithms, a similar recursion can be developed for any prime radix. For example, to obtain a radix 3 Fast Fourier Transform recursion $N/2$ can be replaced everywhere with $N/3$, and the i-th matrix in $\{B\}_{N/3}$ can be defined as

$$B_j = \begin{pmatrix} 1 & W_N^{\ll j \gg} & W_N^{2\ll j \gg} \\ 1 & \exp(i4\pi/3)W_N^{\ll j \gg} & \exp(i2\pi/3)W_N^{2\ll j \gg} \\ 1 & \exp(i2\pi/3)W_N^{\ll j \gg} & \exp(i4\pi/3)W_N^{2\ll j \gg} \end{pmatrix}, \qquad j = 0, 1, \ldots, N/3 - 1$$

where now $\ll j \gg$ is the decimal number obtained from a trit reversal of a $\log_3(n/3)$-trit ternary representation of i, and so on. Nonprime radices can be split into prime factors, each of which may be treated in a manner analogous to that above.

Modified Walsh-Hadamard transform matrices may also be expressed with the recursion given above. Here the ith matrix of $\{B\}_{N/2}$ is given by

$$B_j = \begin{cases} \begin{pmatrix} 1 & 1 \\ 1 & -1 \end{pmatrix}, & j = 0 \\ \sqrt{2}I_2, & \text{otherwise.} \end{cases}$$

The Haar transform may be derived from the modified Walsh-Hadamard transform through zonal bit-reversed ordering permutations of the input and output data.

Hadamard matrices are known to satisfy the recursion

$$R_N = \begin{pmatrix} 1 & 1 \\ 1 & -1 \end{pmatrix} \otimes R_{N/2}, \qquad R_1 = 1$$

which may be understood as a special case of the recursion given above in which each B_j is identical.

With the recursion given above, the matrix B_N in all cases defined above satisfies $\widetilde{R}_N R_N = R_N \widetilde{R}_N = N I_N$ so that $R_N/\sqrt{N}$ is unitary. Therefore fast transform algorithms in all cases are immediately available.

Regalia and Mitra [49] also describe filter bank applications using the definitions and recursions given above. Moreover, they describe matrix factorization of a polyadic permutation matrix with application to a programmable polyadic shift structure.

3.16 Clebsch-Gordan Series

The Clebsch-Gordan series plays a central role in quantum mechanics (Biedenharn and Louck [10], Bohm [11]). Let

$$j \in \{0, 1/2, 1, 3/2, \dots\}$$

and $m = -j, -j+1, \dots, +j$. Let V_1 be the $2j+1$ dimensional vector space over $\mathbb{C}$ with the orthonormal basis $|j, m\rangle$, i.e.

$$\langle j', m'|j, m \rangle = \delta_{j'j}\delta_{m'm}$$

where $\langle \,|\, \rangle$ denotes the scalar product. We define linear operators j_+, j_-, j_3 with

$$j_+|j, m\rangle := C_+(j, m)|j, m+1\rangle, \quad j_-|j, m\rangle := C_-(j, m)|j, m-1\rangle$$

and (eigenvalue equation)

$$j_3|j,m\rangle := m|j,m\rangle$$

where

$$C_\pm = \sqrt{j(j+1) - m(m \pm 1)}.$$

We also define

$$j_1 := \frac{1}{2}(j_+ + j_-), \qquad j_2 := \frac{1}{2i}(j_+ - j_-)$$

and $j^2 := j_1^2 + j_2^2 + j_3^2$. We have the commutation relations

$$[j_k, j_\ell] = i \sum_{m=1}^{3} \epsilon_{k\ell m} j_m, \qquad k, \ell, m = 1, 2, 3$$

where $\epsilon_{k\ell m}$ is the *Levi-Civita symbol*. Now let $j_1 \in \{0, 1/2, 1, 3/2, \ldots\}$ and $m_1 = -j_1, -j_1+1, \ldots, +j_1$. Let V_1 be the $2j_1+1$ dimensional vector space over $\mathbb{C}$ with the orthonormal basis $|j_1, m_1\rangle$. Let $j_2 \in \{0, 1/2, 1, 3/2, \ldots\}$ and $m_2 = -j_2, -j_2+1, \ldots, +j_2$. Let V_2 be the $2j_2+1$ dimensional vector space over $\mathbb{C}$ with the orthonormal basis $|j_2, m_2\rangle$. The tensor product space $V_1 \otimes V_2$ has the orthonormal basis $(2j_1+1) \times (2j_2+1)$ dimensional (product) basis

$$\{ |j_1, m_1\rangle \otimes |j_2, m_2\rangle : m_1 = -j_1, -j_1+1, \ldots, +j_1, \ m_2 = -j_2, j_2+1, \ldots, +j_2 \}.$$

Let I_1 and I_2 be the identity operator in the vector space V_1 and V_2, respectively. Let $\hat{O}_1$ and $\hat{O}_2$ be linear operators in the vector space V_1 and V_2 respectively. Then we have

$$(\hat{O}_1 \otimes I_2 + I_1 \otimes \hat{O}_2)(|j_1, m_1\rangle \otimes |j_2, m_2\rangle)$$

$$= (\hat{O}_1|j_1, m_1\rangle) \otimes |j_2, m_2\rangle + |j_1, m_1\rangle \otimes (\hat{O}_2|j_2, m_2\rangle).$$

Now we define the total angular momentum operators

$$J_1 = j_{1,1} \otimes I_2 + I_1 \otimes j_{2,1}, \quad J_2 = j_{1,2} \otimes I_2 + I_1 \otimes j_{2,2}, \quad J_3 = j_{1,3} \otimes I_2 + I_1 \otimes j_{2,3}.$$

Thus we have the eigenvalue equations

$$J_3(|j_1, m_1\rangle \otimes |j_2, m_2\rangle) = (m_1 + m_2)(|j_1, m_1\rangle \otimes |j_2, m_2\rangle).$$

The linear operators J_1, J_2, J_3 satisfy the same commutation relation as j_1^I, j_2^I, j_3^I and $j_1^{II}, j_2^{II}, j_3^{II}$. Therefore total angular momentum eigenstates $|(j_1 j_2) J M\rangle$ in the tensor product space $V_1 \otimes V_2$ exist

$$J^2|(j_1 j_2) J M\rangle = J(J+1)|(j_1 j_2) J M\rangle, \quad J_z|(j_1 j_2) J M\rangle = M|(j_1 j_2) J M\rangle$$

for $M = -J, -J+1, \ldots, +j$. J satisfies the triangular relation

$$|j_1 - j_2| \leq J \leq j_1 + j_2.$$

The total number of total momentum eigenstates is equal to the dimension of the tensor product space

$$\sum_{J=|j_1-j_2|}^{j_1+j_2} (2J + 1) = (2j_1 + 1)(2j_2 + 1).$$

The total angular momentum states in the tensor product state can be expand using the product space basis

$$|(j_1 j_2)JM\rangle = \sum_{m_1=-j_1}^{j_1} \sum_{m_2=-j_2}^{j_2} (|j_1, m_1\rangle \otimes |j_2, m_2\rangle)\langle j_1, m_1; j_2, m_2|JM\rangle.$$

The complex expansion coefficients $\langle j_1, m_1; j_2, m_2|JM\rangle$ are called Clebsch-Gordan coefficients.

Consider the case $V_1 = V_2$ with $j_1 = 1/2$ and $j_2 = 1/2$. Then $m_1 = -1/2, 1/2$ and $m_2 = -1/2, 1/2$. We have the two dimensional Hilbert space $\mathbb{C}^2$ and the linear operators

$$j_1^I = j_1^{II} = \frac{1}{2}\begin{pmatrix} 0 & 1 \\ 1 & 0 \end{pmatrix}, \quad j_2^I = j_2^{II} = \frac{1}{2}\begin{pmatrix} 0 & -i \\ i & 0 \end{pmatrix}, \quad j_3^I = j_3^{II} = \frac{1}{2}\begin{pmatrix} 1 & 0 \\ 0 & -1 \end{pmatrix}.$$

Therefore

$$j_+^I = j_+^{II} = \begin{pmatrix} 0 & 1 \\ 0 & 0 \end{pmatrix}, \quad j_-^I = j_-^{II} = \begin{pmatrix} 0 & 0 \\ 1 & 0 \end{pmatrix}.$$

We also have

$$|1/2, 1/2\rangle = \begin{pmatrix} 1 \\ 0 \end{pmatrix}, \quad |1/2, -1/2\rangle = \begin{pmatrix} 0 \\ 1 \end{pmatrix}.$$

Consider the case with $j_1 = j_2 = 1$. Then $m_1 = -1, 0, 1$ and $m_2 = -1, 0, 1$ with the matrices

$$j_+ = \begin{pmatrix} 0 & \sqrt{2} & 0 \\ 0 & 0 & \sqrt{2} \\ 0 & 0 & 0 \end{pmatrix}, \quad j_- = \begin{pmatrix} 0 & 0 & 0 \\ \sqrt{2} & 0 & 0 \\ 0 & \sqrt{2} & 0 \end{pmatrix}, \quad j_3 = \begin{pmatrix} 1 & 0 & 0 \\ 0 & 0 & 0 \\ 0 & 0 & -1 \end{pmatrix}.$$

Exercises. (1) Calculate the eigenvalues and eigenvectors of

$$J_3 = j_3 \otimes I + I \otimes j_3.$$

(2) Let $j = 0, 1/2, 1, 3/2, 2, \ldots$ be the spin. The *Clebsch-Gordan decomposition* is

$$V_j \otimes V_j \cong \sum_{s=0}^{2j} \oplus V_s.$$

For $j = 1/2$ we have

$$V_{1/2} \otimes V_{1/2} \cong V_0 \oplus V_1.$$

Consider the spin matrices S_1 and S_3 for spin $\frac{1}{2}$

$$S_1 = \frac{1}{2} \begin{pmatrix} 0 & 1 \\ 1 & 0 \end{pmatrix}, \quad S_3 = \frac{1}{2} \begin{pmatrix} 1 & 0 \\ 0 & -1 \end{pmatrix}.$$

Find the 4×4 unitary matrix U such that

$$U(S_1 \otimes S_3)U^* = \frac{1}{4} \begin{pmatrix} -1 & 0 & 0 & 0 \\ 0 & 0 & 0 & 1 \\ 0 & 0 & 1 & 0 \\ 0 & 1 & 0 & 0 \end{pmatrix} \equiv (-1/4) \oplus \begin{pmatrix} 0 & 0 & 1/4 \\ 0 & 1/4 & 0 \\ 1/4 & 0 & 0 \end{pmatrix}.$$

3.17 Braid-like Relations and Yang-Baxter Relations

Let A, B be nonzero $n \times n$ matrices over $\mathbb{C}$ and $A \neq B$. Assume that

$$ABA = BAB.$$

Then A, B satisfy a *braid-like relation*.

If $ABA = BAB$, then it can easily be proved that

$$(A \otimes A)(B \otimes B)(A \otimes A) = (B \otimes B)(A \otimes A)(B \otimes B)$$
$$(A \otimes I_n)(B \otimes I_n)(A \otimes I_n) = (B \otimes I_n)(A \otimes I_n)(B \otimes I_n)$$

and

$$(A \oplus A)(B \oplus B)(A \oplus A) = (B \oplus B)(A \oplus A)(B \oplus B)$$
$$(A \oplus I_n)(B \oplus I_n)(A \oplus I_n) = (B \oplus I_n)(A \oplus I_n)(B \oplus I_n).$$

Example 3.14. Consider the invertible 2×2 matrices

$$B = \begin{pmatrix} 1 & 1 \\ 0 & 1 \end{pmatrix}, \quad C = \begin{pmatrix} 1 & 0 \\ -1 & 1 \end{pmatrix}.$$

Then B and C are linearly independent and $BCB = CBC$. We also have

$$(B \otimes B)(C \otimes C)(B \otimes B) = (C \otimes C)(B \otimes B)(C \otimes C).$$

Note that $C = ABA^{-1}$, where

$$A = \begin{pmatrix} 0 & 1 \\ 1 & 0 \end{pmatrix}.$$

For the left-hand side we have

$$(B \otimes B)(C \otimes C)(B \otimes B) = (BC) \otimes (BC)(B \otimes B) = (BCB) \otimes (BCB).$$

For the right-hand side we have

$$(C \otimes C)(B \otimes B)(C \otimes C) = (CB) \otimes (CB)(C \otimes C) = (CBC) \otimes (CBC).$$

Since $BCB = CBC$ the identity holds. In general we have

$$B(k)C(k)B(k) = C(k)B(k)C(k)$$

with $B(k) = B(1)^{\otimes k}$, $C(k) = C(1)^{\otimes k}$ and $B(1) = B$, $C(1) = C$.   ♣

Consider the 2×2 matrix

$$B(1) = \begin{pmatrix} 1 & 1 \\ 0 & 1 \end{pmatrix}$$

and $B(k)$ the $2^k \times 2^k$ matrix recursively defined by

$$B(k) := \begin{pmatrix} B(k-1) & B(k-1) \\ 0 & B(k-1) \end{pmatrix} = B(1) \otimes B(k-1), \qquad k = 2, 3, \ldots$$

We find

$$B(1)^{-1} = \begin{pmatrix} 1 & -1 \\ 0 & 1 \end{pmatrix}, \qquad B(k)^{-1} = (B(1)^{-1})^{\otimes k}.$$

Let $A(k)$ be the $2^k \times 2^k$ anti-diagonal matrix defined by $A_{ij}(k) = 1$ if $i = 2^k + 1 - j$ and 0 otherwise with $j = 1, 2, \ldots, 2^k$. Let

$$C(k) = A(k)B(k)^{-1}A(k).$$

We have

$$C(1) = \begin{pmatrix} 1 & 0 \\ -1 & 1 \end{pmatrix}.$$

The matrices $B(k)$ and $C(k)$ satisfy the braid-like relation

$$B(k)C(k)B(k) = C(k)B(k)C(k).$$

The relation is true for $k = 1$ by a direct calculation. Since

$$B(k) = B(1)^{\otimes k}, \qquad C(k) = C(1)^{\otimes k}$$

and using the properties of the Kronecker product we find that the relation is true for every k.

A representation $D(B_n)$ of a Braid group B_n $(n \geq 3)$ can be obtained as

$$g_j = D(b_j, n) = I_N \otimes \cdots \otimes I_N \otimes \sigma \otimes I_N \otimes \cdots \otimes I_N$$

where I_N is the $N \times N$ unit matrix and σ is the j-th position. σ is an invertible $N^2 \times N^2$ matrix satisfying

$$(\sigma \otimes I_N)(I_N \otimes \sigma)(\sigma \otimes I_N) = (I_N \otimes \sigma)(\sigma \otimes I_N)(I_N \otimes \sigma).$$

Let $N = 2$. Find an invertible 4×4 matrix σ such that this condition is satisfied.

If V and W are matrices of the same order, then their *entrywise product* $V \bullet W$ is defined by

$$(V \bullet W)_{j,k} := V_{j,k} W_{j,k}.$$

If all entries of V are nonzero, then we say that X is *Schur invertible* and define its Schur inverse, $V^{(-)}$, by $V^{(-)} \bullet V = J$, where J is the matrix with all 1's.

The vector space $M_n(\mathbb{F})$ of $n \times n$ matrices acts on itself in three distinct ways: if $C \in M_n(\mathbb{F})$ we can define endomorphisms X_C, Δ_C and Y_C by

$$X_C M := CM, \qquad \Delta_C M := C \bullet M, \qquad Y_C := MC^T.$$

Let A, B be $n \times n$ matrices. Assume that X_A is invertible and Δ_B is invertible in the sense of Schur. Note that X_A is invertible if and only if A is, and Δ_B is invertible if and only if the Schur inverse $B^{(-)}$ is defined. We say that (A, B) is a *one-sided Jones pair* if

$$X_A \Delta_B X_A = \Delta_B X_A \Delta_B.$$

We call this the *braid relation*. We give an example for a one-sided Jones pair. A trivial example is the pair (I_n, J_n), where J_n is the $n \times n$ matrix with all ones. For the left-hand side we have

$$(X_A \Delta_B X_A)(M) = (X_A \Delta_B)(X_A(M)) = X_A \Delta_B(AM) = X_A(B \bullet (AM))$$
$$= A(B \bullet (AM)).$$

Thus with $A = I_n$ and $B = J_n$ we obtain M. For the right-hand side we have

$$(\Delta_B X_A \Delta_B)M = \Delta_B X_A(B \bullet M) = \Delta_B(A(B \bullet M)) = B \bullet (A(B \bullet M)).$$

Thus with $A = I_n$ and $B = J_n$ we also obtain M.

Exercises. (1) Find all invertible 2×2 matrices A, B with $AB \neq BA$ such that (*braid-like relation*) $ABA = BAB$. Taking the determinant of both sides and using the assumptions that A and B are invertible provides $\det(A) = \det(B)$. A solution is

$$A = \begin{pmatrix} 1 & 1 \\ 0 & 1 \end{pmatrix}, \qquad B = \begin{pmatrix} 1 & 0 \\ -1 & 1 \end{pmatrix}$$

with $AB \neq BA$.

(2) Given the 2×2 unitary matrices

$$A = \begin{pmatrix} e^{-i\pi/4} & 0 \\ 0 & ie^{-i\pi/4} \end{pmatrix}, \qquad B = \frac{1}{\sqrt{2}} \begin{pmatrix} 1 & i \\ i & 1 \end{pmatrix}.$$

(i) Is the braid-like relation $ABA = BAB$ satisfied?
(ii) Find the smallest $n \in \mathbb{N}$ such that $A^n = I_2$.
(iii) Find the smallest $m \in \mathbb{N}$ such that $B^m = I_2$.

(3) Find all nonzero 2×2 matrices A and B with $[A, B] \neq 0_2$ that satisfy the braid-like relation $ABBA = BAAB$. Assume that $ABBA = BAAB$. Is

$$(A \otimes A)(B \otimes B)(B \otimes B)(A \otimes A) = (B \otimes B)(A \otimes A)(A \otimes A)(B \otimes B) ?$$

(4) Consider the simplest noncommutative braid group B_3. This group is generated by two generators, σ_1 and σ_2, and the unique relation

$$\sigma_1 \sigma_2 \sigma_1 = \sigma_2 \sigma_1 \sigma_2.$$

Let $a := \sigma_1 \sigma_2 \sigma_1$ and $b := \sigma_1 \sigma_2$. Show that $a^2 = b^3$. This implies that $a^2 = (\sigma_1 \sigma_2 \sigma_1)^2$ lies in the center of B_3.

(5) Let $A = (a_{jk})$ be a 2×2 matrix. Find the condition on A such that

$$(A \otimes I_3)(I_3 \otimes A)(A \otimes I_3) = (I_3 \otimes A)(A \otimes I_3)(I_3 \otimes A).$$

3.18   Fast Fourier Transform

The discrete Fourier transform on $\mathbb{C}^n$ is the invertible transformation F_n : $\mathbb{C}^n \to \mathbb{C}^n$ described by the $n \times n$ matrix

$$F_n = \begin{pmatrix} 1 & 1 & 1 & \cdots & 1 \\ 1 & \omega_n & \omega_n^2 & \cdots & \omega_n^{n-1} \\ 1 & \omega_n^2 & \omega_n^4 & \cdots & \omega_n^{2(n-1)} \\ \vdots & \vdots & \vdots & \ddots & \vdots \\ 1 & \omega_n^{n-1} & \omega_n^{2(n-1)} & \cdots & \omega_n^{(n-1)(n-1)} \end{pmatrix}$$

where $\omega_n := e^{-i2\pi/n}$. In other words $(F_n)_{j,k} = \omega_n^{(j-1)(k-1)}$. It follows that

$$F_{2n} = \begin{pmatrix} 1 & 1 & 1 & \cdots & 1 \\ 1 & \omega_{2n} & \omega_{2n}^2 & \cdots & \omega_{2n}^{2n-1} \\ 1 & \omega_{2n}^2 & \omega_{2n}^4 & \cdots & \omega_{2n}^{2(2n-1)} \\ \vdots & \vdots & \vdots & \ddots & \vdots \\ 1 & \omega_{2n}^{2n-1} & \omega_{2n}^{2(2n-1)} & \cdots & \omega_{2n}^{(2n-1)(2n-1)} \end{pmatrix}.$$

Here

$$(F_{2n})_{j,k} = \omega_{2n}^{(j-1)(k-1)} = \omega_n^{(j-1)(k-1)/2}.$$

Since

$$\omega_{2n}^{2(j-1)(k-1)} = \omega_n^{(j-1)(k-1)}, \quad \omega_n^{jn} = 1, \quad \omega_{2n}^{k+n} = -\omega_{2n}^k$$

we find

$$(F_{2n})_{2j-1,k} = (F_{2n})_{2j-1,k+n} = (F_n)_{j,k},$$
$$(F_{2n})_{2j,k} = \omega_{2n}^{k-1}(F_n)_{j,k}, \quad (F_{2n})_{2j,k+n} = -\omega_{2n}^{k-1}(F_n)_{j,k}$$

for $j, k \in \{1, 2, \ldots, n\}$. Consequently

$$F_{2n} = \sum_{j,k=1}^{2n} (F_{2n})_{j,k}\, \mathbf{e}_{j,2n} \otimes \mathbf{e}_{k,2n}^T$$

$$= \sum_{j,k=1}^{n} \Big[(F_{2n})_{2j-1,k}\, \mathbf{e}_{2j-1,2n} \otimes \mathbf{e}_{k,2n}^T + (F_{2n})_{2j-1,k+n}\, \mathbf{e}_{2j-1,2n} \otimes \mathbf{e}_{k+n,2n}^T$$

$$+ (F_{2n})_{2j,k}\mathbf{e}_{2j,2n} \otimes \mathbf{e}_{k,2n}^T + (F_{2n})_{2j,k+n}\mathbf{e}_{2j,2n} \otimes \mathbf{e}_{k+n,2n}^T \Big].$$

Using

$$(\mathbf{e}_{j,n} \otimes \mathbf{e}_{1,2}) \otimes (\mathbf{e}_{1,2} \otimes \mathbf{e}_{k,n})^T = \mathbf{e}_{2j-1,2n} \otimes \mathbf{e}_{k,2n}^T,$$
$$(\mathbf{e}_{j,n} \otimes \mathbf{e}_{1,2}) \otimes (\mathbf{e}_{2,2} \otimes \mathbf{e}_{k,n})^T = \mathbf{e}_{2j-1,2n} \otimes \mathbf{e}_{k+n,2n}^T,$$
$$(\mathbf{e}_{j,n} \otimes \mathbf{e}_{2,2}) \otimes (\mathbf{e}_{1,2} \otimes \mathbf{e}_{k,n})^T = \mathbf{e}_{2j,2n} \otimes \mathbf{e}_{k,2n}^T,$$
$$(\mathbf{e}_{j,n} \otimes \mathbf{e}_{2,2}) \otimes (\mathbf{e}_{2,2} \otimes \mathbf{e}_{k,n})^T = \mathbf{e}_{2j,2n} \otimes \mathbf{e}_{k+n,2n}^T$$

we find

$$F_{2n} = \sum_{j,k=1}^{n} \Big[(F_n)_{j,k}(\mathbf{e}_{j,n} \otimes \mathbf{e}_{1,2}) \otimes (\mathbf{e}_{1,2} \otimes \mathbf{e}_{k,n})^T$$

$$+ (F_n)_{j,k}(\mathbf{e}_{j,n} \otimes \mathbf{e}_{1,2}) \otimes (\mathbf{e}_{2,2} \otimes \mathbf{e}_{k,n})^T$$

$$+ \omega_{2n}^{k-1}(F_n)_{j,k}(\mathbf{e}_{j,n} \otimes \mathbf{e}_{2,2}) \otimes (\mathbf{e}_{1,2} \otimes \mathbf{e}_{k,n})^T$$

$$- \omega_{2n}^{k-1}(F_n)_{j,k}(\mathbf{e}_{j,n} \otimes \mathbf{e}_{2,2}) \otimes (\mathbf{e}_{2,2} \otimes \mathbf{e}_{k,n})^T \Big].$$

Now from

$$P_{n,2} = \sum_{u=1}^{n} \sum_{v=1}^{2} (\mathbf{e}_{v,2} \otimes \mathbf{e}_{u,n})(\mathbf{e}_{u,n} \otimes \mathbf{e}_{v,2})^T$$

and $P_{n,2}^{-1} = P_{2,n}$ we obtain

$$F_{2n} = P_{2,n} P_{n,2} F_{2n}$$

$$= P_{2,n} \sum_{j,k=1}^{n} \Big[(F_n)_{j,k}(\mathbf{e}_{1,2} \otimes \mathbf{e}_{j,n}) \otimes (\mathbf{e}_{1,2} \otimes \mathbf{e}_{k,n})^T$$

$$+ (F_n)_{j,k}(\mathbf{e}_{1,2} \otimes \mathbf{e}_{j,n}) \otimes (\mathbf{e}_{2,2} \otimes \mathbf{e}_{k,n})^T$$

$$+ \omega_{2n}^{k-1}(F_n)_{j,k}(\mathbf{e}_{2,2} \otimes \mathbf{e}_{j,n}) \otimes (\mathbf{e}_{1,2} \otimes \mathbf{e}_{k,n})^T$$

$$- \omega_{2n}^{k-1}(F_n)_{j,k}(\mathbf{e}_{2,2} \otimes \mathbf{e}_{j,n}) \otimes (\mathbf{e}_{2,2} \otimes \mathbf{e}_{k,n})^T \Big].$$

Factoring out common terms on the left-hand side of the Kronecker product yields

$$F_{2n} = P_{2,n} \Bigg[(\mathbf{e}_{1,2}\mathbf{e}_{1,2}^T + \mathbf{e}_{1,2}\mathbf{e}_{2,2}^T) \otimes \left(\sum_{j,k=1}^{n} (F_n)_{j,k}\mathbf{e}_{j,n}\mathbf{e}_{k,n}^T \right)$$

$$+ (\mathbf{e}_{2,2}\mathbf{e}_{1,2}^T - \mathbf{e}_{2,2}\mathbf{e}_{2,2}^T) \otimes \left(\sum_{k=1}^{n} \omega_{2n}^{k-1} \sum_{j=1}^{n} \Big[(F_n)_{j,k}\mathbf{e}_{j,n} \otimes \mathbf{e}_{k,n}^T \Big] \right) \Bigg].$$

Now we use the fact that

$$\left(\sum_{k=1}^{n} \omega_{2n}^{k-1} \sum_{j=1}^{n} \left[(F_n)_{j,k} \mathbf{e}_{j,n} \otimes \mathbf{e}_{k,n}^T \right] \right)$$

$$= \left(\sum_{k=1}^{n} \sum_{j=1}^{n} \left[(F_n)_{j,k} \mathbf{e}_{j,n} \otimes \mathbf{e}_{k,n}^T \right] \right) \left(\sum_{m=1}^{n} \omega_{2n}^{m-1} \mathbf{e}_{m,n} \mathbf{e}_{m,n}^T \right)$$

$$= F_n D_n$$

where D_n is the diagonal matrix

$$D_n := \sum_{m=1}^{n} \omega_{2n}^{m-1} \mathbf{e}_{m,n} \mathbf{e}_{m,n}^T.$$

Inserting this result in F_{2n} yields

$$F_{2n} = P_{2,n} \left[\begin{pmatrix} 1 & 1 \\ 0 & 0 \end{pmatrix} \otimes F_n + \begin{pmatrix} 0 & 0 \\ 1 & -1 \end{pmatrix} \otimes (F_n D_n) \right]$$

$$= P_{2,n} \left[I_2 \otimes F_n \right] \left[\begin{pmatrix} 1 & 1 \\ 0 & 0 \end{pmatrix} \otimes I_n + \begin{pmatrix} 0 & 0 \\ 1 & -1 \end{pmatrix} \otimes D_n \right]$$

$$= P_{2,n} \left[I_2 \otimes F_n \right] \begin{pmatrix} I_n & I_n \\ D_n & -D_n \end{pmatrix}.$$

The matrix multiplications above have very simple implementations. If n is even then we can apply this decomposition again. Thus if $n = 2^l$ is a power of 2 we can apply this process $\log_2(n)$ times to obtain a relatively simple transformation procedure. This is the fast (discrete) Fourier transform.

Let $\mathbf{x} \in \mathbb{C}^n$. Then $F_n \mathbf{x}$ involves n^2 multiplications. On the other hand

$$\begin{pmatrix} I_n & I_n \\ D_n & -D_n \end{pmatrix} \mathbf{x}$$

involves $2n$ multiplications (the I_n result in a multiplication by 1, i.e. we can just add). The matrix P^T swaps rows, so no multiplications are involved. The remaining multiplications follow from $I_{n/2} \otimes F_{n/2}$ which is twice the number of multiplications when calculating $F_{n/2} \mathbf{y}$ for some $\mathbf{y} \in \mathbb{C}^{n/2}$. In other words, let $m(F_n)$ be the least number of nontrivial multiplications performed when applying F_n to an element of $\mathbb{C}^n$ then with $m(F_1) = 0$

(multiplication by 1) we find

$$m(F_n) \leq 2n + 2m(F_{n/2})$$
$$\leq 2n + 2(2 \cdot n/2 + 2m(F_{n/4}))$$
$$\vdots$$
$$\leq \sum_{j=1}^{\log_2(n)} 2n = 2n \log_2(n).$$

For example conventional matrix multiplication for F_{128} involves 16384 multiplications (16129 nontrivial multiplications), while the fast Fourier transform involves 1792 nontrivial multiplications.

The number of additions to calculate $F_n\mathbf{x}$ directly is $n^2 - n$. Let $a(F_n)$ be the least number of additions performed in the fast Fourier transform. Obviously $a(F_1) = 0$. The product

$$\begin{pmatrix} I_n & I_n \\ D_n & -D_n \end{pmatrix} \mathbf{x}$$

requires n additions (one for each row). Applying P^T requires no additions. Thus we have

$$a(F_n) \leq n + 2a(F_{n/2}) \leq n + 2(n/2 + 2a(F_{n/4})) \leq \ldots \leq \sum_{j=1}^{\log_2(n)} n = n \log_2(n).$$

3.19   Entanglement

Consider the Hilbert spaces $\mathbb{C}^2$ and $\mathbb{C}^4$. Given $\mathbf{v} \in \mathbb{C}^4$. We could ask the question whether $\mathbf{v}$ can be written as the Kronecker product of two vectors $\mathbf{x}, \mathbf{y} \in \mathbb{C}^2$, i.e.

$$\begin{pmatrix} v_1 \\ v_2 \\ v_3 \\ v_4 \end{pmatrix} = \begin{pmatrix} x_1 \\ x_2 \end{pmatrix} \otimes \begin{pmatrix} y_1 \\ y_2 \end{pmatrix}.$$

If one cannot find an $\mathbf{x}$ and $\mathbf{y}$ we call the vector $\mathbf{v}$ entangled. If we can find an $\mathbf{x}$ and $\mathbf{y}$ we call the vector $\mathbf{v}$ unentangled, i.e. $\mathbf{v}$ can be written as a product state.

An example for an unentangled state is

$$\begin{pmatrix} 1 \\ -1 \\ 1 \\ -1 \end{pmatrix} = \begin{pmatrix} 1 \\ 1 \end{pmatrix} \otimes \begin{pmatrix} 1 \\ -1 \end{pmatrix}.$$

Examples for entangled states are the *Bell states* given by

$$\frac{1}{\sqrt{2}} \begin{pmatrix} 1 \\ 0 \\ 0 \\ 1 \end{pmatrix}, \quad \frac{1}{\sqrt{2}} \begin{pmatrix} 1 \\ 0 \\ 0 \\ -1 \end{pmatrix}, \quad \frac{1}{\sqrt{2}} \begin{pmatrix} 0 \\ 1 \\ 1 \\ 0 \end{pmatrix}, \quad \frac{1}{\sqrt{2}} \begin{pmatrix} 0 \\ 1 \\ -1 \\ 0 \end{pmatrix}.$$

Let $|0\rangle$, $|1\rangle$ be an orthonormal basis in $\mathbb{C}^2$. Then the four states

$$\frac{1}{\sqrt{2}}(|0\rangle \otimes |0\rangle + |1\rangle \otimes |1\rangle), \quad \frac{1}{\sqrt{2}}(|0\rangle \otimes |0\rangle - |1\rangle \otimes |1\rangle),$$

$$\frac{1}{\sqrt{2}}(|0\rangle \otimes |1\rangle + |1\rangle \otimes |0\rangle), \quad \frac{1}{\sqrt{2}}(|0\rangle \otimes |1\rangle - |1\rangle \otimes |0\rangle)$$

are entangled. They also form an orthonormal basis in $\mathbb{C}^4$ and contain the Bell states.

Another orthonormal basis in $\mathbb{C}^4$ with all vectors entangled is given by

$$\frac{1}{2} \begin{pmatrix} -1 \\ 1 \\ 1 \\ 1 \end{pmatrix}, \quad \frac{1}{2} \begin{pmatrix} 1 \\ -1 \\ 1 \\ 1 \end{pmatrix}, \quad \frac{1}{2} \begin{pmatrix} 1 \\ 1 \\ -1 \\ 1 \end{pmatrix}, \quad \frac{1}{2} \begin{pmatrix} 1 \\ 1 \\ 1 \\ -1 \end{pmatrix}.$$

In the Hilbert space $\mathbb{C}^6$ we can ask two questions. Can the vector $\mathbf{v} \in \mathbb{C}^6$ be written as the Kronecker product of a vector $\mathbf{x} \in \mathbb{C}^2$ and a vector $\mathbf{y} \in \mathbb{C}^3$, i.e. $\mathbf{v} = \mathbf{x} \otimes \mathbf{y}$? Can the vector $\mathbf{v} \in \mathbb{C}^6$ be written as the Kronecker product of a vector $\mathbf{y} \in \mathbb{C}^3$ and a vector $\mathbf{x} \in \mathbb{C}^2$, i.e. $\mathbf{v} = \mathbf{y} \otimes \mathbf{x}$? This two cases do not imply each other.

The vector in $\mathbb{C}^6$

$$\mathbf{v} = \begin{pmatrix} 1 \\ 0 \\ 0 \\ 0 \\ 0 \\ 1 \end{pmatrix}$$

cannot be written as the Kronecker product of 2×3 or 3×2. An example where the vector $\mathbf{v}$ can be written in the 2×3 form is

$$\frac{1}{2}\begin{pmatrix} 1 \\ 0 \\ 1 \\ -1 \\ 0 \\ -1 \end{pmatrix} = \frac{1}{\sqrt{2}}\begin{pmatrix} 1 \\ -1 \end{pmatrix} \otimes \frac{1}{\sqrt{2}}\begin{pmatrix} 1 \\ 0 \\ 1 \end{pmatrix}.$$

But the vector cannot be written in 3×2 form. Finally a case where the vector $\mathbf{v}$ can be written in 2×3 form and 3×2 form is

$$\frac{1}{\sqrt{6}}\begin{pmatrix} 1 \\ 1 \\ 1 \\ 1 \\ 1 \\ 1 \end{pmatrix} = \frac{1}{\sqrt{2}}\begin{pmatrix} 1 \\ 1 \end{pmatrix} \otimes \frac{1}{\sqrt{3}}\begin{pmatrix} 1 \\ 1 \\ 1 \end{pmatrix} = \frac{1}{\sqrt{3}}\begin{pmatrix} 1 \\ 1 \\ 1 \end{pmatrix} \otimes \frac{1}{\sqrt{2}}\begin{pmatrix} 1 \\ 1 \end{pmatrix}.$$

In the Hilbert space $\mathbb{C}^8$ we ask 4×2, 2×4 or even $2 \times 2 \times 2$.

A vector $\mathbf{v}$ in $\mathbb{C}^8$ can be written as

$$\mathbf{v} = \sum_{j,k,\ell=0}^{2} t_{jk\ell}\mathbf{e}_j \otimes \mathbf{e}_k \otimes \mathbf{e}_\ell$$

where $\mathbf{e}_j$ $(j = 0, 1, 2)$ denotes the standard basis in $\mathbb{C}^2$. The expansion coefficients $t_{jk\ell}$ $(j, k, \ell = 0, 1, 2)$ can be considered as the entries of a tensor $T = (t_{jk\ell})$. Consider the GHZ-state in $\mathbb{C}^8$

$$\mathbf{v} = \frac{1}{\sqrt{2}}\begin{pmatrix} 1\,0\,0\,0\,0\,0\,0\,1 \end{pmatrix}^T$$

Then $t_{000} = 1/\sqrt{2}$, $t_{111} = 1/\sqrt{2}$ with all other expansion coefficients equal to 0.

Exercises. (1) Consider the vectors in Hilbert spaces $\mathbb{R}^2$ and $\mathbb{R}^3$, respectively

$$\mathbf{v} = \begin{pmatrix} v_1 \\ v_2 \end{pmatrix}, \quad \mathbf{u} = \begin{pmatrix} u_1 \\ u_2 \\ u_3 \end{pmatrix}.$$

Find the conditions such that $\mathbf{u} \otimes \mathbf{v} = \mathbf{v} \otimes \mathbf{u}$. Find solutions to these conditions.

(2) Can the 4×4 matrix

$$C = \begin{pmatrix} 1 & 0 & 0 & 1 \\ 0 & 1 & 1 & 0 \\ 0 & 1 & -1 & 0 \\ 1 & 0 & 0 & -1 \end{pmatrix}$$

be written as the Kronecker product of two 2×2 matrices A and B, i.e. $C = A \otimes B$?

(3) Show that the normalized vector

$$\begin{pmatrix} 1 & 0 & 0 & 0 \\ 0 & 1 & 0 & 0 \\ 0 & 0 & 0 & 1 \\ 0 & 0 & 1 & 0 \end{pmatrix} \begin{pmatrix} 1/\sqrt{2} \\ 0 \\ 0 \\ 1/\sqrt{2} \end{pmatrix}$$

is not entangled, i.e. it can be written as the Kronecker product of two normalized vectors in $\mathbb{C}^2$. The 4×4 matrix is the CNOT gate.

(4) Show that

$$(U_H \otimes I_2)U_{CNOT} \begin{pmatrix} a_1 \\ a_2 \\ a_3 \\ a_4 \end{pmatrix} = \frac{1}{\sqrt{2}} \begin{pmatrix} a_1 + a_4 \\ a_2 + a_3 \\ a_1 - a_4 \\ a_2 - a_3 \end{pmatrix}$$

where

$$U_H = \frac{1}{\sqrt{2}} \begin{pmatrix} 1 & 1 \\ 1 & -1 \end{pmatrix}, \quad U_{CNOT} = \begin{pmatrix} 1 & 0 & 0 & 0 \\ 0 & 1 & 0 & 0 \\ 0 & 0 & 0 & 1 \\ 0 & 0 & 1 & 0 \end{pmatrix}.$$

(5) Let e_1, e_2 be the standard basis in $\mathbb{C}^2$. Can the vector in $\mathbb{C}^8$

$$e_1 \otimes e_2 \otimes e_1 + e_2 \otimes e_1 \otimes e_2$$

be written as the Kronecker product of a vector in $\mathbb{C}^2$ and a vector in $\mathbb{C}^4$ or a vector in $\mathbb{C}^4$ and a vector in $\mathbb{C}^2$?

3.20 Hyperdeterminant

Cayley [8] in 1845 introduced the hyperdeterminant. Gelfand et al [9] give an in debt discussion of the hyperdeterminant. The hyperdeterminant arises as entanglement measure for three qubits [10, 11, 12], in black hole entropy [13]. The Nambu-Goto action in string theory can be expressed in terms of the hyperdeterminant [14]. A computer algebra program for the hyperdeterminant is given by Steeb and Hardy [11]

Let $\epsilon_{00} = \epsilon_{11} = 0$, $\epsilon_{01} = 1$, $\epsilon_{10} = -1$, i.e. we consider the 2×2 matrix

$$\epsilon = \begin{pmatrix} 0 & 1 \\ -1 & 0 \end{pmatrix}.$$

Then the determinant of a 2×2 matrix $A_2 = (a_{ij})$ with $i, j = 0, 1$ can be defined as

$$\det(A_2) := \frac{1}{2} \sum_{i=0}^{1} \sum_{j=0}^{1} \sum_{\ell=0}^{1} \sum_{m=0}^{1} \epsilon_{ij} \epsilon_{\ell m} a_{i\ell} a_{jm}.$$

Thus $\det(A_2) = a_{00}a_{11} - a_{01}a_{10}$. In analogy the hyperdeterminant of the $2 \times 2 \times 2$ array $A_3 = (a_{ijk})$ with $i, j, k = 0, 1$ is defined as

$$\mathrm{Det}(A_3) :=$$

$$-\frac{1}{2} \sum_{ii'=0}^{1} \sum_{jj'=0}^{1} \sum_{kk'=0}^{1} \sum_{mm'=0}^{1} \sum_{nn'=0}^{1} \sum_{pp'=0}^{1} \epsilon_{ii'} \epsilon_{jj'} \epsilon_{kk'} \epsilon_{mm'} \epsilon_{nn'} \epsilon_{pp'}$$

$$a_{ijk} a_{i'j'm} a_{npk'} a_{n'p'm'}.$$

There are $2^8 = 256$ terms, but only 24 are nonzero. We find

$$\begin{aligned}
\mathrm{Det}(A_3) = {}& a_{000}^2 a_{111}^2 + a_{001}^2 a_{110}^2 + a_{001}^2 a_{101}^2 + a_{100}^2 a_{011}^2 \\
& - 2(a_{000}a_{001}a_{110}a_{111} + a_{000}a_{010}a_{101}a_{111} \\
& + a_{000}a_{100}a_{011}a_{111} + a_{001}a_{010}a_{101}a_{110} \\
& + a_{001}a_{100}a_{011}a_{110} + a_{010}a_{100}a_{011}a_{101}) \\
& + 4(a_{000}a_{011}a_{101}a_{110} + a_{001}a_{010}a_{100}a_{111}).
\end{aligned}$$

The hyperdeterminant $\mathrm{Det}(A)$ of the three-dimensional array $A_3 = (a_{ijk}) \in \mathbb{R}^{2 \times 2 \times 2}$ can also be calculated as follows

$$\begin{aligned}
\mathrm{Det}(A_3) = {}& \frac{1}{4} \left(\det \left(\begin{pmatrix} a_{000} & a_{010} \\ a_{001} & a_{011} \end{pmatrix} + \begin{pmatrix} a_{100} & a_{110} \\ a_{101} & a_{111} \end{pmatrix} \right) \right. \\
& \left. - \det \left(\begin{pmatrix} a_{000} & a_{010} \\ a_{001} & a_{011} \end{pmatrix} - \begin{pmatrix} a_{100} & a_{110} \\ a_{101} & a_{111} \end{pmatrix} \right) \right)^2 \\
& - 4 \det \begin{pmatrix} a_{000} & a_{010} \\ a_{001} & a_{011} \end{pmatrix} \det \begin{pmatrix} a_{100} & a_{110} \\ a_{101} & a_{111} \end{pmatrix}.
\end{aligned}$$

If only one of the coefficients a_{ijk} is nonzero we find that the hyperdeterminant of A_3 is 0.

Given a $2 \times 2 \times 2$ hypermatrix $A_3 = (a_{jk\ell})$, $j, k, \ell = 0, 1$ and the 2×2 matrix

$$S = \begin{pmatrix} s_{00} & s_{01} \\ s_{10} & s_{11} \end{pmatrix}.$$

The multiplication $A_3 S$ which is again a 2×2 hypermatrix is defined by

$$(A_3 S)_{jk\ell} := \sum_{r=0}^{1} a_{jkr} s_{r\ell}.$$

If $\det(S) = 1$, i.e. $S \in SL(2, \mathbb{C})$, then $\mathrm{Det}(A_3 S) = \mathrm{Det}(A_3)$.

3.21 Tensor Eigenvalue Problem

Let $n \geq 1$ and $M \geq 1$. Consider the $T = (t_{j_1, \ldots, j_m})$ order-m tensor of size $(n \times \cdots \times n)$ (m times), $(j_1, \ldots, j_m = 1, \ldots, n)$. One defines the operator on $\mathbf{v} \in \mathbb{C}^n$ written as

$$(T\mathbf{v}^{m-1})_k := \sum_{j_2=1}^{n} \cdots \sum_{j_m=1}^{n} t_{k, j_2, \ldots, j_m} \mathbf{v}_{j_2} \cdots \mathbf{v}_{j_m}, \quad k = 1, \ldots, n.$$

The $(E-)$ eigenvector of T (Cartwright and Sturmfels [13]) are the fixed points (up to scaling) of this operator

$$T\mathbf{v}^{m-1} = \lambda \mathbf{v} \quad \text{where} \quad \mathbf{v} \neq \mathbf{0}.$$

Example. Let $m = 3$, $n = 2$ with $t_{1,2,2} = 1$, $t_{2,1,1} = 1$ and all other entries are 0. For this case we obtain

$$(T\mathbf{v}^2)_1 = t_{1,2,2} v_2^2 = v_2^2 = \lambda v_1$$
$$(T\mathbf{v}^2)_2 = t_{2,1,1} v_1^2 = v_1^2 = \lambda v_2.$$

Now $\lambda = 0$ is not a solution, since $v_1^2 = v_2^2 = 0$ implies $v_1 = v_2 = 0$. If $\lambda \neq 0$, then $v_1 \neq 0$, $v_2 \neq 0$. We also have $v_1^3 = v_2^3$.

3.22 Carleman Matrix and Bell Matrix

Consider the vector space of all analytic functions $f : \mathbb{R} \to \mathbb{R}$. Function composition is denoted by $\circ$. If f and g are analytic functions, then $f \circ g$ is an analytic function again.

Let $j, k = 0, 1, 2, \ldots$. The *Carleman matrix* $C[f]$ of an analytic function f is defined as

$$C[f]_{j,k} := \frac{1}{k!} \left(\frac{d^k}{dx^k} (f(x))^j \right)_{x=0}.$$

It follows that

$$(f(x))^j = \sum_{k=0}^{\infty} C[f]_{j,k} x^k.$$

The *Bell matrix* of an analytic function f is defined as

$$B[f]_{j,k} = \frac{1}{j!} \left(\frac{d^j}{dx^j} (f(x))^k \right)_{x=0}.$$

It follows that

$$(f(x))^k = \sum_{j=0}^{\infty} B[f]_{j,k} x^j.$$

We have the properties

$$C[f \circ g] = C[f]C[g], \quad B[f \circ g] = B[g]B[f]$$

for two analytic functions f and g and $C[f]C[g]$ denotes matrix multiplication. The functions $f(x) = x$ maps into the infinite dimensional identity matrix. Consider the analytic function

$$f(x) = \sum_{k=0}^{\infty} c_k x^k.$$

The Carleman matrix is

$$C[f] = \begin{pmatrix} 1 & 0 & 0 & 0 & \cdots \\ c_0 & c_1 & c_2 & c_3 & \cdots \\ c_0^2 & 2c_0 c_1 & c_1^2 & 2(c_0 c_3 + c_1 c_2) & \cdots \\ c_0^3 & 2c_0^2 c_1 & 6(c_0^2 c_2 + c_0 c_1^2) & & \cdots \\ \vdots & \vdots & \vdots & & \ddots \end{pmatrix}$$

Let `SetAnalytic` be the category of sets with analytic functions as *morphism* and `VecInfinite` be the category of vector spaces with infinite dimensional matrices as morphism between them. Then the Carleman matrix C is a (covariant) functor from `SetAnalytic` to `VecInfinite` and the Bell matrix B is a covariant functor from `SetAnalytic` to `VecInfinite`.

Exercise. (i) Find the Carleman matrix and Bell matrix for the analytic function $f(x) = \exp(-x)$.

(ii) Find the Carleman matrix and Bell matrix for the analytic function $\exp(-x^2/2)$.

Chapter 4

Tensor Product

4.1 Hilbert Spaces

In this section we introduce the concept of a Hilbert space (Young [74], Steeb [55], Halmos [27], Berberian [9], Weidmann [73]). Hilbert spaces play the central rôle in quantum mechanics. The proofs of the theorems given in this chapter can be found in Prugovečki [48]. We assume that the reader is familiar with the notation of a linear space. First we introduce the pre-Hilbert space.

Definition 4.1. A linear space L is called a *pre-Hilbert space* if there is defined a numerical function called the *scalar product* (or *inner product*) which assigns to every f, g of "vectors" of L ($f, g \in L$) a complex number. The scalar product $\langle , \rangle$ satisfies the conditions

(a) $\qquad \langle f, f \rangle \geq 0; \qquad \langle f, f \rangle = 0 \quad \text{iff} \quad f = 0$

(b) $\qquad \langle f, g \rangle = \overline{\langle g, f \rangle}$

(c) $\quad \langle cf, g \rangle = c\langle f, g \rangle$ where c is an arbitrary complex number

(d) $\qquad \langle f_1 + f_2, g \rangle = \langle f_1, g \rangle + \langle f_2, g \rangle$

where $\overline{\langle g, f \rangle}$ denotes the complex conjugate of $\langle g, f \rangle$.

It follows that $\langle f, g_1 + g_2 \rangle = \langle f, g_1 \rangle + \langle f, g_2 \rangle$ and $\langle f, cg \rangle = \bar{c}\langle f, g \rangle$.

Definition 4.2. A linear space E is called *normed space*, if for every $f \in E$ there is associated a real number $\|f\|$, the norm of the vector f such that

(a) $\qquad \|f\| \geq 0, \qquad \|f\| = 0 \quad \text{iff} \quad f = 0$

(b) $\quad \|cf\| = |c|\|f\|$ where c is an arbitrary complex number

(c) $\qquad \|f + g\| \leq \|f\| + \|g\|.$

The topology of a normed linear space E is thus defined by the distance
$$d(f,g) := \|f - g\|.$$
If a scalar product is given we can introduce a norm. The norm of f is defined by
$$\|f\| := \sqrt{\langle f, f \rangle}.$$
A vector $f \in L$ is called normalized if $\|f\| = 1$.

Let $f, g \in L$. The following identity holds (*parallelogram identity*)
$$\|f + g\|^2 + \|f - g\|^2 \equiv 2(\|f\|^2 + \|g\|^2).$$

Definition 4.3. Two functions $f \in L$ and $g \in L$ are called *orthogonal* if
$$\langle f, g \rangle = 0.$$

Definition 4.4. A sequence $\{f_n\}$ ($n \in \mathbb{N}$) of elements in a normed space E is called a *Cauchy sequence* if, for every $\epsilon > 0$, there exists a number M_ϵ such that $\|f_p - f_q\| < \epsilon$ for $p, q > M_\epsilon$.

Definition 4.5. A normed space E is said to be *complete* if every Cauchy sequence of elements in E converges to an element in E.

Example 4.1. Let $\mathbb{Q}$ be the rational numbers. Since the sum and product of two rational numbers are again rational numbers we obviously have a pre-Hilbert space with the scalar product $\langle q_1, q_2 \rangle := q_1 q_2$. However, the pre-Hilbert space is not complete. Consider the sequence
$$f_n = 1 + \frac{1}{1!} + \frac{1}{2!} + \cdots + \frac{1}{(n-1)!}$$
with $n = 1, 2, \ldots$. The sequence f_n is a Cauchy sequence. However
$$\lim_{n \to \infty} f_n \to e$$
with $e \notin \mathbb{Q}$.   ♣

Definition 4.6. A complete pre-Hilbert space is called a *Hilbert space*.

Example 4.2. Let $\mathcal{H}$ be a Hilbert space, $u, v \in \mathcal{H}$ and $\alpha \in \mathbb{C}$. Then
$$\langle u + \alpha v, u + \alpha v \rangle = \langle u, u \rangle + \langle u, \alpha v \rangle \langle \alpha v, u \rangle + \langle \alpha v, \alpha v \rangle$$
$$= \langle u, u \rangle + \alpha \langle u, v \rangle + \overline{\alpha} \langle v, u \rangle + \alpha \overline{\alpha} \langle v, v \rangle$$
$$= \langle u, u \rangle + \alpha \langle u, v \rangle + \overline{\alpha} \overline{\langle u, v \rangle} + \alpha \overline{\alpha} \langle v, v \rangle$$
$$\geq 0.$$
                                                                    ♣

Definition 4.7. A complete normed space is called a *Banach space*.

A Hilbert space will be denoted by $\mathcal{H}$ and a Banach space will be denoted by $\mathcal{B}$.

Theorem 4.1. *Every pre-Hilbert space L admits a completion $\mathcal{H}$ which is a Hilbert space.*

Example 4.3. Let $L = \mathbb{Q}$. Then $\mathcal{H} = \mathbb{R}$. ♣

Before we discuss some examples of Hilbert spaces we give the definitions of strong and weak convergence in Hilbert spaces.

Definition 4.8. A sequence $\{f_n\}$ of vectors in a Hilbert space $\mathcal{H}$ is said to *converge strongly* to f if $\|f_n - f\| \to 0$ as $n \to \infty$. We write $s-\lim_{n\to\infty} f_n \to f$.

Definition 4.9. A sequence $\{f_n\}$ of vectors in a Hilbert space $\mathcal{H}$ is said to *converge weakly* to f if $\langle f_n, g \rangle \to \langle f, g \rangle$ as $n \to \infty$, for any vector g in $\mathcal{H}$. We write $w - \lim_{n\to\infty} f_n \to f$.

It can be shown that strong convergence implies weak convergence. The converse is not generally true, however.

Let us now give several examples of Hilbert spaces which are important in quantum mechanics.

Example 4.4. Every finite dimensional vector space with an inner product is a Hilbert space. Let $\mathbb{C}^n$ be the linear space of n-tuples of complex numbers with the scalar product

$$\langle \mathbf{u}, \mathbf{v} \rangle := \sum_{j=1}^{n} u_j \bar{v}_j.$$

Then $\mathbb{C}^n$ is a Hilbert space. Let $\mathbf{u} \in \mathbb{C}^n$. We write the vector $\mathbf{u}$ as a column vector

$$\mathbf{u} = \begin{pmatrix} u_1 & u_2 & \dots & u_n \end{pmatrix}^T$$

where ^{T} denotes transpose. Thus we can write the scalar product in matrix notation $\langle \mathbf{u}, \mathbf{v} \rangle = \mathbf{u}^T \bar{\mathbf{v}}$, where $\mathbf{u}^T$ is the transpose of $\mathbf{u}$. ♣

Example 4.5. By $\ell_2(\mathbb{N})$ we mean the set of all infinite dimensional vectors (sequences) $\mathbf{u} = (u_1, u_2, \dots)^T$ of complex numbers u_j such that

$$\sum_{j=1}^{\infty} |u_j|^2 < \infty.$$

Here $\ell_2(\mathbb{N})$ is a linear space with operations $(a \in \mathbb{C})$

$$a\mathbf{u} = (au_1, au_2, \ldots)^T, \qquad \mathbf{u} + \mathbf{v} = (u_1 + v_1, u_2 + v_2, \ldots)^T$$

with $\mathbf{v} = (v_1, v_2, \ldots)^T$ and $\sum_{j=1}^{\infty} |v_j|^2 < \infty$. One has

$$\sum_{j=1}^{\infty} |u_j + v_j|^2 \leq \sum_{j=1}^{\infty}(|u_j|^2 + |v_j|^2 + 2|u_j v_j|) \leq 2\sum_{j=1}^{\infty}(|u_j|^2 + |v_j|^2) < \infty.$$

The scalar product is defined as

$$\langle \mathbf{u}, \mathbf{v} \rangle := \sum_{j=1}^{\infty} u_j \bar{v}_j = \mathbf{u}^T \bar{\mathbf{v}}.$$

It can also be proved that this pre-Hilbert space is complete. Therefore $\ell_2(\mathbb{N})$ is a Hilbert space. As an example, consider

$$\mathbf{u} = (1, 1/2, 1/3, \ldots, 1/j, \ldots)^T.$$

Since $\sum_{j=1}^{\infty} 1/j^2 < \infty$ we find that $\mathbf{u} \in \ell_2(\mathbb{N})$. Let

$$\mathbf{u} = \left(1, 1/\sqrt{2}, 1/\sqrt{3}, \ldots, 1/\sqrt{j}, \ldots\right)^T.$$

Then $\mathbf{u} \notin \ell_2(\mathbb{N})$.   ♣

Example 4.6. $L_2(M)$ is the space of Lebesgue square-integrable functions on M, where M is a Lebesgue measurable subset of $\mathbb{R}^n$, where $n \in \mathbb{N}$. If $f \in L_2(M)$, then

$$\int_M |f|^2 \, dm < \infty.$$

The integration is performed in the Lebesgue sense. The scalar product in $L_2(M)$ is defined as

$$\langle f, g \rangle := \int_M f(x) \bar{g}(x) \, dm$$

where $\bar{g}$ denotes the complex conjugate of g. It can be shown that this pre-Hilbert space is complete. Therefore $L_2(M)$ is a Hilbert space. Instead of dm we also write dx in the following. If the Riemann integral exists then it is equal to the Lebesgue integral. However, the Lebesgue integral exists also in cases in which the Riemann integral does not exist.   ♣

Example 4.7. Consider the linear space $M^n(\mathbb{C})$ of all $n \times n$ matrices over $\mathbb{C}$. The *trace* of an $n \times n$ matrix $A = (a_{jk})$ is given by

$$\text{tr}(A) := \sum_{j=1}^{n} a_{jj}.$$

We define a *scalar product* by

$$\langle A, B \rangle := \text{tr}(AB^*)$$

where B^* denotes the conjugate transpose matrix of B. We recall that

$$\text{tr}(C + D) = \text{tr}(C) + \text{tr}(D)$$

where C and D are $n \times n$ matrices.   ♣

Consider the linear space of all infinite dimensional matrices $A = (a_{jk})$ over $\mathbb{C}$ such that

$$\sum_{j=1}^{\infty} \sum_{k=1}^{\infty} |a_{jk}|^2 < \infty.$$

We define a scalar product by $\langle A, B \rangle := \text{tr}(AB^*)$, where tr denotes the trace and B^* denotes the conjugate transpose matrix of B. Then

$$\text{tr}(C + D) = \text{tr}(C) + \text{tr}(D)$$

where C and D are infinite dimensional matrices. The infinite dimensional unit matrix does not belong to this Hilbert space.

Example 4.8. Let D be an open set of the Euclidean space $\mathbb{R}^n$. Now $L_2(D)^{pq}$ denotes the space of all $q \times p$ matrix functions Lebesgue measurable on D such that

$$\int_D \text{tr}(f(x)f(x)^*)dm < \infty$$

where m denotes the Lebesgue measure, $*$ denotes the conjugate transpose, and tr is the trace of the $q \times q$ matrix. We define the scalar product as

$$\langle f, g \rangle := \int_D \text{tr}(f(x)g(x)^*)dm.$$

Then $L_2(D)^{pq}$ is a Hilbert space.   ♣

Theorem 4.2. *All complex infinite dimensional Hilbert spaces are isomorphic to $\ell_2(\mathbb{N})$ and consequently are mutually isomorphic.*

Definition 4.10. Let S be a subset of the Hilbert space $\mathcal{H}$. The subset S is *dense* in $\mathcal{H}$ if for every $f \in \mathcal{H}$ there exists a Cauchy sequence $\{f_j\}$ in S such that $f_j \to f$ as $j \to \infty$.

Definition 4.11. A Hilbert space is *separable* if it contains a countable dense subset $\{f_1, f_2, \ldots\}$.

Example 4.9. The set of all $\mathbf{u} = (u_1, u_2, \ldots)^T$ in $\ell_2(\mathbb{N})$ with only finitely many nonzero components u_j is dense in $\ell_2(\mathbb{N})$. ♣

Example 4.10. Let $C^1_{(2)}(\mathbb{R})$ be the linear space of the once continuously differentiable functions that vanish at infinity together with their first derivative and are square integrable. Then $C^1_{(2)}(\mathbb{R})$ is dense in $L_2(\mathbb{R})$. ♣

In almost all applications in quantum mechanics the underlying Hilbert space is separable.

Definition 4.12. A *subspace* $\mathcal{K}$ of a Hilbert space $\mathcal{H}$ is a subset of vectors which themselves forms a Hilbert space.

It follows from this definition that, if $\mathcal{K}$ is a subspace of $\mathcal{H}$, then so too is the set $\mathcal{K}^\perp$ of vectors orthogonal to all those in $\mathcal{K}$. The subspace $\mathcal{K}^\perp$ is termed the *orthogonal complement* of $\mathcal{K}$ in $\mathcal{H}$. Moreover, any vector f in $\mathcal{H}$ may be uniquely decomposed into components $f_{\mathcal{K}}$ and $f_{\mathcal{K}^\perp}$, lying in $\mathcal{K}$ and $\mathcal{K}^\perp$, respectively, i.e. $f = f_{\mathcal{K}} + f_{\mathcal{K}^\perp}$.

Example 4.11. Consider the Hilbert space $\mathcal{H} = \ell_2(\mathbb{N})$. Then the vectors

$$\mathbf{u}^T = (u_1, u_2, \ldots, u_N, 0, \ldots)$$

with $u_n = 0$ for $n > N$, form a subspace $\mathcal{K}$. The orthogonal complement $\mathcal{K}^\perp$ of $\mathcal{K}$ then consists of the vectors $(0, \ldots, 0, u_{N+1}, u_{N+2}, \ldots)$ with $u_n = 0$ for $n \leq N$. ♣

Definition 4.13. A sequence $\{\phi_j\}$, $j \in I$ and $\phi_j \in \mathcal{H}$ is called an *orthonormal sequence* if

$$\langle \phi_j, \phi_k \rangle = \delta_{jk}$$

where I is a countable index set and δ_{jk} denotes the *Kronecker delta*, i.e.

$$\delta_{jk} := \begin{cases} 1 \text{ for } j = k \\ 0 \text{ for } j \neq k. \end{cases}$$

Definition 4.14. An orthonormal sequence $\{\phi_j \;:\; j \in \mathbb{I}\}$ in $\mathcal{H}$ is an *orthonormal basis* if every $f \in \mathcal{H}$ can be expressed as

$$f = \sum_{j \in \mathbb{I}} a_j \phi_j \qquad \mathbb{I}: \text{ Index set}$$

for some constants $a_j \in \mathbb{C}$. The expansion coefficients a_j are given by

$$a_j := \langle f, \phi_j \rangle.$$

Example 4.12. Consider the Hilbert space $\mathcal{H} = \mathbb{C}^2$. The scalar product is defined as

$$\langle \mathbf{u}, \mathbf{v} \rangle := \sum_{j=1}^{2} u_j \bar{v}_j.$$

An orthonormal basis in $\mathcal{H}$ is given by

$$\mathbf{e}_1 = \frac{1}{\sqrt{2}} \begin{pmatrix} 1 \\ i \end{pmatrix}, \qquad \mathbf{e}_2 = \frac{1}{\sqrt{2}} \begin{pmatrix} 1 \\ -i \end{pmatrix}.$$

Let $\mathbf{u} = \begin{pmatrix} 1 & 2 \end{pmatrix}^T$. Then the expansion coefficients are given by

$$a_1 = \langle \mathbf{u}, \mathbf{e}_1 \rangle = \frac{1}{\sqrt{2}}(1 - 2i), \qquad a_2 = \langle \mathbf{u}, \mathbf{e}_2 \rangle = \frac{1}{\sqrt{2}}(1 + 2i).$$

Consequently

$$\begin{pmatrix} 1 \\ 2 \end{pmatrix} = \frac{1}{\sqrt{2}}(1 - 2i)\mathbf{e}_1 + \frac{1}{\sqrt{2}}(1 + 2i)\mathbf{e}_2.$$

♣

The *Bell basis*

$$\frac{1}{\sqrt{2}} \begin{pmatrix} 1 \\ 0 \\ 0 \\ 1 \end{pmatrix}, \quad \frac{1}{\sqrt{2}} \begin{pmatrix} 1 \\ 0 \\ 0 \\ -1 \end{pmatrix}, \quad \frac{1}{\sqrt{2}} \begin{pmatrix} 0 \\ 1 \\ 1 \\ 0 \end{pmatrix}, \quad \frac{1}{\sqrt{2}} \begin{pmatrix} 0 \\ 1 \\ -1 \\ 0 \end{pmatrix}$$

forms an orthonormal basis in the Hilbert space $\mathbb{C}^4$. This basis can be built from the standard basis in $\mathbb{C}^2$ and the Kronecker product $\otimes$ as follows

$$\frac{1}{\sqrt{2}} \left(\begin{pmatrix} 1 \\ 0 \end{pmatrix} \otimes \begin{pmatrix} 1 \\ 0 \end{pmatrix} + \begin{pmatrix} 0 \\ 1 \end{pmatrix} \otimes \begin{pmatrix} 0 \\ 1 \end{pmatrix} \right), \quad \frac{1}{\sqrt{2}} \left(\begin{pmatrix} 1 \\ 0 \end{pmatrix} \otimes \begin{pmatrix} 1 \\ 0 \end{pmatrix} - \begin{pmatrix} 0 \\ 1 \end{pmatrix} \otimes \begin{pmatrix} 0 \\ 1 \end{pmatrix} \right),$$

$$\frac{1}{\sqrt{2}} \left(\begin{pmatrix} 1 \\ 0 \end{pmatrix} \otimes \begin{pmatrix} 0 \\ 1 \end{pmatrix} + \begin{pmatrix} 0 \\ 1 \end{pmatrix} \otimes \begin{pmatrix} 1 \\ 0 \end{pmatrix} \right), \quad \frac{1}{\sqrt{2}} \left(\begin{pmatrix} 1 \\ 0 \end{pmatrix} \otimes \begin{pmatrix} 0 \\ 1 \end{pmatrix} - \begin{pmatrix} 0 \\ 1 \end{pmatrix} \otimes \begin{pmatrix} 1 \\ 0 \end{pmatrix} \right).$$

Example 4.13. Let $\mathcal{H} = L_2(-\pi, \pi)$. Then an orthonormal basis is given by

$$\left\{ \phi_k(x) = \frac{1}{\sqrt{2\pi}} \exp(ikx) \ : \ k \in \mathbb{Z} \right\}.$$

Let $f \in L_2(-\pi, \pi)$ with $f(x) = x$. Then

$$a_k = \langle f, \phi_k \rangle = \int_{-\pi}^{\pi} f(x)\bar{\phi}_k(x)dx = \frac{1}{\sqrt{2\pi}} \int_{-\pi}^{\pi} x \exp(-ikx)dx.$$

♣

We call the expansion

$$f = \sum_{k \in \mathbb{Z}} \langle f, \phi_k \rangle \phi_k$$

the *Fourier expansion* of f.

Example 4.14. The *Fock space* $\mathcal{F}$ is the Hilbert space of entire functions with inner product given by

$$\langle f | g \rangle := \frac{1}{\pi} \int_{\mathbb{C}} f(z)\overline{g(z)}e^{-|z|^2} dxdy, \qquad z = x + iy$$

where $\mathbb{C}$ denotes the complex numbers. Therefore the growth of functions in the Hilbert space $\mathcal{F}$ is dominated by $\exp(|z|^2/2)$. ♣

Theorem 4.3. *Every separable Hilbert space has at least one orthonormal basis.*

Inequality of Schwarz. Let $f, g \in \mathcal{H}$. Then $|\langle f, g \rangle| \leq \|f\| \cdot \|g\|$.

Triangle inequality. Let $f, g \in \mathcal{H}$. Then $\|f + g\| \leq \|f\| + \|g\|$.

We also have the inequality

$$\big| \|f\| - \|g\| \big| \leq \|f + g\|$$

and the identity

$$\|f + g\|^2 + \|f - g\|^2 = 2(\|f\|^2 + \|g\|^2).$$

Let $B = \{ \phi_n \; : \; n \in \mathbb{I} \}$ be an orthonormal basis in a Hilbert space $\mathcal{H}$. $\mathbb{I}$ is the countable index set. Then

$$(1) \qquad \langle \phi_n, \phi_m \rangle = \delta_{nm}$$

$$(2) \qquad \bigwedge_{f \in \mathcal{H}} \quad f = \sum_{n \in \mathbb{I}} \langle f, \phi_n \rangle \phi_n$$

$$(3) \qquad \bigwedge_{f,g \in \mathcal{H}} \quad \langle f, g \rangle = \sum_{n \in \mathbb{I}} \overline{\langle f, \phi_n \rangle} \langle g, \phi_n \rangle$$

$$(4) \qquad \left(\bigwedge_{\phi_n \in B} \langle f, \phi_n \rangle = 0 \right) \Rightarrow f = 0$$

$$(5) \qquad \bigwedge_{f \in \mathcal{H}} \| f \|^2 = \sum_{n \in \mathbb{I}} |\langle f, \phi_n \rangle|^2$$

Equation (3) is called *Parseval's relation*.

Exercises. (1) Let $|0\rangle$, $|1\rangle$, $|2\rangle$, $|3\rangle$ be an orthonormal basis in the Hilbert space $\mathbb{C}^4$. Is

$$V = \sum_{j,k=0}^{3} (-1)^{jk} |j\rangle \langle k|$$

a unitary matrix? Is the matrix hermitian?

(2) Let $|0\rangle$, $|1\rangle$, $|2\rangle$ be an orthonormal basis in the Hilbert space $\mathbb{C}^3$. Show that the linear operator $(3 \times 3$ matrix)

$$T = |0\rangle\langle 1| + |1\rangle\langle 2| + |2\rangle\langle 0|$$

is invertible and the inverse is given by

$$T^{-1} = |1\rangle\langle 0| + |2\rangle\langle 1| + |0\rangle\langle 2|.$$

(3) Let $d \geq 2$. Consider the orthonormal basis $|0\rangle$, $|1\rangle$, $\ldots$, $|d-1\rangle$ in the Hilbert space $\mathbb{C}^d$. Is the $d \times d$ matrix

$$T = \sum_{j=0}^{d-1} (-1)^j |j\rangle \langle j|$$

invertible?

(4) The vectors

$$\mathbf{v}_1 = \frac{1}{\sqrt{2}} \begin{pmatrix} 1 \\ 0 \\ 1 \end{pmatrix}, \quad \mathbf{v}_2 = \begin{pmatrix} 0 \\ 1 \\ 0 \end{pmatrix}, \quad \mathbf{v}_3 = \frac{1}{\sqrt{2}} \begin{pmatrix} 1 \\ 0 \\ -1 \end{pmatrix}$$

form an orthonormal basis in the Hilbert space $\mathbb{C}^3$. Find the unitary matrices U_{12}, U_{23}, U_{31} such that

$$U_{12}\mathbf{v}_1 = \mathbf{v}_2, \quad U_{23}\mathbf{v}_2 = \mathbf{v}_3, \quad U_{31}\mathbf{v}_3 = \mathbf{v}_1.$$

Show that

$$U_{31}U_{23}U_{12} = I_3.$$

Calculate (spectral decomposition)

$$V = \lambda_1\mathbf{v}_1\mathbf{v}_1^* + \lambda_2\mathbf{v}_2\mathbf{v}_2^* + \lambda_3\mathbf{v}_3\mathbf{v}_3^*$$

where the complex numbers λ_1, λ_2, λ_3 satisfy $\lambda_1\overline{\lambda}_1 = 1$, $\lambda_2\overline{\lambda}_2 = 1$, $\lambda_3\overline{\lambda}_3 = 1$.

4.2   Hilbert Tensor Products of Hilbert Spaces

In this section we introduce the concept of the tensor product of Hilbert spaces (Prugovečki [48]). Then we provide some applications.

Definition 4.15. Let $\mathcal{H}_1, \ldots, \mathcal{H}_n$ be Hilbert spaces with inner products $\langle \cdot, \cdot \rangle_1, \ldots, \langle \cdot, \cdot \rangle_n$, respectively. Let $\mathcal{H}_1 \otimes_a \cdots \otimes_a \mathcal{H}_n$ be the algebraic tensor product of $\mathcal{H}_1, \ldots, \mathcal{H}_n$. We denote the inner product in the algebraic tensor product space by $(\cdot, \cdot)$ and define

$$\langle f, g \rangle := \prod_{k=1}^{n} \langle f^{(k)}, g^{(k)} \rangle_k$$

for f and g of the form

$$f = f^{(1)} \otimes \cdots \otimes f^{(n)}, \qquad g = g^{(1)} \otimes \cdots \otimes g^{(n)}.$$

The equation defines a unique inner product. The *Hilbert tensor product* $\mathcal{H}_1 \otimes \cdots \otimes \mathcal{H}_n$ of the Hilbert spaces $\mathcal{H}_1, \ldots, \mathcal{H}_n$ is the Hilbert space which is the completion of the pre-Hilbert space $\mathcal{E}$ with the inner product.

Consider now two Hilbert spaces $L_2(\Omega_1, \mu_1)$ and $L_2(\Omega_2, \mu_2)$. Denote by $\mu_1 \times \mu_2$ the product of the measures μ_1 and μ_2 on the Borel subsets of $\Omega_1 \times \Omega_2$. If $f \in L_2(\Omega_1, \mu_1)$ and $g \in L_2(\Omega_2, \mu_2)$, then $f(x_1)g(x_2)$ represents an element of $L_2(\Omega_1 \times \Omega_2, \mu_1 \times \mu_2)$, which we denote by $f \cdot g$

$$(f \cdot g)(x_1, x_2) = f(x_1)g(x_2).$$

Theorem 4.4. *The linear mapping*

$$h \mapsto \hat{h}, \qquad h \in \mathcal{H}_1 \otimes_a \mathcal{H}_2, \qquad \hat{h} \in \mathcal{H}_3$$

$$h = \sum_{k=1}^{n} a_k f_k \otimes g_k, \qquad \hat{h} = \sum_{k=1}^{n} a_k f_k \cdot g_k$$

of the algebraic tensor product of $\mathcal{H}_1 = L_2(\Omega_1, \mu_1)$ *and* $\mathcal{H}_2 = L_2(\Omega_2, \mu_2)$ *into*

$$\mathcal{H}_3 = L_2(\Omega_1 \times \Omega_2, \mu_1 \times \mu_2)$$

can be extended uniquely to a unitary transformation of $\mathcal{H}_1 \otimes \mathcal{H}_2$ *onto* $\mathcal{H}_3$.

For the proof we refer to Prugovečki [48].

If $\mathcal{H}_k$ is separable, then there is a countable orthogonal basis

$$\{\, e_i^{(k)} : i \in \mathcal{U}_k \,\}$$

in $\mathcal{H}_k$.

Theorem 4.5. *The Hilbert tensor product* $\mathcal{H}_1 \otimes \cdots \otimes \mathcal{H}_n$ *of separable Hilbert spaces* $\mathcal{H}_1, \ldots, \mathcal{H}_n$ *is separable; if* $\{\, e_i^{(k)} : i \in \mathcal{U}_k \,\}$ *is an orthonormal basis in the* $\mathcal{H}_k$, *then*

$$\{\, e_{i_1}^{(1)} \otimes \cdots \otimes e_{i_n}^{(n)} : i_1 \in \mathcal{U}_1, \ldots, i_n \in \mathcal{U}_n \,\}$$

is an orthonormal basis in $\mathcal{H}_1 \otimes \cdots \otimes \mathcal{H}_n$. *Consequently, the set*

$$\mathbf{T} := \{\, e_{i_1}^{(1)} \otimes \cdots \otimes e_{i_n}^{(n)} : i_1 \in \mathcal{U}_1, \ldots, i_n \in \mathcal{U}_n \,\}$$

is also countable. In addition, $\mathbf{T}$ *is an orthonormal system*

$$\langle e_{i_1}^{(1)} \otimes \cdots \otimes e_{i_n}^{(n)}, e_{j_1}^{(1)} \otimes \cdots \otimes e_{j_n}^{(n)} \rangle = \langle e_{i_1}^{(n)}, e_{j_1}^{(1)} \rangle_1 \cdots \langle e_{i_n}^{(n)}, e_{j_n}^{(n)} \rangle_n = \delta_{i_1 j_1} \cdots \delta_{i_n j_n}.$$

Let the operator $\hat{A}_k$ be an observable of the system $\mathcal{G}_k$. In the case where $\mathcal{G}_k$ is an independent part of the system $\mathcal{G}$, we can measure $\hat{A}_k$. Hence, this question arises: which mathematical entity related to the product Hilbert space $\mathcal{H} = \mathcal{H}_1 \otimes \cdots \otimes \mathcal{H}_n$ represents this observable. We define the following concept.

Definition 4.16. Let $\hat{A}_1, \ldots, \hat{A}_n$ be n bounded linear operators acting on the Hilbert spaces $\mathcal{H}_1, \ldots, \mathcal{H}_n$, respectively. The tensor product

$$\hat{A}_1 \otimes \cdots \otimes \hat{A}_n$$

of these n operators is that bounded linear operator on $\mathcal{H}_1 \otimes \cdots \otimes \mathcal{H}_n$ which acts on a vector $f_1 \otimes \cdots \otimes f_n$, $f_1 \in \mathcal{H}_1, \ldots, f_n \in \mathcal{H}_n$ as follows

$$(\hat{A}_1 \otimes \cdots \otimes \hat{A}_n)(f_1 \otimes \cdots \otimes f_n) = (\hat{A}_1 f_1) \otimes \cdots \otimes (\hat{A}_n f_n).$$

The above relation determines the operator $\hat{A}_1 \otimes \cdots \otimes \hat{A}_n$ on the set of all vectors of the form $f_1 \otimes \cdots \otimes f_n$, and therefore, due to the presupposed linearity of $\hat{A}_1 \otimes \cdots \otimes \hat{A}_n$, on the linear manifold spanned by all such vectors. Since this linear manifold is dense in $\mathcal{H}_1 \otimes \cdots \otimes \mathcal{H}_n$ and the operator defined above in this manifold is bounded, it has a unique extension to $\mathcal{H}_1 \otimes \cdots \otimes \mathcal{H}_n$. Consequently, the above definition is consistent.

How do we represent in $\mathcal{H} = \mathcal{H}_1 \otimes \cdots \otimes \mathcal{H}_n$ an observable of $\mathcal{S}_k$ representable in $\mathcal{H}_k$ by the bounded operator $\hat{A}_k$? We have

$$\widetilde{A}_k := I \otimes \cdots \otimes I \otimes \hat{A}_k \otimes I \otimes \cdots \otimes I$$

where I is the identity operator. We can verify that the physically meaningful quantities, i.e. the expectation values are the same for $\hat{A}_k$ when $\mathcal{S}_k$ is an isolated system in the state Ψ_k at t, or when $\mathcal{S}_k$ is at the instant t an independent part of $\mathcal{G}$, which is at t in the state

$$\Psi_1 \otimes \cdots \otimes \Psi_k \otimes \cdots \otimes \Psi_n.$$

If we take the normalized state vectors $\Psi_1, \ldots, \Psi_n$, then

$$\langle \Psi_1 \otimes \cdots \otimes \Psi_n, (I \otimes \cdots \otimes I \otimes \hat{A}_k \otimes I \otimes \cdots \otimes I)\Psi_1 \otimes \cdots \otimes \Psi_n \rangle$$

$$= \langle \Psi_1, \Psi_1 \rangle_1 \cdots \langle \Psi_k, A_k \Psi_k \rangle_k \cdots \langle \Psi_n, \Psi_n \rangle_n = \langle \Psi_k, \hat{A}_k \Psi_k \rangle_k.$$

Example 4.15. Consider the case of n different particles without spin. In that case $\mathcal{H}$ is $L_2(\mathbb{R}^{3n})$, and we can take $\mathcal{H}_1 = \cdots = \mathcal{H}_n = L_2(\mathbb{R}^3)$. The Hilbert space $L_2(\mathbb{R}^{3n})$ is isomorphic to $L_2(\mathbb{R}^3) \otimes \cdots \otimes L_2(\mathbb{R}^3)$ (n factors), and the mapping

$$\psi_1(\mathbf{r}_1) \otimes \cdots \otimes \psi_n(\mathbf{r}_n) \to \psi_1(\mathbf{r}_1) \cdots \psi_n(\mathbf{r}_n)$$

induces a unitary transformation U between these two Hilbert spaces. The Hamilton operator $\hat{H}_k$ is in this case given by the differential operator

$$-(\hbar^2/(2m_k))\Delta_k$$

and therefore $\hat{H}_1 + \cdots + \hat{H}_n$ is given by

$$-((\hbar^2/(2m_1))\Delta_1 + \cdots + (\hbar^2/(2m_n))\Delta_n)$$

and represents in this case the kinetic energy of the system $\mathcal{S}$. The interaction term V is determined by the potential $V(\mathbf{r}_1, \ldots, \mathbf{r}_n)$,

$$(V\psi)(\mathbf{r}_1, \ldots, \mathbf{r}_n) = V(\mathbf{r}_1, \ldots, \mathbf{r}_n)\psi(\mathbf{r}_1, \ldots, \mathbf{r}_n).$$

♣

Definition 4.17. The closed linear subspace of the Hilbert space

$$\mathcal{H} = \mathcal{H}_1 \otimes \cdots \otimes \mathcal{H}_n, \qquad \mathcal{H}_1 \equiv \cdots \equiv \mathcal{H}_n$$

spanned by all the vectors of the form

$$f_1 \otimes^S \cdots \otimes^S f_n := \frac{1}{\sqrt{n!}} \sum_{(k_1,\ldots,k_n)} f_{k_1} \otimes \cdots \otimes f_{k_n}, \qquad f_1,\ldots,f_n \in \mathcal{H}_1$$

where the sum is taken over all the permutations $(k_1,\ldots,k_n)$ of $(1,\ldots,n)$, is called the *symmetric tensor product* of $\mathcal{H}_1,\ldots,\mathcal{H}_n$, and it is denoted by

$$\mathcal{H}_1 \otimes^S \cdots \otimes^S \mathcal{H}_n \quad \text{or} \quad \mathcal{H}_1^{\otimes^S n}.$$

Similarly, in the case of the set of all vectors in $\mathcal{H}$ of the form

$$f_1 \otimes^A \cdots \otimes^A f_n := \frac{1}{\sqrt{n!}} \sum_{(k_1,\ldots,k_n)} \pi(k_1,\ldots,k_n) f_{k_1} \otimes \cdots \otimes f_{k_n}, \quad f_1,\ldots,f_n \in \mathcal{H}_1$$

where

$$\pi(k_1,\ldots,k_n) := \begin{cases} +1 \text{ if } (k_1,\ldots,k_n) \text{ is even} \\ -1 \text{ if } (k_1,\ldots,k_n) \text{ is odd} \end{cases}$$

the closed linear subspace spanned by this set is called the *antisymmetric tensor product* of $\mathcal{H}_1,\ldots,\mathcal{H}_n$, and it is denoted by

$$\mathcal{H}_1 \otimes^A \cdots \otimes^A \mathcal{H}_n \quad \text{or} \quad \mathcal{H}_1^{\otimes^A n}.$$

The factor $(n!)^{-1/2}$ has been introduced for the sake of convenience in dealing with orthonormal bases. In case that $f_1,\ldots,f_n$ are orthogonal to each other and normalized in $\mathcal{H}_1$, then the symmetric and antisymmetric tensor products of these vectors will be also normalized.

Thus if we take the inner product

$$\langle f_1 \otimes^S \cdots \otimes^S f_n, f_1 \otimes^A \cdots \otimes^A f_n \rangle$$

we obtain zero. This implies that $\mathcal{H}_1^{\otimes^S n}$ is orthogonal to $\mathcal{H}_1^{\otimes^A n}$, when these spaces are treated as subspaces of $\mathcal{H}_1^{\otimes n}$.

If $\hat{A}_1,\ldots,\hat{A}_n$ are bounded linear operators on $\mathcal{H}_1$, then $\hat{A}_1 \otimes \cdots \otimes \hat{A}_n$ will not leave, in general, $\mathcal{H}_1^{\otimes^S n}$ or $\mathcal{H}_1^{\otimes^A n}$ invariant. We can define, however, the symmetric tensor product

$$\hat{A}_1 \otimes^S \cdots \otimes^S \hat{A}_n := \sum_{(k_1,\ldots,k_n)} \hat{A}_{k_1} \otimes \cdots \otimes \hat{A}_{k_n}$$

which leaves $\mathcal{H}_1^{\otimes^A n}$ and $\mathcal{H}_1^{\otimes^S n}$ invariant, and which can be considered therefore as defining linear operators on these respective spaces, by restricting the domain of definition of this operator to these spaces.

Example 4.16. Consider the Hilbert space $\mathcal{H} = \mathbb{R}^2$ and

$$\begin{pmatrix} a \\ b \end{pmatrix}, \ \begin{pmatrix} c \\ d \end{pmatrix} \in \mathbb{R}^2.$$

Let

$$\mathbf{u} = \begin{pmatrix} a \\ b \end{pmatrix} \otimes \begin{pmatrix} c \\ d \end{pmatrix} + \begin{pmatrix} c \\ d \end{pmatrix} \otimes \begin{pmatrix} a \\ b \end{pmatrix}, \quad \mathbf{v} = \begin{pmatrix} a \\ b \end{pmatrix} \otimes \begin{pmatrix} c \\ d \end{pmatrix} - \begin{pmatrix} c \\ d \end{pmatrix} \otimes \begin{pmatrix} a \\ b \end{pmatrix}.$$

Then $\mathbf{u}, \mathbf{v} \in \mathbb{R}^4$. We find the vectors

$$\mathbf{u} = \begin{pmatrix} 2ac \\ ad + bc \\ bc + ad \\ 2bd \end{pmatrix}, \qquad \mathbf{v} = \begin{pmatrix} 0 \\ ad - bc \\ bc - ad \\ 0 \end{pmatrix}$$

and $\langle \mathbf{u}, \mathbf{v} \rangle = 0$.   ♣

Exercise. Let $\mathbb{Z}$ be the set of integers and $n, m \in \mathbb{Z}$. We consider the Hilbert space $\ell_2(\mathbb{Z})$ and denotes the standard basis by $|n\rangle$ $(n \in \mathbb{Z})$ and the dual one by $\langle m|$ $(m \in \mathbb{Z})$ with $\langle m|n \rangle = \delta_{m,n}$. Let $\mathbb{C}^2$ be the two-dimensional Hilbert space with the standard basis

$$|+\rangle = \begin{pmatrix} 1 \\ 0 \end{pmatrix}, \quad |-\rangle = \begin{pmatrix} 0 \\ 1 \end{pmatrix}, \quad \langle +| = (1 \ 0), \quad \langle -| = (0 \ 1).$$

We consider the product Hilbert space $\ell_2(\mathbb{Z}) \otimes \mathbb{C}^2$ and define the linear operator $\hat{S}$ acting on $\ell_2(\mathbb{Z}) \otimes \mathbb{C}^2$ as

$$\hat{S}(|n\rangle \otimes |\pm\rangle) = |n \pm 1\rangle \otimes |\pm\rangle$$

and the unitary operator $U(\theta_n)$ $(\theta_n \in \mathbb{R}), n \in \mathbb{Z}$ defined by

$$\hat{U}(\theta_n) = \sum_{n \in \mathbb{Z}} ((|n\rangle\langle n|) \otimes ((\cos(\theta_n)|+\rangle\langle +| + (\sin(\theta_n)|+\rangle\langle -|)$$

$$+ (\sin(\theta_n)|-\rangle\langle +| - (\cos(\theta_n)|-\rangle\langle -|)).$$

The underlying matrix is

$$\begin{pmatrix} \cos(\theta_n) & \sin(\theta_n) \\ \sin(\theta_n) & -\cos(\theta_n) \end{pmatrix}$$

with determinant -1. Setting $\theta_n = \pi/4$ for all $n \in \mathbb{Z}$ we obtain the Hadamard matrix. We can consider now the linear operator $\hat{S}\hat{U}(\theta_n)$. Assume that θ_n does not depend on n and we set $\theta_n = \theta$. Consider the normalized state $|0\rangle \otimes |+\rangle$ in the product Hilbert space $\ell_2(\mathbb{Z}) \otimes \mathbb{C}^2$. Find the state $(\hat{S}\hat{U}(\theta))(|0\rangle \otimes |+\rangle)$ and

$$(\langle 0| \otimes \langle +|)(\hat{S}\hat{U}(\theta))(|0\rangle \otimes |+\rangle)), \quad (\langle 0| \otimes \langle -|)(\hat{S}\hat{U}(\theta))(|0\rangle \otimes |+\rangle)).$$

4.3 Spin and Statistics for the n-Body Problem

As an illustration of the above consideration on the connection between spin and statistics we formulate the wave mechanics for n identical particles of spin σ. A state vector will be represented in this case by a wavefunction

$$\psi(\mathbf{r}_1, s_1, \ldots, \mathbf{r}_n, s_n), \quad \mathbf{r}_k \in \mathbb{R}^3, \qquad s_k = -\sigma, -\sigma+1, \ldots, +\sigma$$

$(k = 1, \ldots, n)$ which is Lebesgue square integrable and symmetric

$$\psi(\ldots, \mathbf{r}_i, s_i, \ldots, \mathbf{r}_j, s_j, \ldots) = -\psi(\ldots, \mathbf{r}_j, s_j, \ldots, \mathbf{r}_i, s_i, \ldots)$$

if σ is half-integer. In most cases we have $\sigma = 1/2$. Thus the Hilbert spaces of functions in which the inner product is taken to be

$$(f, g) = \sum_{s_1=-\sigma}^{+\sigma} \cdots \sum_{s_n=-\sigma}^{+\sigma} \int_{\mathbb{R}^{3n}} f^*(\mathbf{r}_1, s_1, \ldots, \mathbf{r}_n, s_n) g(\mathbf{r}_1, s_1, \ldots, \mathbf{r}_n, s_n) d\mathbf{r}_1 \ldots d\mathbf{r}_n$$

are unitarily equivalent under the unitary transformation induced by the mapping

$$\psi(\mathbf{r}_1, s_1) \otimes \cdots \otimes \psi(\mathbf{r}_n, s_n) \to \psi(\mathbf{r}_1, s_1) \cdots \psi(\mathbf{r}_n, s_n)$$

to the spaces $\mathcal{H}_1^{\otimes^S n}$ and $\mathcal{H}_1^{\otimes^A n}$, respectively, where

$$\mathcal{H}_1 = L_2(\mathbb{R}^3) \oplus \cdots \oplus L_2(\mathbb{R}^3), \qquad (2\sigma + 1 \text{ terms}).$$

The Hamilton operator $\hat{H}$ of the system is taken to be of the form

$$\hat{H} = \hat{T} + \hat{V}$$

where $\hat{T}$ is the kinetic energy operator given by

$$(-\hbar^2/(2m))(\Delta_1 + \cdots + \Delta_n).$$

Thus, $\hat{T}$ is already symmetric with respect to the n particles. If $\hat{V}$ is the potential energy given by a potential $\hat{V}(\mathbf{r}_1, s_1, \ldots, \mathbf{r}_n, s_n)$, then the principle of indistinguishability of identical particles requires that this function is symmetric under any permutation of the indices $1, \ldots, n$.

Example 4.17. We calculate the eigenvalues of the Hamilton operator

$$\hat{H} := \lambda(S_1 \otimes \hat{L}_1 + S_2 \otimes \hat{L}_2 + S_3 \otimes \hat{L}_3)$$

where we consider a subspace G_1 of the Hilbert space $L_2(\mathbb{S}^2)$ with

$$\mathbb{S}^2 := \{ (x_1, x_2, x_3) : x_1^2 + x_2^2 + x_3^2 = 1 \}.$$

Here $\otimes$ denotes the *tensor product*. The linear operators $\hat{L}_1$, $\hat{L}_2$, $\hat{L}_3$ act in the subspace G_1. A basis of subspace G_1 is

$$Y_{1,0} = \sqrt{\frac{3}{4\pi}}\cos(\theta), \quad Y_{1,1} = -\sqrt{\frac{3}{8\pi}}\sin(\theta)e^{i\phi}, \quad Y_{1,-1} = \sqrt{\frac{3}{8\pi}}\sin(\theta)e^{-i\phi}.$$

The operators (matrices) S_1, S_2, S_3 act in the Hilbert space $\mathbb{C}^2$ with the standard basis

$$\begin{pmatrix} 1 \\ 0 \end{pmatrix}, \quad \begin{pmatrix} 0 \\ 1 \end{pmatrix}.$$

The spin matrices S_3, S_+ and S_- are given by

$$S_3 := \frac{1}{2}\hbar\begin{pmatrix} 1 & 0 \\ 0 & -1 \end{pmatrix}, \qquad S_+ := \hbar\begin{pmatrix} 0 & 1 \\ 0 & 0 \end{pmatrix}, \qquad S_- := \hbar\begin{pmatrix} 0 & 0 \\ 1 & 0 \end{pmatrix}$$

where $S_\pm := S_1 \pm iS_2$. The operators $\hat{L}_1$, $\hat{L}_2$ and $\hat{L}_3$ take the form

$$\hat{L}_+ := \hbar e^{i\phi}\left(\frac{\partial}{\partial\theta} + i\cot(\theta)\frac{\partial}{\partial\phi}\right), \quad \hat{L}_- := \hbar e^{-i\phi}\left(-\frac{\partial}{\partial\theta} + i\cot(\theta)\frac{\partial}{\partial\phi}\right),$$

$$\hat{L}_z := -i\hbar\frac{\partial}{\partial\phi}, \quad \hat{L}_\pm := \hat{L}_1 \pm i\hat{L}_2.$$

We give an interpretation of the Hamilton operator $\hat{H}$ and of the subspace G_1. The Hamilton operator can be written as

$$\hat{H} = \lambda(S_3 \otimes \hat{L}_3) + \frac{\lambda}{2}(S_+ \otimes \hat{L}_- + S_- \otimes \hat{L}_+).$$

In the tensor product space $\mathbb{C}^2 \otimes G_1$ a basis is given by

$$|1\rangle = \begin{pmatrix} 1 \\ 0 \end{pmatrix} \otimes Y_{1,0}, \qquad |2\rangle = \begin{pmatrix} 1 \\ 0 \end{pmatrix} \otimes Y_{1,-1}, \qquad |3\rangle = \begin{pmatrix} 1 \\ 0 \end{pmatrix} \otimes Y_{1,1}$$

$$|4\rangle = \begin{pmatrix} 0 \\ 1 \end{pmatrix} \otimes Y_{1,0}, \qquad |5\rangle = \begin{pmatrix} 0 \\ 1 \end{pmatrix} \otimes Y_{1,-1}, \qquad |6\rangle = \begin{pmatrix} 0 \\ 1 \end{pmatrix} \otimes Y_{1,1}.$$

In the following we use

$$\hat{L}_+Y_{1,1} = 0, \qquad \hat{L}_+Y_{1,0} = \hbar\sqrt{2}Y_{1,1}, \qquad \hat{L}_+Y_{1,-1} = \hbar\sqrt{2}Y_{1,0}$$

$$\hat{L}_-Y_{1,1} = \hbar\sqrt{2}Y_{1,0}, \qquad \hat{L}_-Y_{1,0} = \hbar\sqrt{2}Y_{1,-1}, \qquad \hat{L}_-Y_{1,-1} = 0.$$

For the state $|1\rangle$ we find

$$\hat{H}|1\rangle = [\lambda(S_3 \otimes \hat{L}_3) + \frac{\lambda}{2}(S_+ \otimes \hat{L}_- + S_- \otimes \hat{L}_+)]\begin{pmatrix} 1 \\ 0 \end{pmatrix} \otimes Y_{1,0}.$$

Thus

$$\hat{H}|1\rangle = \lambda[S_3 \begin{pmatrix} 1 \\ 0 \end{pmatrix} \otimes \hat{L}_3 Y_{1,0}] + \frac{\lambda}{2}[S_+ \begin{pmatrix} 1 \\ 0 \end{pmatrix} \otimes \hat{L}_- Y_{1,0} + S_- \begin{pmatrix} 1 \\ 0 \end{pmatrix} \otimes \hat{L}_+ Y_{1,0}].$$

Finally

$$\hat{H}|1\rangle = \frac{\lambda}{2} S_- \begin{pmatrix} 1 \\ 0 \end{pmatrix} \otimes \hat{L}_+ Y_{1,0} = \frac{\lambda}{\sqrt{2}} \hbar^2 |6\rangle.$$

Analogously, we find

$$\hat{H}|2\rangle = -\frac{\lambda\hbar^2}{2}|2\rangle + \frac{\lambda\hbar^2}{\sqrt{2}}|4\rangle,$$

$$\hat{H}|3\rangle = \frac{\lambda\hbar^2}{2}|3\rangle, \quad \hat{H}|4\rangle = \frac{\lambda\hbar^2}{2}|2\rangle, \quad \hat{H}|5\rangle = \frac{\lambda\hbar^2}{2}|5\rangle,$$

$$\hat{H}|6\rangle = -\frac{\lambda\hbar^2}{2}|6\rangle + \frac{\lambda\hbar^2}{\sqrt{2}}|1\rangle.$$

Hence the states $|3\rangle$ and $|5\rangle$ are eigenstates with the eigenvalues $E_{1,2} = \lambda\hbar^2/2$. The states $|1\rangle$ and $|6\rangle$ form a two dimensional subspace. The matrix representation is given by

$$\begin{pmatrix} 0 & \dfrac{\lambda\hbar^2}{\sqrt{2}} \\ \dfrac{\lambda\hbar^2}{\sqrt{2}} & -\dfrac{\lambda\hbar^2}{2} \end{pmatrix}.$$

The eigenvalues are

$$E_{3,4} = -\frac{\lambda\hbar^2}{2} \pm \frac{3\lambda\hbar^2}{4}.$$

Analogously, the states $|2\rangle$ and $|4\rangle$ form a two dimensional subspace. The matrix representation is given by

$$\begin{pmatrix} -\dfrac{\lambda\hbar^2}{2} & \dfrac{\lambda\hbar^2}{\sqrt{2}} \\ \dfrac{\lambda\hbar^2}{\sqrt{2}} & 0 \end{pmatrix}.$$

The eigenvalues are

$$E_{5,6} = -\frac{\lambda\hbar^2}{2} \pm \frac{3\lambda\hbar^2}{4}.$$

The Hamilton operator describes the *spin-orbit coupling*. We also find the notation $\hat{H} = \lambda \mathbf{S} \cdot \mathbf{L}$. ♣

4.4   Exciton-Phonon Systems

In this section we consider Bose-spin systems which can model exciton-phonon systems (Steeb [60]). The first model under investigation is given by

$$\hat{H} = -\Delta I_B \otimes \sigma_1 - k(b^\dagger + b) \otimes \sigma_3 + \Omega b^\dagger b \otimes I_2$$

where Δ, k, and Ω are constants and $\otimes$ denotes the tensor product. The first term describes transitions between the excitonic sites, the second term represents the exciton-phonon interaction, and the third term is the oscillatory energy. The quantity I_2 is the 2×2 matrix and I_B is the unit operator in the vector space in which the *Bose operators* act. In matrix representation I_B is the infinite unit matrix. The *commutation relations* for the *Bose operators* $b^\dagger, b$ are given by

$$[b, b^\dagger] = I_B, \qquad [b, b] = 0, \qquad [b^\dagger, b^\dagger] = 0.$$

It follows that

$$[b^\dagger b, b] = -b, \qquad [b^\dagger b, b^\dagger] = b^\dagger, \qquad [b^\dagger b, b^\dagger + b] = b^\dagger - b.$$

Thus the operators $b^\dagger$, b, I_B, $b^\dagger b$ form a Lie algebra. Note that the operators $b^\dagger$, b, $b^\dagger b$ are unbounded. For the matrix representation of $b^\dagger b$ we have

$$b^\dagger b = \mathrm{diag}(0, 1, 2, \dots).$$

If σ_j is one of the Pauli spin matrices, then

$$[b \otimes \sigma_j, b^\dagger \otimes \sigma_j] = I_B \otimes I_2$$

since $\sigma_j^2 = I_2$ for $j = 1, 2, 3$. For the mathematical treatment it is more convenient to cast the Hamilton operator $\hat{H}$ into a new form. With the help of the unitary transformation

$$\tilde{H} = \exp(S)\hat{H}\exp(-S), \qquad S = \left(\frac{i\pi}{4}\right) I_B \otimes \sigma_2$$

the Hamilton operator $\hat{H}$ takes the form

$$\tilde{H} = -\Delta I_B \otimes \sigma_3 + k(b^\dagger + b) \otimes \sigma_1 + \Omega b^\dagger b \otimes I_2$$

where we used that

$$\sigma_3 = \exp(i\pi\sigma_2/4)\sigma_1\exp(-i\pi\sigma_2/4)$$
$$\sigma_1 = \exp(i\pi\sigma_2/4)\sigma_3\exp(-i\pi\sigma_2/4).$$

Since $\sigma_+ := \sigma_1 + i\sigma_2$, $\sigma_- := \sigma_1 - i\sigma_2$ and $\sigma_1 \equiv (\sigma_+ + \sigma_-)/2$ we can write

$$\tilde{H} = -\Delta I_B \otimes \sigma_3 + (k/2)(b^\dagger + b) \otimes (\sigma_+ + \sigma_-) + \Omega b^\dagger b \otimes I_2.$$

We now determine the constants of motion for the Hamilton operator $\widetilde{H}$. For the Hamilton operator $\widetilde{H}$ we find that

$$\hat{P} = \exp(i\pi(b^\dagger b \otimes I_2) + I_B \otimes \sigma_3/2 + I_B \otimes I_2/2)$$

(*parity operator*), and

$$I_B \otimes (\sigma_1^2 + \sigma_2^2 + \sigma_3^2) \equiv I_B \otimes (\sigma_3^2 + (1/2)(\sigma_+\sigma_- + \sigma_-\sigma_+))$$

are constants of motion, i.e. $[\widetilde{H}, \widetilde{H}] = 0$, $[\widetilde{H}, \hat{P}] = 0$, $[\widetilde{H}, I_B \otimes \sigma^2] = 0$. Notice that $\widetilde{H}$ does not commute with the operator

$$\hat{N} := b^\dagger b \otimes I_2 + I_B \otimes \sigma_3/2.$$

The constant of motion $\hat{P}$ enables us to simplify the eigenvalue problem for $\widetilde{H}$. The operator $\hat{P}$ has a discrete spectrum, namely two eigenvalues (infinitely degenerate). With the help of $\hat{P}$ we can decompose the product Hilbert space into two invariant subspaces. In both subspaces we cannot determine the eigenvalues exactly. Next we look at the matrix representation of $\widetilde{H}$. Owing to the constant of motion $\hat{P}$ we can decompose the set of basis vectors

$$\left\{ |n\rangle \otimes \begin{pmatrix} 1 \\ 0 \end{pmatrix}, \ |n\rangle \otimes \begin{pmatrix} 0 \\ 1 \end{pmatrix} \ : \ n = 0, 1, 2, \dots \right\}$$

into two subspaces S_1 and S_2, each of them belonging to an $\hat{H}$-invariant subspace

$$S_1 := \left\{ |0\rangle \otimes \begin{pmatrix} 0 \\ 1 \end{pmatrix}, \ |1\rangle \otimes \begin{pmatrix} 1 \\ 0 \end{pmatrix}, \ |2\rangle \otimes \begin{pmatrix} 0 \\ 1 \end{pmatrix}, \dots \right\}$$

$$S_2 := \left\{ |0\rangle \otimes \begin{pmatrix} 1 \\ 0 \end{pmatrix}, \ |1\rangle \otimes \begin{pmatrix} 0 \\ 1 \end{pmatrix}, \ |2\rangle \otimes \begin{pmatrix} 1 \\ 0 \end{pmatrix}, \dots \right\}.$$

The corresponding infinite symmetric matrices are tridiagonal. For the subspace S_1 we have

$$\widetilde{H}_{n,n} = (-1)^n \Delta + n\Omega, \qquad \widetilde{H}_{n+1,n} = \widetilde{H}_{n,n+1} = k(n+1)^{1/2}$$

where $n = 0, 1, 2, \dots$. For the subspace S_2 we find the matrix representation

$$\widetilde{H}_{n,n} = -(-1)^n \Delta + n\Omega, \qquad \widetilde{H}_{n+1,n} = \widetilde{H}_{n,n+1} = k(n+1)^{1/2}.$$

Closely related to the Hamilton operator $\widetilde{H}$ is the Hamilton operator

$$\hat{H}_R = -\Delta I_B \otimes \sigma_3 + (k/2) \otimes (b^\dagger \otimes \sigma_- + b \otimes \sigma_+) + \Omega b^\dagger b \otimes I_2.$$

The Hamilton operator $\hat{H}_R$ commutes with the operators $\hat{H}_R$, $\hat{P}$, $I_B \otimes \sigma^2$, and $\hat{N}$. Notice that

$$[\hat{N}, \hat{P}] = 0 \otimes 0_2.$$

Owing to the constants of motion $\hat{N}$, we can write the infinite matrix representation of $\hat{H}_R$ as the *direct sum* of 2×2 matrices. Thus we can solve the eigenvalue problem exactly.

Consider the matrix representation of $\hat{H}_R$. For the subspace S_1 we find the matrix representation

$$\begin{pmatrix} -\Delta & k \\ k & \Delta + \Omega \end{pmatrix} \oplus \begin{pmatrix} -\Delta + 2\Omega & \sqrt{3}k \\ \sqrt{3}k & \Delta + 3\Omega \end{pmatrix} \oplus \cdots \oplus \begin{pmatrix} -\Delta + 2n\Omega & \sqrt{2n+1}k \\ \sqrt{2n+1}k & \Delta + (2n+1)\Omega \end{pmatrix} \oplus \cdots$$

where $\oplus$ denotes the direct sum. For the subspace S_2 we obtain

$$\Delta \oplus \begin{pmatrix} -\Delta + \Omega & \sqrt{2}k \\ \sqrt{2}k & \Delta + 2\Omega \end{pmatrix} \oplus \cdots \oplus \begin{pmatrix} -\Delta + (2n+1)\Omega & \sqrt{2n+2}k \\ \sqrt{2n+2}k & \Delta + (2n+2)\Omega \end{pmatrix} \oplus \cdots.$$

Consequently, we can solve the eigenvalue problem exactly. For example the eigenvalues in the subspace S_1 are given by

$$E_{n\pm} = \left(2n + \frac{1}{2}\right)\Omega \pm \left(\left(\Delta + \frac{\Omega}{2}\right)^2 + (2n+1)k^2\right)^{1/2}$$

where $n = 0, 1, 2, \ldots$.

Exercise. Let b, $b^\dagger$ be Bose annihilation and creation operators, respectively. Let c, $c^\dagger$ be Fermi annihilation and creation operators, respectively. Let I_B be the identity operator in the vector space of the Bose operators. Let I_F be the identity operator in the vector space of the Fermi operators. Calculate the commutators between all the operators in the set

$$\left\{ b \otimes I_F, \ b^\dagger \otimes I_F, \ I_B \otimes c, \ I_B \otimes c^\dagger, \ b \otimes c, \ b \otimes c^\dagger, \ b^\dagger \otimes c, \ b^\dagger \otimes c^\dagger \right\}.$$

4.5   Interpretation of Quantum Mechanics

In this section we discuss the interpretation of quantum mechanics and the application of the Kronecker and tensor product. We follow the articles of Elby and Bub [19] and Kent [38].

Quantum mechanics gives us well-understood rules for calculating probabilities, but no precise mathematical criterion for a unique sample space of events (or physical properties, or propositions) over which a probability distribution is defined. This has motivated many attempts to find mathematical structures which could be used to define a distribution.

Consider two finite dimensional Hilbert spaces $\mathcal{H}_A$ and $\mathcal{H}_B$ with

$$\dim(\mathcal{H}_A) = d_A, \qquad \dim(\mathcal{H}_B) = d_B.$$

The *Schmidt decomposition theorem* tells us that any bi-partite normalized pure state $|\Psi\rangle \in \mathcal{H}_A \otimes \mathcal{H}_B$ can be written in terms of some orthonormal bases $|\psi_k\rangle_A \in \mathcal{H}_A$ $(k = 0, 1, \ldots, d_A - 1)$, $|\phi_\ell\rangle_B$ $(\ell = 0, 1, \ldots, d_B - 1)$

$$|\Psi\rangle = \sum_{k=0}^{\min(d_A-1, d_B-1)} \lambda_k |\psi_k\rangle \otimes |\phi_k\rangle.$$

The positive coefficients λ_k satisfy

$$\sum_{k=0}^{\min(d_A-1, d_B-1)} \lambda_k = 1.$$

The orthonormal bases $|\psi_k\rangle_A$ and $|\phi_\ell\rangle_B$ dependent on the given normalized state $|\Psi\rangle$. Let $|j\rangle_A$, $|\ell\rangle_B$ $(j = 0, 1, \ldots, d_A - 1; \ell = 0, 1, \ldots, d_B - 1)$ be state independent orthonormal bases. Then there are $d_A \times d_A$ and $d_B \times d_B$ unitary matrices $U_A(\alpha)$ and $U_B(\beta)$ such that

$$|\psi_j\rangle_A = U_A(\alpha)|j\rangle_A, \qquad |\phi_k\rangle_B = U_B(\beta)|k\rangle_B$$

where α, β are parameters to specify the unitary matrices. The quantum entanglement of the joint state $|\Psi\rangle$ resides only in the coefficients λ_k and not in the parameters α and β.

Example 4.18. Consider a spin-$\frac{1}{2}$ particle initially described by a superposition of eigenstates of S_3, the z component of spin

$$|\Phi\rangle = c_1|S_3 = +\rangle + c_2|S_3 = -\rangle, \qquad |c_1|^2 + |c_2|^2 = 1.$$

Let $|R = +\rangle$ and $|R = -\rangle$ denote the "up" and "down" pointer-reading eigenstates of an S_3-measuring apparatus. Owing to quantum mechanics (with no wave-function collapse), if the apparatus ideally measures the particle, the combined system evolves into an entangled state

$$|\varphi\rangle = c_1|S_3 = +\rangle \otimes |R = +\rangle + c_2|S_3 = -\rangle \otimes |R = -\rangle.$$

For $c_1 = c_2 = 1/\sqrt{2}$ we have a *Bell state* which is maximally entangled. Thus after the measurement, the pointer reading should be definite. Owing to the "orthodox" value-assignment rule, however, the pointer reading is definite only if the quantum state is an eigenstate of $\hat{R}$, the pointer-reading operator. Since $|\varphi\rangle$ is not an eigenstate of $\hat{R}$, the pointer reading is indefinite. The interpretations of quantum mechanics attempt to deal with this aspect of the measurement problem.   ♣

Example 4.19. Consider the entangled state

$$|a_1\rangle \otimes |b_1\rangle + |a_2\rangle \otimes |b_2\rangle$$

where the states $|a_1\rangle$ and $|a_2\rangle$ are orthonormal and $|b_1\rangle$ and $|b_2\rangle$ are orthonormal. Then if system A is found in $|a_1\rangle$, then B must be found in $|b_1\rangle$. If system A is found in $|a_2\rangle$, then system B must be found in $|b_2\rangle$. ♣

If the Hilbert space $\mathcal{H}$ is written as the tensor product of spaces $\mathcal{H}_1$ and $\mathcal{H}_2$, then a generic state in $\mathcal{H}$ has a unique decomposition into orthonormal basis vectors of a particular type. The *Schmidt decomposition* is for studying models of system-apparatus measurement. The Schmidt decomposition contains the complete description of the physical properties of a system state. This is called the *modal interpretation*. The modalists have so far been unable to produce a natural joint probability distribution for the physical properties of a system at several times, even when these properties all correspond to a single Hilbert space on an equal footing. We assume that a particular splitting is somehow specified. If a set of possible statements about physics can be represented by sets of projective decompositions at various times, and if these projections decohere, then the decoherence functional defines a natural joint probability distribution. Since the Schmidt decompositions are projective, this gives a natural condition and any hypothetical joint probability distribution satisfying the modalists' single-time axiom: when a subset of the Schmidt projections is consistent, the marginal distribution for that subset should be defined by the decoherence functional. One asks what the consistency condition implies when the sets of projections are defined by the Schmidt decompositions at each point in time. While the condition does restrict the possible probability joint distribution, it is possible to find joint distributions satisfying both the consistency condition and the modalist' single-time axiom.

Consider the Heisenberg picture and a closed quantum system with Hilbert space $\mathcal{H}$, in the state ψ, evolving under the Hamilton operator $\hat{H}$ from time $t = 0$ onwards. We suppose that an isomorphism

$$\mathcal{H} \simeq \mathcal{H}_1 \otimes \mathcal{H}_2$$

is given at $t = 0$. We write $\dim(\mathcal{H}_j) = n_j$ and suppose that $n_1 \leq n_2$. Let $\{v_k^j : 1 \leq k \leq n_j\}$ be orthonormal bases of $\mathcal{H}_j$ for $j = 1, 2$, so that $\{v_k^1 \otimes v_l^2\}$ forms an orthonormal basis for $\mathcal{H}$. With respect to this isomorphism, the *Schmidt decomposition* of ψ at time t is an expression of the form

$$|\psi(t)\rangle = \sum_{k=1}^{n_1} (p_k(t))^{1/2} \exp(-i\hat{H}t/\hbar)(|w_k^1(t)\rangle \otimes |w_k^2(t)\rangle)$$

where $\{w_k^1\}$ form an orthonormal basis of $\mathcal{H}_1$ and $\{w_k^2\}$ form part of an orthonormal basis of $\mathcal{H}_2$. We can take the function $p_k(t)$ to be real and positive, and we take the positive square root. For fixed time t, any decomposition of the form given above then has the same list of weights $\{p_k(t)\}$, and the decomposition is unique provided that this list is nondegenerate. We write

$$|\psi_k(t)\rangle = \exp(-i\hat{H}t/\hbar)(|w_k^1(t)\rangle \otimes |w_k^2(t)\rangle).$$

Let $W(t)$ be the set of Schmidt weights at time t: that is, the list of $\{p_k(t)\}$ with repetitions deleted. The *Schmidt projections* at time t are defined to be the set of projections $P_p(t)$ onto subspaces of the form span $\{\psi_k(t) : p_k(T) = p\}$, for $p \in W(t)$, together with the projection

$$P_0(t) = (1 - \sum_{p \in W(t)} P_p(t)).$$

These define a projective decomposition of the identity

$$\sigma(t) = \{ P_p(t) : p = 0 \quad \text{or} \quad p \in W(t) \}.$$

Kent [38] gives a consistency criterion for the probability distribution on time-dependent Schmidt trajectories and shows that it can be satisfied.

Elby and Bub [19] showed that when a quantum state vector can be written in the *triorthogonal form*

$$|\Psi\rangle = \sum_j c_j |u_j\rangle \otimes |v_j\rangle \otimes |w_j\rangle$$

then there exists no other triorthogonal basis in terms of which $|\Psi\rangle$ can be expanded. Several interpretations of quantum mechanics can make use of this special basis. Many-world adherents claim that a branching of worlds occurs in the preferred basis picked out by the unique triorthogonal decomposition. Modal interpreters can postulate that the triorthogonal basis helps to pick out which observables possess definite values at a given time.

Decoherence theorists can cite the uniqueness of the triorthogonal decomposition as a principled reason for asserting that pointer readings become "classical" upon interacting with the environment. Many-world, decoherence, and modal interpretations of quantum mechanics suffer from a basis degeneracy problem arising from the nonuniqueness of some biorthogonal decompositions.

The $|GHZ\rangle$ state can be written in triorthogonal form

$$|GHZ\rangle = \frac{1}{\sqrt{2}}(|\mathbf{e}_1\rangle \otimes |\mathbf{e}_1\rangle \otimes |\mathbf{e}_1\rangle + |\mathbf{e}_2\rangle \otimes |\mathbf{e}_2\rangle \otimes |\mathbf{e}_2\rangle)$$

where $|\mathbf{e}_1\rangle$, $|\mathbf{e}_2\rangle$ is the standard basis in $\mathbb{C}^2$.

Many-world interpretations address the measurement problem by hypothesizing that when the combined system occupies state $|\varphi\rangle$, the two branches of the superposition split into separate worlds, in some sense. The pointer reading becomes definite relative to its branch. For instance, in the "up" world, the particle has spin up and the apparatus possesses the corresponding pointer reading. In this way, many-world interpreters explain why we always find definite pointer readings, instead of superpositions. This approach suffers from a technical problem, the basis degeneracy problem, which arises from the nonuniqueness of some biorthogonal decompositions. Owing to the biorthogonal decomposition theorem, any quantum state vector describing two systems can, for a certain choice of bases, be expanded in the simple form

$$\sum_j c_j |\mathbf{u}_j\rangle \otimes |\mathbf{v}_j\rangle$$

where the $\{|\mathbf{u}_j\rangle\}$ and $\{|\mathbf{v}_j\rangle\}$ vectors are orthonormal, and are therefore eigenstates of Hermitian operators (observables) $\hat{A}$ and $\hat{B}$ associated with systems 1 and 2, respectively. This biorthogonal expansion picks out the Schmidt basis. The basis degeneracy problem arises because the biorthogonal decomposition is unique just in case all of the nonzero $|c_j|$'s are different. When $|c_1| = |c_2|$, we can biorthogonally expand $|\varphi\rangle$ in an infinite number of bases. For instance, we can construct $\hat{S}_1$ eigenstates out of linear combinations of $\hat{S}_3$ eigenstates. Similarly, one can introduce a new apparatus observable $\hat{R}'$, whose eigenstates are superpositions of pointer-reading eigenstates

$$|S_1 = \pm\rangle = \frac{1}{\sqrt{2}}(|S_3 = +\rangle \pm |S_3 = -\rangle)$$

$$|R' = \pm\rangle = \frac{1}{\sqrt{2}}(|R = +\rangle \pm |R = -\rangle).$$

When $c_1 = c_2 = 1/\sqrt{2}$, we can obtain the fully entangled state (*Bell state*)

$$|\varphi\rangle = \frac{1}{\sqrt{2}}(|S_3 = +\rangle \otimes |R' = +\rangle + |S_1 = -\rangle \otimes |R' = -\rangle).$$

These two ways of writing $|\varphi\rangle$ correspond to two ways of writing the reduced density operator ρ_a that describes the apparatus. Taking the partial trace over the particle's states we obtain the density matrix

$$\rho_a = \frac{1}{2}(|R' = +\rangle\langle R' = +| + |R' = -\rangle\langle R' = -|)$$

where the index a indicates the apparatus. The basis degeneracy problem leaves many-world interpreters without a purely formal algorithm for deciding how splitting occurs. As the decoherence theorists show, when the environment interacts with the combined particle-apparatus system, the following state results

$$|\Psi\rangle = c_1|S_3 = +\rangle \otimes |R = +\rangle \otimes |E_+\rangle + c_2|S_3 = -\rangle \otimes |R = -\rangle \otimes |E_-\rangle$$

where $|E_\pm\rangle$ is the state of the rest of the universe after the environment interacts with the apparatus. As time passes, these environmental states quickly approach orthogonality: $\langle E_+|E_-\rangle \to 0$. In this limit, we have a triorthogonal decomposition of $|\Psi\rangle$. Even if $c_1 = c_2$, the triorthogonal decomposition is unique. In other words, no transformed bases exist such that $|\Psi\rangle$ can be expanded as

$$d_1|S' = +\rangle \otimes |R' = +\rangle \otimes |E'_+\rangle + d_2|S' = -\rangle \otimes |R' = -\rangle \otimes |E'_-\rangle.$$

Therefore, a preferred basis is chosen. Many-world interpreters can postulate that this basis determines the branches into which the universe splits.

Decoherence theorists claim that the environment picks out the pointer-reading basis. An existential interpretation relies on the environment to select the "correct" basis. This interpretation suffers from a version of the basis degeneracy problem. If $|\Psi\rangle$ describes the universe, and if $\langle E_+|E_-\rangle = 0$, then the reduced density operator $\rho_{p\&a}$ describing the particle and apparatus (found by tracing over the environmental degrees of freedom) is

$$\rho_{p\&a} = |c_1|^2|S_3 = +\rangle\langle S_3 = +||R = +\rangle\langle R = +|$$
$$+ |c_2|^2|S_3 = -\rangle\langle S_3 = -||R = -\rangle\langle R = -|$$

the same mixture as would be obtained upon wave-function collapse. If $c_1 = c_2$, however, then we can decompose this mixture into another basis, in which case the pointer reading loses its "special" status. For example, define the Bell states

$$|q_\pm\rangle := \frac{1}{\sqrt{2}}(|S_3 = +\rangle \otimes |R = +\rangle \pm |S_3 = -\rangle \otimes |R = -\rangle).$$

If $c_1 = c_2 = 1/\sqrt{2}$, then we can write the density matrix $\rho_{p\&a}$ (mixed state) as

$$\rho_{p\&a} = \frac{1}{2}(|q_+\rangle\langle q_+| + |q_-\rangle\langle q_-|).$$

Although decoherence-based interpretations can deal with their basis degeneracy problem in many ways, a particularly clean formal solution is to invoke the uniqueness of the triorthogonal decomposition of $|\Psi\rangle$. Uniqueness holds even when $c_1 = c_2$.

A third kind of interpretation is modal interpretation that relies on the biorthogonal decomposition theorem. According to most modal interpretations, if

$$\sum_j c_j |\mathbf{u}_j\rangle \otimes |\mathbf{v}_j\rangle$$

is the unique biorthogonal decomposition of the quantum state, then system 1 has a definite value for observable $\hat{A}$, and system 2 has a definite value for $\hat{B}$. Consider the example given above. According to modal interpretations, if $|c_1| \neq |c_2|$, then the particle has a definite z component of spin, and the apparatus has a definite pointer reading. These possessed values result not for a world splitting; the entangled wave function still exists entirely in our world, and continues to determine the dynamical evolution of the system. Rather, these modally possessed values are a kind of hidden variable. According to modal interpretations, an observable can possess a definite value even when the quantum state is not an eigenstate of that observable. The unique biorthogonal decomposition determines which observables take on definite values. Thus modal interpretations suffer from the basis degeneracy problem, when $|c_1| = |c_2|$. When the particle-apparatus system interacts with its environment, it evolves into the state $|\Psi\rangle$, which is (uniquely) triorthogonally decomposed. By allowing unique triorthogonal decompositions – as well as unique biorthogonal decompositions – to pick out which observables receive definite values, modal interpreters can explain why all ideal measurements have definite results. The basis selected by a triorthogonal decomposition never conflicts with the basis picked out by the unique biorthogonal decomposition, when one exists. Elby and Bub [19] proved that when a quantum state can be written in the triorthogonal form

$$|\Psi\rangle = \sum_j c_j |\mathbf{u}_j\rangle \otimes |\mathbf{v}_j\rangle \otimes |\mathbf{w}_j\rangle$$

then, even if some of the c_i's are equal, no alternative bases exist such that $|\Psi\rangle$ can be rewritten

$$\sum_j d_j |\mathbf{u}'_j\rangle \otimes |\mathbf{v}'_j\rangle \otimes |\mathbf{w}'_j\rangle.$$

They used this preferred basis to address the basis degeneracy problem.

Exercise. (1) Consider the tripartite states in the Hilbert space $\mathbb{C}^8$

$$|GHZ\rangle = \frac{1}{\sqrt{2}}(|000\rangle + |111\rangle), \qquad |W\rangle = \frac{1}{\sqrt{3}}(|001\rangle + |010\rangle + |100\rangle).$$

Find the probability $p = |\langle W|GHZ\rangle|^2$.

(2) In the following A stands for Alice and B stands for Bob. Alice and Bob share two qubits in the maximally entangled Bell state

$$|\psi\rangle = \frac{1}{\sqrt{2}}(|0\rangle_A \otimes |0\rangle_B + |1\rangle_A \otimes |1\rangle_B)$$

where $|0\rangle$, $|1\rangle$ denotes the standard basis in $\mathbb{C}^2$. Let σ_1, σ_2, σ_3 be the Pauli spin matrices. Alice's observables correspond to the operators (4×4 hermitian matrices)

$$A_1 = \sigma_3 \otimes I_2, \quad A_2 = \sigma_1 \otimes I_2, \quad A_3 = \sigma_2 \otimes I_2.$$

Bob's observables correspond to

$$B_1 = I_2 \otimes \frac{1}{\sqrt{3}}(\sigma_1 - \sigma_2 + \sigma_3), \quad B_2 = I_2 \otimes \frac{1}{\sqrt{3}}(-\sigma_1 + \sigma_2 + \sigma_3),$$

$$B_3 = I_2 \otimes \frac{1}{\sqrt{3}}(\sigma_1 + \sigma_2 - \sigma_3), \quad B_4 = I_2 \otimes \frac{1}{\sqrt{3}}(-\sigma_1 - \sigma_2 - \sigma_3).$$

Find $A_j|\psi\rangle$ ($j = 1, 2, 3$) and $B_k|\psi\rangle$ ($k = 1, 2, 3, 4$). Then calculate

$$|\langle\psi|A_j|\psi\rangle|^2, \qquad |\langle\psi|B_k|\psi\rangle|^2$$

with $j = 1, 2, 3$; $k = 1, 2, 3, 4$.

(3) Consider the density matrices (pure states)

$$\rho_1 = \frac{1}{2}\begin{pmatrix} 1 & 1 \\ 1 & 1 \end{pmatrix} \equiv \frac{1}{\sqrt{2}}\begin{pmatrix} 1 \\ 1 \end{pmatrix}\frac{1}{\sqrt{2}}\begin{pmatrix} 1 & 1 \end{pmatrix},$$

$$\rho_2 = \frac{1}{2}\begin{pmatrix} 1 & -1 \\ -1 & 1 \end{pmatrix} \equiv \frac{1}{\sqrt{2}}\begin{pmatrix} 1 \\ -1 \end{pmatrix}\frac{1}{\sqrt{2}}\begin{pmatrix} 1 & -1 \end{pmatrix}$$

and the Hamilton operator

$$\hat{H} = \frac{1}{\sqrt{2}}\hbar\omega\begin{pmatrix} 1 & 0 & 0 & 1 \\ 0 & 1 & 1 & 0 \\ 0 & 1 & -1 & 0 \\ 1 & 0 & 0 & -1 \end{pmatrix}.$$

Find $\text{tr}((\rho_1 \otimes \rho_2)\hat{H})$.

4.6   Universal Enveloping Algebra

Here we introduce the concept of a universal enveloping algebra (Bäuerle and de Kerf [7], Humphreys [34]). Let V be a complex vector space, for example $\mathbb{C}^n$. We denote the multiple tensor products by

$$T^0V := \mathbb{C}, \quad T^1V := V, \quad T^2V := V \otimes V, \ \ldots, T^nV := V \otimes T^{n-1}V.$$

Then one defines the vector space $T(V)$ as the direct sum

$$T(V) := \bigoplus_{j=0}^{\infty} T^jV.$$

Thus elements t in the vector space $T(V)$ can be written as

$$t = \sum_{k=0}^{\infty} \sum_{j_1 \cdots j_k} c_{j_1 \cdots j_k} v_{j_1} \otimes \cdots \otimes v_{j_k}$$

where $v_{j_1}, \ldots, v_{j_k}$ are vectors in V and the expansion coefficients $c_{j_1 \cdots j_k} \in \mathbb{C}$. By definition only a finite number of terms in this expansion are unequal to 0. The vector space $T(V)$ can be given the structure of an associative algebra by defining a product between vectors in $T(V)$. Let

$$v_{j_1} \otimes \cdots \otimes v_{j_m} \in T^mV, \quad w_{k_1} \otimes \cdots \otimes w_{k_n} \in T^nV$$

and $c \in \mathbb{C}$. Then

$$(v_{j_1} \otimes \cdots \otimes v_{j_m}) \otimes (w_{k_1} \otimes \cdots \otimes w_{k_n}) := (v_{j_1} \otimes \cdots \otimes v_{j_m} \otimes w_{k_1} \otimes \cdots \otimes w_{k_n}) \in T^{m+n}V$$

and

$$c \otimes (v_{j_1} \otimes \cdots \otimes v_{j_m}) := (cv_{j_1}) \otimes (v_{j_2} \otimes \cdots \otimes v_{j_m})$$

with $v_{j_1}, \ldots, v_{j_m}, w_{j_1}, \ldots, w_{j_n}$ vectors in V and $c \in T^0 V \equiv \mathbb{C}$. This is then extended linearly to all of the vector space $T(V)$. The vector space $T(V)$ equipped with this multiplication is called the *tensor algebra* generated by the given vector space V. Obviously this multiplication in $T(V)$ is associative. Since $1 \in \mathbb{C}$ we have a unit element in $T(V)$. Thus we have an associative algebra with unit element. The tensor algebra $T(V)$ is universal. Let $\Phi : V \to U$ be a linear map from the vector space V to an associative algebra U. Then there exists a unique associative algebra homomorphism $\Psi : T(V) \to U$ such that $\Phi = \Psi \circ \iota$, where $\iota : V \to T(V)$ is the canonical embedding of V in $T(V)$ and $\iota(t^1) = t^1$ ($t^1 \in V$). For $t \in T(V)$ we have

$$t = t^0 + t^1 + \cdots t^k + \cdots, \qquad t^k \in T^k V.$$

The map Ψ is defined by

$$\Psi \left(\sum_{k=0}^{\infty} \sum_{j_1 \cdots j_k} c_{j_1 \ldots j_k} v_{j_1} \otimes \cdots \otimes v_{j_k} \right) := \sum_{k=0}^{\infty} \sum_{j_1 \cdots j_k} c_{j_1 \ldots j_k} \Phi(v_{j_1}) \otimes \cdots \otimes \Phi(v_{j_k}).$$

Let L be a Lie algebra. A universal enveloping algebra of the Lie algebra L is a pair $(U(L), i)$ with $U(L)$ an associative algebra with unit element and i a Lie algebra homomorphism $i : L \to U(L)$, where $U(L)$ is considered as a Lie algebra. This means that the linear map i satisfies

$$i([x, y]) = i(x)i(y) - i(y)i(x), \qquad x, y \in L.$$

A universal algebra is unique up to isomorphisms. We obtain $U(L)$ and i from the Lie algebra L as follows: Since L is a vector space we can construct the tensor algebra $T(L)$ with unit element

$$T(L) = \mathbb{C} \oplus L \oplus T^2 L \oplus T^3 L \oplus \cdots.$$

Using the commutator in the Lie algebra L we consider in the vector space $L \oplus T^2 L \subset T(L)$ elements of the form

$$I_{x,y} = [x, y] \oplus (y \otimes x - x \otimes y), \qquad x, y \in L.$$

Now we define

$$I := \left\{ \sum_{x,y \in L} t \otimes I_{x,y} \otimes t' : t, t' \in T(L) : x, y \in L \right\}.$$

With this two-sided ideal I one finds the quotient algebra $U(L) := T(L)/I$ and the canonical projection

$$\Pi : t \in T(L) \to \Pi(t) \in U(L) = T(L)/I.$$

Thus $\Pi(I) = 0 \in U(L)$ and

$$
\begin{aligned}
\Pi((x \otimes y - y \otimes x) - [x,y]) &= \Pi(x \otimes y - y \otimes x) - \Pi([x,y]) \\
&= \Pi(x)\Pi(y) - \Pi(y)\Pi(x) - \Pi([x,y]) \\
&= 0.
\end{aligned}
$$

The restriction of Π to the subspace L of $T(L)$ provides the homomorphism $i = \Pi|_L : L \to U(L)$. We have

$$
i([x,y]) = \Pi([x,y]) = \Pi(x)\Pi(y) - \Pi(y)\Pi(x) = i(x)i(y) - i(y)i(x).
$$

Exercises. (1) Let A, B be $n \times n$ matrices over $\mathbb{C}$. Show that

$$
\operatorname{tr}((A \otimes B - B \otimes A) \oplus (-[A,B])) = 0.
$$

Find $\det((A \otimes B - B \otimes A) \oplus (-[A,B]))$.

(2) Let A, B be $n \times n$ matrices. Assume that $[A,B] = 0_n$. Can we conclude that $A \otimes B - B \otimes A = 0_{n^2}$?

(3) Let A, B be $n \times n$ matrices. Assume that $A \otimes B - B \otimes A = 0_{n^2}$. Can we conclude that $[A,B] = 0_n$?

(4) Let A, B be nonzero $n \times n$ matrices. Assume that $[A \otimes A, B \otimes B] = 0_{n^2}$. Can we conclude that $[A,B] = 0_n$?

4.7   Tensor Fields, Metric Tensor Fields and Ricci Tensors

Let M be a smooth manifold with $\dim(M) = m$ and $\mathbf{x} \in M$. A smooth manifold M (Choquet-Bruhat and DeWitt-Morette [15]) looks locally like an Euclidean space, i.e. it can be covered by a smooth atlas made out of local coordinate charts. A tangent vector at $\mathbf{x} \in M$ is a given by $d\mathbf{c}(\tau)/d\tau|_{\tau=0}$, where $\mathbf{c}(\tau) \in M$ is a smooth curve with $\mathbf{c}(0) = \mathbf{x}$. The linear space of all tangents at $\mathbf{x}$ is the tangent space $TM|_{\mathbf{x}}$ while

$$
TM = \cup_{\mathbf{x} \in M} TM|_{\mathbf{x}}
$$

is the tangent bundle. The cotangent space $T^*M|_{\mathbf{x}}$ consists of all linear functionals acting on elements of $TM|_{\mathbf{x}}$.

The local representation of tensor fields leads to the classical notation of tensors. For the local coordinates $x_1, \ldots, x_m$ with basis

$$\{\partial/\partial x_i\}_{1 \leq i \leq m}$$

of $TM|_{\mathbf{x}}$, the basis

$$\{dx_i\}_{1 \leq i \leq m}$$

yields a basis for $T_s^r M|_{\mathbf{x}}$ by r-fold tensor product

$$\frac{\partial}{\partial x_{i_1}} \otimes \cdots \otimes \frac{\partial}{\partial x_{i_r}}$$

and s-fold tensor product

$$dx_{j_1} \otimes \cdots \otimes dx_{j_s}.$$

Thus

$$T = \sum_{i_1,\ldots,i_r=1}^{m} \sum_{j_1,\ldots,j_s=1}^{m} t_{j_1,\ldots,j_s}^{i_1,\ldots,i_r}(\mathbf{x}) \frac{\partial}{\partial x_{i_1}} \otimes \cdots \otimes \frac{\partial}{x_{i_r}} \otimes dx_{j_1} \otimes \cdots \otimes dx_{j_s}$$

so that the local field of type (r, s), $T_s^r M|_{\mathbf{x}}$, has local basis

$$\left\{ \frac{\partial}{\partial x_{i_1}} \otimes \cdots \otimes \frac{\partial}{\partial x_{i_r}} \otimes dx_{j_1} \otimes \cdots \otimes dx_{j_s} \right\}_{1 \leq i_k \leq m, \ 1 \leq j_k \leq m}.$$

Covariant tensor fields of order s are given by

$$T = \sum_{j_1,\ldots,j_s=1}^{m} t_{j_1,\ldots,j_s}(\mathbf{x}) dx_{j_1} \otimes \cdots \otimes dx_{j_s}.$$

For $s = 1$ we obtain smooth differential one-forms.

Contravariant tensor fields of order r are given by

$$T = \sum_{i_1,\ldots,i_r=1}^{m} t^{i_1,\ldots,i_r}(\mathbf{x}) \frac{\partial}{\partial x_{i_1}} \otimes \cdots \otimes \frac{\partial}{\partial x_{i_r}}.$$

For $r = 1$ we obtain smooth vector fields.

Example 4.20. The *Riemann curvature tensor*

$$R(\mathbf{x}) := \sum_{i,j,k,l=1}^{m} R_{jkl}^{i}(\mathbf{x}) dx_j \otimes dx_k \otimes dx_l \otimes \frac{\partial}{\partial x_i}$$

is a tensor field of type $(1, 3)$ on M.

♣

Definition 4.18. An m-dimensional manifold M is said to be *orientable* if and only if there is a volume-form on M that is an m-form

$$\Omega \in \bigwedge_m T^*M|_{\mathbf{x}}$$

such that $\Omega|_{\mathbf{x}} \neq 0$ for all $\mathbf{x} \in M$.

Definition 4.19. A *Riemannian manifold* is a differentiable manifold M with dimension m on which there is given, in any local coordinate system, a *metric tensor field* which is a covariant tensor field of type $(0, 2)$, denoted by

$$g = \sum_{i=1}^{m} \sum_{j=1}^{m} g_{ij}(\mathbf{x}) dx_i \otimes dx_j.$$

The function $g_{ij}(\mathbf{x})$ of $\mathbf{x} \in M$ determines a Riemannian metric on M. The volume form Ω on M determined by this Riemannian metric is then given in local coordinates $(x_1, \ldots, x_m)$ by

$$\Omega := \sqrt{|\det(g_{ij})|} dx_1 \wedge \cdots \wedge dx_m$$

and is called the *Riemannian volume form*. If the determinant of the matrix (g_{ij}) is negative, the manifold is called *pseudo-Riemannian*.

Example 4.21. Consider the metric tensor field $g = dx \otimes dx$ and the map $f : \mathbb{R} \to \mathbb{R}$, $f(x) = 4x(1 - x)$. Then with $df = 4dx - 8xdx$ we obtain

$$f^*(g) = df \otimes df = 16(1 - 4x + 4x^2)dx \otimes dx$$

and the solution of $f^*(g) = 0$ is given by $x = 1/2$ which provides the maximum of f with $f(1/2) = 1$. ♣

The metric tensor field in terms of a Lorentz-orthonormal frame of differential one-forms α^j $(j = 0, 1, 2, 3)$ is given by

$$g = \sum_{j,k=0}^{3} g_{j,k} \alpha^j \otimes \alpha^k$$

where $g_{j,k} = \text{diag}(-1, +1, +1, +1)$ (*signature*). The geometric content of this metric tensor field is then expressed in the *structure equations*

$$d\alpha^j = \sum_{k=0}^{3} \alpha^k \wedge \alpha^j_k, \quad \alpha_{jk} + \alpha_{kj} = 0,$$

$$d\alpha^j_k + \sum_{\ell=0}^{3} \alpha^j_\ell \wedge \alpha^\ell_k = \frac{1}{2} \sum_{\ell,m=0}^{3} R^j_{k\ell m} \alpha^\ell \wedge \alpha^m.$$

The first equation defines the *connection forms* α_k^j. The second equation provides the components of the Riemann curvature tensor field.

Exercises. (1) Consider the metric tensor field of $\mathbb{E}^3$

$$g = dx_1 \otimes dx_1 + dx_2 \otimes dx_2 + dx_3 \otimes dx_3$$

and let $\mathbb{S}^2(R)$ be the two-dimensional sphere of radius R. Introducing *spherical coordinates*

$$x_1(\phi, \theta) = R\sin(\phi)\cos(\theta)$$
$$x_2(\phi, \theta) = R\sin(\phi)\sin(\theta)$$
$$x_3(\phi, \theta) = R\cos(\phi)$$

show that

$$g_{\mathbb{S}^2(R)} = R^2 d\phi \otimes d\phi + R^2 \sin(\phi) d\theta \otimes d\theta.$$

Show that the non-zero Christoffel symbols are $\Gamma^\theta_{\phi\theta} = \cot(\phi)$, $\Gamma^\phi_{\theta\theta} = -\sin(\phi)\cos(\phi)$.

(2) Consider the metric tensor field of $\mathbb{E}^3$

$$g = dx_1 \otimes dx_1 + dx_2 \otimes dx_2 + dx_3 \otimes dx_3$$

and let $(R > 0)$

$$\mathbb{H}^2(R) := \{ (x_1, x_2, x_3) \in \mathbb{R}^3 : x_1^2 + x_2^2 - x_3^2 = -R^2 \}.$$

Given

$$x_1(\alpha, \beta) = R\cosh(\alpha)\sinh(\beta)$$
$$x_2(\alpha, \beta) = R\sinh(\alpha)$$
$$x_3(\alpha, \beta) = R\cosh(\alpha)\cosh(\beta)$$

with $\alpha \in \mathbb{R}$, $\beta \in \mathbb{R}$. Show that

$$g_{\mathbb{H}^2(R)} = R^2 d\alpha \otimes d\alpha + R^2 \sinh^2(\alpha) d\beta \otimes d\beta.$$

Show that the non-zero Christoffel symbols are $\Gamma^\beta_{\alpha\beta} = \coth(\alpha)$, $\Gamma^\alpha_{\beta\beta} = -\sinh(\alpha)\cosh(\beta)$.

(3) Consider the metric tensor field

$$g = \frac{1}{x_2^2}(dx_1 \otimes dx_1 + dx_2 \otimes dx_2)$$

for the hyperbolic space in the Poincaré upper half plane

$$\{ (x_1, x_2) : x_2 > 0 \}.$$

Show that the Killing vector fields are given by

$$V_1 = \frac{\partial}{\partial x_1}, \quad V_2 = x_1 \frac{\partial}{\partial x_1} + x_2 \frac{\partial}{\partial x_2}, \quad V_3 = (x_1^2 - x_2^2) \frac{\partial}{\partial x_1} + 2 x_1 x_2 \frac{\partial}{\partial x_2}$$

i.e.

$$L_{V_j} g = 0, \quad j = 1, 2, 3$$

where $L_V(.)$ denotes the Lie derivative. The Lie derivative is linear and obeys the product rule. For example one has

$$L_V(dx_j \otimes dx_k) = (L_V dx_j) \otimes dx_k + dx_j \otimes (L_V dx_k).$$

Let

$$V = \sum_{j=1}^{n} V_j(\mathbf{x}) \frac{\partial}{\partial x_j}$$

be a smooth vector field. Then

$$L_V dx_j = dV_j = \sum_{k=1}^{n} \frac{\partial V_j}{\partial x_k} dx_k$$

and

$$L_V \left(\frac{\partial}{\partial x_j} \right) = \left[V, \frac{\partial}{\partial x_j} \right] = - \sum_{k=1}^{n} \frac{\partial V_k}{\partial x_j} \frac{\partial}{\partial x_k}$$

where $j = 1, \dots, n$.

Chapter 5

Software Implementations

Most of the available computer algebra packages have implemented with linear algebra the Kronecker product. Here we consider the two packages. SymbolicC++ (Hardy, Tan and Steeb [28]) is a computer algebra system written for C++ and Maxima [76] is a computer algebra system implemented in LISP. We give a number of applications in SymbolicC++ and Maxima.

In SymbolicC++ the command for the Kronecker product is `kron()`. The command for the Kronecker product in Maxima is `kronecker_product()`. In Maxima matrix multiplication is indicated by . and in SymbolicC++ matrix multiplication is indicated by $*$. In Maxima for the trace we have the command `mattrace()`, the determinant command is `determinant()` and to find the inverse of a square matrix M the command `invert(M)` is used. In SymbolicC++ we have `tr()` for the trace, `det()` for the determinant and `inverse()` for the inverse. For the transpose of an $n \times m$ matrix the command `transpose()` is used in Maxima and also in SymbolicC++.

The *Pauli spin matrices* are given by

$$\sigma_1 := \begin{pmatrix} 0 & 1 \\ 1 & 0 \end{pmatrix}, \quad \sigma_2 := \begin{pmatrix} 0 & -i \\ i & 0 \end{pmatrix}, \quad \sigma_3 := \begin{pmatrix} 1 & 0 \\ 0 & -1 \end{pmatrix}.$$

Together with the 2×2 identity matrix the matrices σ_1, σ_2, σ_3 form a basis in the Hilbert space of the 2×2 matrices over $\mathbb{C}$ with the scalar product $\langle A, B \rangle := \text{tr}(AB^*)$. In the programs we use the notation `sig1, sig2, sig3`.

Example 5.1. In the Maxima program we consider the vector in $\mathbb{C}^2$

$$\mathbf{v} = \begin{pmatrix} i \\ i \end{pmatrix} \in \mathbb{C}^2 \;\Rightarrow\; \mathbf{v}^* = \begin{pmatrix} -i & -i \end{pmatrix}.$$

We obtain the transpose and conjugate complex and then find the scalar product. Then we normalize the vector $\mathbf{v}$. Furthermore we calculate the 2×2 matrix $\mathbf{vv}^*$ for the normalized eigenvectors and obtain the eigenvalues. Next we consider the 2×3 matrix

$$A = \begin{pmatrix} 0 & 1 & 0 \\ 1 & 0 & 1 \end{pmatrix}.$$

We find the matrices AA^T and $A^T A$ and the eigenvalues and eigenvectors of it.

```
/* Transpose.mac */
v: matrix([%i],[%i]);  /* %i = sqrt(-1) */
vT: transpose(v);      /* transpose */
vTC: conjugate(vT);    /* conjugate complex */
scp: vTC . vT;         /* scalar product */
vn: v/sqrt(scp);       /* sqrt: square root */
vTn: transpose(vn);    /* transpose */
vTCn: conjugate(vTn);  /* conjugate complex */
scpn: vTCn . vn;
M: vn . vTCn;
EM: eigenvalues(M);    /* eigenvalues of M */
E11: M[1,1];           /* entries of matrix */
E12: M[1,2];           /* entries of matrix */
E21: M[2,1];           /* entries of matrix */
E22: M[2,2];           /* entries of matrix */
DM: determinant(M);    /* determinant */
A: matrix([1,1],[0,1],[1,0]);
AT: transpose(A);
AAT: A . AT;           /* matrix multiplication */
EAAT: eigenvalues(AAT);
ATA: AT . A;
EATA: eigenvalues(ATA);
```

♣

The eigenvalues of the matrix

$$\mathbf{vv}^* = \frac{1}{2}\begin{pmatrix} 1 & 1 \\ 1 & 1 \end{pmatrix}$$

(for the normalized vectors $\mathbf{v}$, $\mathbf{v}^*$) are 0 and 1. The eigenvalues of AA^T are 3, 1, 0 and the eigenvalues of $A^T A$ are 3, 1.

Example 5.2. We consider the 2×2 matrices

$$X = \begin{pmatrix} a & b \\ 0 & c \end{pmatrix}, \qquad Y = \begin{pmatrix} d & e \\ 0 & f \end{pmatrix}$$

and

$$F = X.Y.X - Y.X.Y$$

where . denotes matrix multiplication. Then we find solutions to the coupled system of four equations

$$F = 0_2.$$

```
/* braid22.mac */
X: matrix([a,b],[0,c]);
Y: matrix([d,e],[0,f]);
F: X . Y . X - Y . X . Y;
R: solve([F[1,1]=0,F[1,2]=0,F[2,1]=0,F[2,2]=0],[a,b,c,d,e,f]);
R: expand(R);
```

For example solutions are

$$a = 0, \ b = r_1, \ c = r_2, \ d = r_3, \ e = r_4, \ f = 0$$

and

$$a = r_5, \ b = r_6, \ c = 0, \ d = 0, \ e = r_7, \ f = r_8$$

with r_j arbitrary. ♣

Example 5.3. We consider the spin matrices S_1, S_2, S_3 for spin $1/2$, 1, $3/2$, 2, $5/2$. They satisfy the commutation relations

$$[S_1, S_2] = iS_3, \quad [S_2, S_3] = iS_1, \quad [S_3, S_1] = iS_2.$$

Thus one can write $\mathbf{S} \times \mathbf{S} = i\mathbf{S}$. Note that

$$S_1^2 + S_2^2 + S_3^2 = s(s+1)I_{2s+1}$$

and one has the identity

$$(\mathbf{S} \cdot \mathbf{a})(\mathbf{S} \cdot \mathbf{b}) + (\mathbf{S} \times \mathbf{a}) \cdot (\mathbf{S} \times \mathbf{b}) \equiv \mathbf{S}^2(\mathbf{a} \cdot \mathbf{b}) + i\mathbf{S} \cdot (\mathbf{a} \times \mathbf{b})$$

where $\mathbf{a}, \mathbf{b} \in \mathbb{R}^3$ and $\cdot$ denotes the scalar product. The eigenvalues of S_1, S_2, S_3 for a given spin s are given by $-s, -s+1, \ldots, +s$.

We show that

$$\frac{1}{\sqrt{3}}(S_1 + S_2 + S_3)$$

admit the same eigenvalues as S_1, S_2, S_3. Hence for spin-$1/2$ we obtain $-1/2$, $1/2$. For spin-1 we obtain -1, 0, 1. For spin-$3/2$ we obtain $-3/2$, $-1/2$, $1/2$, $3/2$. For spin-2 we obtain -2, -1, 0, 1, 2. For spin-$5/2$ we obtain $-5/2$, $-3/2$, $-1/2$, $1/2$, $3/2$, $5/2$.

```
/* spinsadd.mac */
load("nchrpl");

/* spin 1/2 */
S123: matrix([1,0],[0,-1])/2;
Sp12: matrix([0,1],[0,0]);
Sm12: transpose(Sp12);
S121: (Sp12 + Sm12)/2;
S122: -%i*(S123 . S121 - S121 . S123);
T12: (S121 + S122 + S123)/sqrt(3);
T12: ratsimp(T12);
D12: determinant(T12);
D12: ratsimp(D12);
E12: eigenvalues(T12);
E12: ratsimp(E12);

/* spin 1 */
S11: matrix([0,1,0],[1,0,1],[0,1,0])/sqrt(2);
S12: matrix([0,-%i,0],[%i,0,-%i],[0,%i,0])/sqrt(2);
S13: matrix([1,0,0],[0,0,0],[0,0,-1]);
T1: (S11 + S12 + S13)/sqrt(3); T1: ratsimp(T1);
D1: determinant(T1); D1: ratsimp(D1);
tr11: mattrace(T1); tr11: ratsimp(tr11);
tr12: mattrace(T1 . T1); tr12: ratsimp(tr12);
tr13: mattrace(T1 . T1 . T1); tr13: ratsimp(tr13);
E1: eigenvalues(T1); E1: ratsimp(E1);

/* spin 3/2 */
S321: matrix([0,sqrt(3),0,0],[sqrt(3),0,2,0],
             [0,2,0,sqrt(3)],[0,0,sqrt(3),0])/2;
S322: matrix([0,-%i*sqrt(3),0,0],[%i*sqrt(3),0,-2*%i,0],
             [0,2*%i,0,-%i*sqrt(3)],[0,0,%i*sqrt(3),0])/2;
S323: matrix([3/2,0,0,0],[0,1/2,0,0],[0,0,-1/2,0],[0,0,0,-3/2]);
T32: (S321 + S322 + S323)/sqrt(3);
D32: determinant(T32); D32: ratsimp(D32);
E32: eigenvalues(T32); E32: ratsimp(E32);

/* spin 2 */
S21: matrix([0,1,0,0,0],[1,0,sqrt(6)/2,0,0],[0,sqrt(6)/2,0,sqrt(6)/2,0],
            [0,0,sqrt(6)/2,0,1],[0,0,0,1,0]);
S22: matrix([0,-%i,0,0,0],[%i,0,-%i*sqrt(6)/2,0,0],
            [0,%i*sqrt(6)/2,0,-%i*sqrt(6)/2,0],
            [0,0,%i*sqrt(6)/2,0,-%i],[0,0,0,%i,0]);
S23: matrix([2,0,0,0,0],[0,1,0,0,0],[0,0,0,0,0],
            [0,0,0,-1,0],[0,0,0,0,-2]);
T2: (S21 + S22 + S23)/sqrt(3); T2: ratsimp(T2);
```

```
D2: determinant(T2); D2: ratsimp(D2);
tr21: mattrace(T2); tr21: ratsimp(tr21);
tr22: mattrace(T2 . T2); tr22: ratsimp(tr22);
tr23: mattrace(T2 . T2 . T2); tr23: ratsimp(tr23);
tr24: mattrace(T2 . T2 . T2 . T2); tr24: ratsimp(tr24);
tr25: mattrace(T2 . T2 . T2 . T2 . T2); tr25: ratsimp(tr25);
E2: eigenvalues(T2);
E2: ratsimp(E2);


/* spin 5/2 */
S523: matrix([5/2,0,0,0,0,0],[0,3/2,0,0,0,0],[0,0,1/2,0,0,0],
[0,0,0,-1/2,0,0],[0,0,0,0,-3/2,0],[0,0,0,0,0,-5/2]);
Sp52: matrix([0,sqrt(5),0,0,0,0],[0,0,sqrt(8),0,0,0],[0,0,0,sqrt(9),0,0],
[0,0,0,0,sqrt(8),0],[0,0,0,0,0,sqrt(5)],[0,0,0,0,0,0]);
Sm52: transpose(Sp52);
S521: (Sp52+Sm52)/2;
S522: -%i*(S523 . S521 - S521 . S523);
T52: (S521 + S522 + S523)/sqrt(3); T52: ratsimp(T52);
E52: eigenvalues(T52); E52: ratsimp(E52);
```
♣

Example 5.4. Consider the four 4×4 matrices

$$A = \begin{pmatrix} 1 & 0 & 0 & 1 \\ 0 & 1 & 1 & 0 \\ 0 & 1 & 1 & 0 \\ 1 & 0 & 0 & 1 \end{pmatrix}, \quad B = \begin{pmatrix} 0 & 1 & 1 & 0 \\ 1 & 0 & 0 & 1 \\ 1 & 0 & 0 & 1 \\ 0 & 1 & 1 & 0 \end{pmatrix},$$

$$C = \begin{pmatrix} -1 & 1 & 1 & -1 \\ 1 & -1 & -1 & 1 \\ 1 & -1 & -1 & 1 \\ -1 & 1 & 1 & -1 \end{pmatrix}, \quad D = \begin{pmatrix} 1 & -1 & -1 & 1 \\ -1 & 1 & 1 & -1 \\ -1 & 1 & 1 & -1 \\ 1 & -1 & -1 & 1 \end{pmatrix}.$$

The eigenvalues of a 4×4 matrix M can be found from the set of equations

$$\text{tr}(M) = \lambda_1 + \lambda_2 + \lambda_3 + \lambda_4$$
$$\text{tr}(M^2) = \lambda_1^2 + \lambda_2^2 + \lambda_3^2 + \lambda_4^2$$
$$\text{tr}(M^3) = \lambda_1^3 + \lambda_2^3 + \lambda_3^3 + \lambda_4^3$$
$$\text{tr}(M^4) = \lambda_1^4 + \lambda_2^4 + \lambda_3^4 + \lambda_4^4.$$

The following Maxima program provides the eigenvalues for A, B, C, D.

```
/* traceigen.mac */
load("nchrpl");
A: matrix([1,0,0,1],[0,1,1,0],[0,1,1,0],[1,0,0,1]);
B: matrix([0,1,1,0],[1,0,0,1],[1,0,0,1],[0,1,1,0]);
```

```
C: matrix([-1,1,1,-1],[1,-1,-1,1],[1,-1,-1,1],[-1,1,1,-1]);
D: matrix([1,-1,-1,1],[-1,1,1,-1],[-1,1,1,-1],[1,-1,-1,1]);
trA: mattrace(A); trA2: mattrace(A . A);
trA3: mattrace(A . A . A); trA4: mattrace(A . A . A . A);
solve([l1+l2+l3+l4=trA,l1*l1+l2*l2+l3*l3+l4*l4=trA2,
       l1*l1*l1+l2*l2*l2+l3*l3*l3+l4*l4*l4=trA3,
       l1*l1*l1*l1+l2*l2*l2*l2+l3*l3*l3*l3+l4*l4*l4*l4=trA4],
       [l1,l2,l3,l4]);
trB: mattrace(B); trB2: mattrace(B . B);
trB3: mattrace(B . B . B); trB4: mattrace(B . B . B . B);
solve([l1+l2+l3+l4=trB,l1*l1+l2*l2+l3*l3+l4*l4=trB2,
       l1*l1*l1+l2*l2*l2+l3*l3*l3+l4*l4*l4=trB3,
       l1*l1*l1*l1+l2*l2*l2*l2+l3*l3*l3*l3+l4*l4*l4*l4=trB4],
       [l1,l2,l3,l4]);
trC: mattrace(C); trC2: mattrace(C . C);
trC3: mattrace(C . C . C); trC4: mattrace(C . C . C . C);
solve([l1+l2+l3+l4=trC,l1*l1+l2*l2+l3*l3+l4*l4=trC2,
       l1*l1*l1+l2*l2*l2+l3*l3*l3+l4*l4*l4=trC3,
       l1*l1*l1*l1+l2*l2*l2*l2+l3*l3*l3*l3+l4*l4*l4*l4=trC4],
       [l1,l2,l3,l4]);
trD: mattrace(D); trD2: mattrace(D . D);
trD3: mattrace(D . D . D); trD4: mattrace(D . D . D . D);
solve([l1+l2+l3+l4=trD,l1*l1+l2*l2+l3*l3+l4*l4=trD2,
       l1*l1*l1+l2*l2*l2+l3*l3*l3+l4*l4*l4=trD3,
       l1*l1*l1*l1+l2*l2*l2*l2+l3*l3*l3*l3+l4*l4*l4*l4=trD4],
       [l1,l2,l3,l4]);
```

For the matrix A we obtain the eigenvalues 0 (twice) and 2 (twice). For the matrix B we obtain the eigenvalues -2, 0 (twice), 2. For the matrix C we obtain the eigenvalues -4 and 0 (three times). For the matrix D we obtain the eigenvalues 4 and 0 (three times).   ♣

Example 5.5. The spin matrices S_1, S_2, S_3 for spin $s = 1/2, 1, 3/2, 2, \ldots$ satisfy the commutation relations

$$[S_1, S_2] = iS_3, \quad [S_2, S_3] = iS_1, \quad [S_3, S_1] = iS_2.$$

The matrices S_+ and S_- are given by

$$S_+ := S_1 + iS_2, \quad S_- := S_1 - iS_2.$$

For the case with spin-$\frac{3}{2}$ we have the 4×4 matrix for S_-

$$S_- = \begin{pmatrix} 0 & 0 & 0 & 0 \\ \sqrt{3} & 0 & 0 & 0 \\ 0 & 2 & 0 & 0 \\ 0 & 0 & \sqrt{3} & 0 \end{pmatrix}.$$

Hence S_- is a nilpotent matrix. We calculate $\exp(zS_-)$ and $\exp(zS_-)v$, where $v = \begin{pmatrix} 1 & 0 & 0 & 0 \end{pmatrix}^T$. The SymbolicC++ program is

```
/* nilpotent.cpp */
#include<iostream>
#include "symbolicc++.h"
using namespace std;

int main(void)
{
Symbolic Sm = ((Symbolic(0),0,0,0),(sqrt(Symbolic(3)),Symbolic(0),0,0),
               (Symbolic(0),2,0,0),(Symbolic(0),0,sqrt(Symbolic(3)),0));
Symbolic sum;
Symbolic T = Sm.identity();
Symbolic ZM = 0*Sm;
Symbolic z("z");
int j=0; int fact=1;
while(T != ZM) { sum += T/fact; T *= z*Sm; fact *= ++j; }
cout << "Sm = " << Sm << endl << endl;
cout << "exp(zA) = " << sum << endl;
Symbolic v("v",4,1);
v(0,0) = 1; v(1,0) = 0; v(2,0) = 0; v(3,0) = 0;
cout << v << endl;
Symbolic w = sum*v;
cout << "w = " << w << endl;
return 0;
}
```

The Maxima program takes the form

```
/* nilpotent.mac */
Sm: matrix([0,0,0,0],[sqrt(3),0,0,0],[0,2,0,0],[0,0,sqrt(3),0]);
sum: 0$
T: identfor(Sm)$
j: 0$
while not(T=0*Sm) do (sum: sum+T/(j!),T: z*(T.Sm),j: j+1)$
exp(z*'Sm) = sum;
v: matrix([1],[0],[0],[0]);
w: sum . v;
```

The output of the programs is

$$
\exp(zS_-) = \begin{pmatrix} 1 & 0 & 0 & 0 \\ \sqrt{3}z & 1 & 0 & 0 \\ \sqrt{3}z^2 & 2z & 1 & 0 \\ z^3 & \sqrt{3}z^2 & \sqrt{3}z & 1 \end{pmatrix}, \quad \exp(zS_-)v = \begin{pmatrix} 1 \\ \sqrt{3}z \\ \sqrt{3}z^2 \\ z^3 \end{pmatrix}.
$$

Obviously the vector $\exp(zS_-)v$ is not normalized.   ♣

Example 5.6. Let A be an $n \times n$ hermitian matrix and I_n be the $n \times n$ identity matrix. Then $(A + iI_n)^{-1}$ exists and

$$U_A = (A - iI_n)(A + iI_n)^{-1}$$

is called the *Cayley transform* of A. Then U_A is a unitary matrix and the inverse transform is given by

$$A = i(I_n + U_A)(I_n - U_A)^{-1}.$$

Let $a \in \mathbb{R}$. A Maxima implementation to find the Cayley transform of the matrix

$$A = a\sigma_2 \equiv \begin{pmatrix} 0 & -ai \\ ai & 0 \end{pmatrix}$$

is given by

```
/* Cayley.mac */
I2: matrix([1,0],[0,1]);
sig2: matrix([0,-%i],[%i,0]);
A: a*sig2;
Ap: A + %i*I2; Am: A - %i*I2;
Api: invert(Ap); Api: ratsimp(Api);
UC: Am . Api; UC: ratsimp(UC);
UCT: transpose(UC); UCT: ratsimp(UCT);
F: UC . UCT;
F: ratsimp(F);
```
                                                                                 ♣

Example 5.7. Given an $n \times n$ matrix H. We find all $n \times n$ permutation matrices P such

$$HP = PH \quad \Leftrightarrow \quad PHP^{-1} = H.$$

Note that $P^{-1} = P^T$ for permutation matrices. In the present program we apply it to the 4×4 matrices

$$H_1 = \begin{pmatrix} 0 & 1 & 1 & 0 \\ 1 & 0 & 0 & 1 \\ 1 & 0 & 0 & 1 \\ 0 & 1 & 1 & 0 \end{pmatrix}, \quad H_2 = \begin{pmatrix} 1 & 0 & 0 & 1 \\ 0 & 1 & 1 & 0 \\ 0 & 1 & 1 & 0 \\ 1 & 0 & 0 & 1 \end{pmatrix}.$$

```
/* permutation.mac */

findperm(n,use) := block([i,j,k,P,c],
  i: 1,
```

```
    array(j,n),
    for k: 1 thru n do j[k]: 0,
    while i > 0 do (
     j[i]: j[i]+1,
     if equal(j[i],n+1) then (j[i]: 0,i: i-1) else (
      if i > 0 then (c: 0,
       for k: 1 thru i-1 do if equal(j[k],j[i]) then c: c+1,
       if equal(c,0) then (
       i: i+1,
        if equal(i,n+1) then (
        P: genmatrix(lambda([r,c],0),n,n),
         for k: 1 thru n do P[k,j[k]]: 1,use(P),i: i-1)
      )
     )
    )
   )
  )$

commutesH(P) :=
 if equal(H.P,P.H) then (print('P[count]=P),count: count+1)$
count: 0$
H: matrix([0,1,1,0],[1,0,0,1],[1,0,0,1],[0,1,1,0]);
print(H);
findperm(length(H),commutesH)$
R: eigenvalues(H); print(R);
cout: 0$
H: matrix([1,0,0,1],[0,1,1,0],[0,1,1,0],[1,0,0,1]); print(H);
findperm(length(H),commutesH);
R: eigenvalues(H); print(R);
```

In both cases we find eight permutation matrices (including the identity matrix) and the permutation matrix

$$P = \begin{pmatrix} 0 & 0 & 0 & 1 \\ 0 & 0 & 1 & 0 \\ 0 & 1 & 0 & 0 \\ 1 & 0 & 0 & 0 \end{pmatrix}.$$

Obviously these permutation matrices form a subgroup of all 24 4×4 permutation matrices. ♣

Example 5.8. We consider the matrices

$$A = \begin{pmatrix} a_1 & 1 \\ 0 & a_2 \end{pmatrix}, \qquad B = \begin{pmatrix} 1 & b_1 \\ b_2 & 0 \end{pmatrix}$$

and calculate $A \otimes B$, $\det(A \otimes B)$, $\mathrm{tr}(A \otimes B)$, $A \otimes B - B \otimes A$, $\det(A \otimes B - B \otimes A)$.

The implementation in SymbolicC++ is

```
#include <iostream>
#include "symbolicc++.h"
using namespace std;

int main(void)
{
 Symbolic a1("a1"), a2("a2"), b1("b1"), b2("b2");
 Symbolic A = ((a1,1),(0,a2));
 Symbolic B = ((1,b1),(b2,0));
 Symbolic KAB = kron(A,B);
 cout << KAB << endl;
 cout << det(KAB) << endl;
 cout << tr(KAB) << endl;
 cout << kron(A,B)-kron(B,A) << endl;
 cout << det(kron(A,B)-kron(B,A));
 return 0;
}
```

The implementation in Maxima is

```
A: matrix([a1,1],[0,a2]);
B: matrix([1,b1],[b2,0]);
KAB: kronecker_product(A,B);
determinant(KAB);
mat_trace(KAB);
T: kronecker_product(A,B)-kronecker_product(B,A);
determinant(T);
```

♣

Example 5.9. Let $\sigma_0 = I_2$ and σ_1 be the Pauli spin matrix. We define

$$T_1 = \sigma_0 \otimes \sigma_0, \quad T_2 = \sigma_0 \otimes \sigma_1, \quad T_3 = \sigma_1 \otimes \sigma_0, \quad T_4 = \sigma_1 \otimes \sigma_1,$$

$R = (T_1 + T_2 + T_3 - T_4)/2$, the *flip matrix*

$$F = (1) \oplus \begin{pmatrix} 0 & 1 \\ 1 & 0 \end{pmatrix} \oplus (1)$$

and $R_T = F.R$, where . is matrix multiplication. We show that

$$(\sigma_0 \otimes R_T)(R_T \otimes \sigma_0)(\sigma_0 \otimes R_T) = (R_T \otimes \sigma_0)(\sigma_0 \otimes R_T)(R_T \otimes \sigma_0)$$

Then we apply R_T to the *Bell states*.

```
/* braidrelation.mac */
sig0: matrix([1,0],[0,1]);
sig1: matrix([0,1],[1,0]);
T1: kronecker_product(sig0,sig0);
```

```
T2: kronecker_product(sig0,sig1);
T3: kronecker_product(sig1,sig0);
T4: kronecker_product(sig1,sig1);
R: (T1 + T2 + T3 - T4)/2;
F: matrix([1,0,0,0],[0,0,1,0],[0,1,0,0],[0,0,0,1]);
RT: F . R;
V1: kronecker_product(sig0,RT);
V2: kronecker_product(RT,sig0);
/* braid relation */
Z: V1 . V2 . V1 - V2 . V1 . V2;
/* Bell states */
B1: matrix([1],[0],[0],[1])/sqrt(2);
B2: matrix([0],[1],[1],[0])/sqrt(2);
B3: matrix([0],[1],[-1],[0])/sqrt(2);
B4: matrix([1],[0],[0],[-1])/sqrt(2);
/* applying RT to the Bell states */
RT . B1; RT . B2; RT . B3; RT . B4;
```

♣

Example 5.10. In the SymbolicC++ program we consider the simple Lie algebra $so(3)$ with the basis

$$X_1 = \begin{pmatrix} 0 & 0 & 0 \\ 0 & 0 & -1 \\ 0 & 1 & 0 \end{pmatrix}, \quad X_2 = \begin{pmatrix} 0 & 0 & 1 \\ 0 & 0 & 0 \\ -1 & 0 & 0 \end{pmatrix}, \quad X_3 = \begin{pmatrix} 0 & -1 & 0 \\ 1 & 0 & 0 \\ 0 & 0 & 0 \end{pmatrix}.$$

We show that

$$[X_1 \otimes I_3 + I_3 \otimes X_1, X_2 \otimes I_3 + I_3 \otimes X_2] = X_3 \otimes I_3 + I_3 \otimes X_3.$$

```
// kronecker1.cpp
#include <iostream>
#include "symbolicc++.h"
using namespace std;

int main(void)
{
int n = 3;
Symbolic I("I",n,n);
I = I.identity();
Symbolic X1("X1",n,n);
X1(0,0) = 0; X1(0,1) = 0; X1(0,2) = 0;
X1(1,0) = 0; X1(1,1) = 0; X1(1,2) = -1;
X1(2,0) = 0; X1(2,1) = 1; X1(2,2) = 0;
Symbolic X2("X2",n,n);
X2(0,0) = 0; X2(0,1) = 0; X2(0,2) = 1;
X2(1,0) = 0; X2(1,1) = 0; X2(1,2) = 0;
```

```
X2(2,0) = -1; X2(2,1) = 0; X2(2,2) = 0;
Symbolic X3("X3",n,n);
X3(0,0) = 0; X3(0,1) = -1; X3(0,2) = 0;
X3(1,0) = 1; X3(1,1) = 0; X3(1,2) = 0;
X3(2,0) = 0; X3(2,1) = 0; X3(2,2) = 0;
Symbolic R1 = kron(X1,I) + kron(I,X1);
Symbolic R2 = kron(X2,I) + kron(I,X2);
Symbolic C = R1*R2-R2*R1;   // commutator
cout << "C = " << C << endl;
Symbolic T = C-(kron(X3,I)+kron(I,X3));
cout << "T = " << T << endl;
return 0;
}
```

♣

Example 5.11. The n-qubit *Pauli group* is defined by

$$\mathcal{P}_n := \{I_2, \sigma_1, \sigma_2, \sigma_3\}^{\otimes n} \otimes \{\pm 1, \pm i\}$$

where σ_1, σ_2, σ_3 are the 2×2 Pauli matrices and I_2 is the 2×2 identity matrix. The dimension of the Hilbert space under consideration is dim $\mathcal{H} = 2^n$. In the SymbolicC++ program we implement the Pauli group for $n = 2$. All 64 elements of $\mathcal{P}_2$ are generated. In the SymbolicC++ program we calculate the product

$$(\sigma_1 \otimes \sigma_1)(\sigma_3 \otimes \sigma_3) = -\sigma_2 \otimes \sigma_2.$$

The group element $\sigma_1 \otimes \sigma_1$ is given by g(5) since

$$5 = 0 \cdot 16 + 3 \cdot 4 + 1$$

i.e. $j = 0$, $k = 1$, $\ell = 1$. The group element g(15) is given by $\sigma_3 \otimes \sigma_3$ since

$$15 = 0 \cdot 16 + 3 \cdot 4 + 3$$

i.e. $j = 0$, $k = 3$, $\ell = 3$. For the element g(42) we have

$$42 = 2 \cdot 16 + 2 \cdot 4 + 2$$

i.e. $j = 2$, $k = 2$, $\ell = 2$.

```
// Pauli_group.cpp
#include <iostream>
#include <map>
#include <sstream>
#include <string>
#include "symbolicc++.h"
using namespace std;

int operator < (const Symbolic &s1,const Symbolic &s2)
```

```
{
ostringstream os1, os2;
os1 << s1; os2 << s2;
return (os1.str() < os2.str());
}

int main(void)
{
int j,k,l;
using SymbolicConstant::i;
map<Symbolic,Symbolic> rep;
Symbolic I2 = Symbolic("",2,2).identity();
Symbolic I4 = Symbolic("",4,4).identity();
Symbolic sig1 = ((Symbolic(0),Symbolic(1)),(Symbolic(1),Symbolic(0)));
Symbolic sig2 = ((Symbolic(0),-i),(i,Symbolic(0)));
Symbolic sig3 = ((Symbolic(1),Symbolic(0)),(Symbolic(0),-Symbolic(1)));
Symbolic g("g",64), gr("gr",64), s("s",4);
s(0) = I2; s(1) = sig1; s(2) = sig2; s(3) = sig3;
for(j=0;j<4;j++)
for(k=0;k<4;k++)
for(l=0;l<4;l++)
{
gr(j*16+k*4+l) = ((i^j)*kron(s(k),s(l)))[(i^3)==-i];
rep[gr(j*16+k*4+l)] = g(j*16+k*4+l);
cout << gr(j*16+k*4+l) << " => " << rep[gr(j*16+k*4+l)] << endl;
}
cout << gr(5)*gr(15) << " => " << rep[gr(5)*gr(15)] << endl;
return 0;
}
```

♣

Example 5.12. Let σ_2, σ_3 the Pauli spin matrices. We show that

$$\sigma_2 \otimes \sigma_3 = UAU^{-1}$$

with

$$A = \begin{pmatrix} -1 & 0 & 0 & 0 \\ 0 & 0 & 0 & 1 \\ 0 & 0 & 1 & 0 \\ 0 & 1 & 0 & 0 \end{pmatrix}, \quad U = \frac{1}{\sqrt{2}} \begin{pmatrix} 0 & 1 & 0 & i \\ 1 & 0 & 1 & 0 \\ 0 & -1 & 0 & i \\ i & 0 & -i & 0 \end{pmatrix}$$

where U is a unitary matrix, i.e. $U^{-1} = U^*$.

```
/* vec.mac */
sig2: matrix([0,-%i],[%i,0]);
sig3: matrix([1,0],[0,-1]);
I2: matrix([1,0],[0,1]);
```

```
I4: kronecker_product(I2,I2);
S23: kronecker_product(sig2,sig3);
A: matrix([-1,0,0,0],[0,0,0,1],[0,0,1,0],[0,1,0,0]);
X: kronecker_product(I4,S23) - kronecker_product(A,I4);
VU: matrix([u11],[u21],[u31],[u41],[u12],[u22],[u32],[u42],
    [u13],[u23],[u33],[u43],[u14],[u24],[u34],[u44]);
R: X . VU;
U: matrix([0,1,0,%i],[1,0,1,0],[0,-1,0,%i],[%i,0,-%i,0])/sqrt(2);
UT: transpose(U);
UTC: conjugate(UT);
F: S23 . U - U . A;                                                    ♣
```

Example 5.13. We prove the identities

$$U_{CNOT}(\sigma_1 \otimes I_2) \equiv (\sigma_1 \otimes \sigma_1)U_{CNOT}, \quad U_{CNOT}(\sigma_3 \otimes I_2) \equiv (\sigma_3 \otimes I_2)U_{CNOT}$$

$$U_{CNOT}(I_2 \otimes \sigma_1) \equiv (I_2 \otimes \sigma_1)U_{CNOT}, \quad U_{CNOT}(I_2 \otimes \sigma_3) \equiv (\sigma_3 \otimes \sigma_3)U_{CNOT}$$

where the U_{CNOT} gate is given by

$$U_{CNOT} = \begin{pmatrix} 1 & 0 & 0 & 0 \\ 0 & 1 & 0 & 0 \\ 0 & 0 & 0 & 1 \\ 0 & 0 & 1 & 0 \end{pmatrix}.$$

```
/* CNOT.mac */
I2: matrix([1,0],[0,1]);
sig1: matrix([0,1],[1,0]);
sig2: matrix([0,-%i],[%i,0]);
sig3: matrix([1,0],[0,-1]);
CNOT: matrix([1,0,0,0],[0,1,0,0],[0,0,0,1],[0,0,1,0]);
R1: CNOT.kronecker_product(sig1,I2)-kronecker_product(sig1,sig1).CNOT;
R2: CNOT.kronecker_product(sig3,I2)-kronecker_product(sig3,I2).CNOT;
R3: CNOT.kronecker_product(I2,sig1)-kronecker_product(I2,sig1).CNOT;
R4: CNOT.kronecker_product(I2,sig3)-kronecker_product(sig3,sig3).CNOT;  ♣
```

Example 5.14. An application for quantum groups is as follows: The quantum Yang-Baxter equation plays an important role both in physics and mathematics. With solutions of the quantum Yang-Baxter equation one can construct exactly solvable models and find their eigenvalues and eigenstates. On the other hand, any solution of the quantum Yang-Baxter equation can be generally used to find the new quasi-triangular Hopf algebra. Many multiparameter solutions (4×4 matrices) of the quantum Yang-Baxter equation have been found. Corresponding to the case of the standard one-parameter R matrix, the algebras related to the standard two-parameter R matrix are considered. Consider the 2×2 matrix

$$T = \begin{pmatrix} a & b \\ c & d \end{pmatrix}$$

where a, b, c and d are noncommutative linear operators. We may consider a, b, c and d as $n \times n$ matrices over the complex or real numbers. Let I be the 2×2 unit matrix

$$I = \begin{pmatrix} 1 & 0 \\ 0 & 1 \end{pmatrix}$$

where 1 is the unit operator (unit matrix). Now we define

$$T_1 := T \otimes I, \qquad T_2 = I \otimes T$$

where $\otimes$ denotes the Kronecker product. Thus T_1 and T_2 are operator valued 4×4 matrices. Thus we find

$$T_1 = T \otimes I = \begin{pmatrix} a & 0 & b & 0 \\ 0 & a & 0 & b \\ c & 0 & d & 0 \\ 0 & c & 0 & d \end{pmatrix}, \quad T_2 = I \otimes T = \begin{pmatrix} a & b & 0 & 0 \\ c & d & 0 & 0 \\ 0 & 0 & a & b \\ 0 & 0 & c & d \end{pmatrix}.$$

Consider now the 4×4 matrix

$$R_q := \begin{pmatrix} 1 & 0 & 0 & 0 \\ 0 & -1 & 0 & 0 \\ 0 & 1+q & q & 0 \\ 0 & 0 & 0 & 1 \end{pmatrix}$$

where q is a nonzero complex number. The algebra related to this quantum matrix is given by the *Yang-Baxter equation*

$$R_q T_1 T_2 = T_2 T_1 R_q.$$

From the Yang-Baxter equation we obtain the relations of the algebra elements a, b, c and d

$$ab = q^{-1}ba, \quad dc = qcd, \quad bc = -qcb, \quad bd = -db, \quad ac = -ca,$$

$$[a, d] = (1 + q^{-1})bc$$

where all the commutative relations have been omitted. There is no nilpotent element in quantum matrix T. We can define an element of the algebra

$$\delta(T) \equiv \delta := ad - q^{-1}bc.$$

Then we can show that δ satisfies the following commutation relations

$$[a, \delta] = 0, \qquad [d, \delta] = 0$$

and

$$\{c, \delta\}_q := qc\delta + \delta c = 0, \qquad \{\delta, b\}_q := q\delta b + b\delta = 0.$$

The element δ commutes only with a and b. Hence it is not the centre of the algebra. If we formally define δ^{-1} to be inverse of δ, then

$$T^{-1} = \begin{pmatrix} d\delta^{-1} & -qb\delta^{-1} \\ -q^{-1}\delta^{-1}c & a\delta^{-1} \end{pmatrix}.$$

In the SymbolicC++ program we implement the Kronecker product. Then we evaluate the matrices T_1 and T_2. Finally we evaluate the Yang-Baxter equation and thus find the commutation relations.

```
// qg.cpp
#include <iostream>
#include "symbolicc++.h"
using namespace std;

int main(void)
{
Symbolic a("a"), b("b"), c("c"), d("d"), q("q");
a = ~a; b = ~b; c = ~c; d = ~d; // noncommutative
// matrices
Symbolic T = ( (a,b),(c,d) );
Symbolic ID = Symbolic("",2,2).identity();
// Kronecker product
Symbolic T1 = kron(T,ID); Symbolic T2 = kron(ID,T);
cout << "T1 = " << T1 << endl; cout << "T2 = " << T2 << endl;

Symbolic R = ((Symbolic(1),0,0,0),(Symbolic(0),-1,0,0),
              (Symbolic(0),1+q,q,0),(Symbolic(0),0,0,1));
Symbolic res = R*T1*T2-T2*T1*R;
cout << "res = " << res << endl;
cout << solve(res(0,1),a*b); cout << solve(res(1,0),a*c);
cout << solve(res(1,2),b*c); cout << solve(res(1,3),b*d);
cout << solve(res(2,2),a*d); cout << solve(res(3,1),d*c);
return 0;
}
```
                                                                    ♣

The output is the matrix T_1, the matrix T_2 and the commutation conditions.

Example 5.15. An application for gamma matrices and spin matrices is as follows. The Pauli spin matrices σ_1, σ_2, σ_3 and the gamma matrices γ_1, γ_2, γ_3, γ_4 play a central role in describing the electron (or more general Fermi particles with spin-$\frac{1}{2}$). The scalar product of the Pauli spin matrices is

$$\langle \sigma_1, \sigma_2 \rangle = 0, \quad \langle \sigma_1, \sigma_3 \rangle = 0, \quad \langle \sigma_2, \sigma_3 \rangle = 0.$$

The *gamma matrices* are given by

$$\gamma_1 := \begin{pmatrix} 0 & 0 & 0 & -i \\ 0 & 0 & -i & 0 \\ 0 & i & 0 & 0 \\ i & 0 & 0 & 0 \end{pmatrix}, \qquad \gamma_2 := \begin{pmatrix} 0 & 0 & 0 & -1 \\ 0 & 0 & 1 & 0 \\ 0 & 1 & 0 & 0 \\ -1 & 0 & 0 & 0 \end{pmatrix},$$

$$\gamma_3 := \begin{pmatrix} 0 & 0 & -i & 0 \\ 0 & 0 & 0 & i \\ i & 0 & 0 & 0 \\ 0 & -i & 0 & 0 \end{pmatrix}, \qquad \gamma_4 := \begin{pmatrix} 1 & 0 & 0 & 0 \\ 0 & 1 & 0 & 0 \\ 0 & 0 & -1 & 0 \\ 0 & 0 & 0 & -1 \end{pmatrix}.$$

Let I_2 be the 2×2 unit matrix. Then the gamma matrices can be expressed as Kronecker products of the Pauli spin matrices and the identity matrix I_2

$$\gamma_1 = \sigma_2 \otimes \sigma_1, \quad \gamma_2 = \sigma_2 \otimes \sigma_2, \quad \gamma_3 = \sigma_2 \otimes \sigma_3, \quad \gamma_4 = \sigma_3 \otimes I_2.$$

Taking all the 16 possible Kronecker products of I_2, σ_1, σ_2, σ_3

$$I_2 \otimes I_2, \quad I_2 \otimes \sigma_1, \quad I_2 \otimes \sigma_2, \quad I_2 \otimes \sigma_3,$$

$$\sigma_1 \otimes I_2, \quad \sigma_1 \otimes \sigma_1, \quad \sigma_1 \otimes \sigma_2, \quad \sigma_1 \otimes \sigma_3,$$

$$\sigma_2 \otimes I_2, \quad \sigma_2 \otimes \sigma_1, \quad \sigma_2 \otimes \sigma_2, \quad \sigma_2 \otimes \sigma_3,$$

$$\sigma_3 \otimes I_2, \quad \sigma_3 \otimes \sigma_1, \quad \sigma_3 \otimes \sigma_2, \quad \sigma_3 \otimes \sigma_3$$

we find a basis for the Hilbert space of the 4×4 matrices. We calculate the gamma matrices from the Pauli spin matrices and the 2×2 unit matrix.

```cpp
// gamma.cpp
#include <iostream>
#include "symbolicc++.h"

int main(void)
{
using SymbolicConstant::i;
Symbolic I = Symbolic("",2,2).identity();
```

```
Symbolic sig1 = ((Symbolic(0),Symbolic(1)),(Symbolic(1),Symbolic(0)));
Symbolic sig2 = ((Symbolic(0),-i),(i,Symbolic(0)));
Symbolic sig3 = ((Symbolic(1),Symbolic(0)),(Symbolic(0),Symbolic(-1)));
Symbolic g1 = kron(sig2,sig3), g2 = kron(sig2,sig2);
Symbolic g3 = kron(sig2,sig3), g4 = kron(sig3,I);
cout << "g1 = " << g1 << endl; cout << "g2 = " << g2 << endl;
cout << "g3 = " << g3 << endl; cout << "g4 = " << g4 << endl;
return 0;
}
```

The output is

```
g1 =
[ 0   0 -i   0]
[ 0   0  0   i]
[ i   0  0   0]
[ 0  -i  0   0]
g2 =
[ 0   0  0  -1]
[ 0   0  1   0]
[ 0   1  0   0]
[-1   0  0   0]
g3 =
[ 0   0 -i   0]
[ 0   0  0   i]
[ i   0  0   0]
[ 0  -i  0   0]
g4 =
[ 1   0  0   0]
[ 0   1  0   0]
[ 0   0 -1   0]
[ 0   0  0  -1]
```

♣

Example 5.16. An application of the Kronecker product in *teleportation* (Steeb and Hardy [59]) is as follows. In quantum teleportation we start with the following state in the Hilbert space $\mathbb{C}^8$

$$|\psi\rangle \otimes |0\rangle \otimes |0\rangle \equiv (a|0\rangle + b|1\rangle) \otimes |0\rangle \otimes |0\rangle \equiv |\psi 00\rangle$$

where $|a|^2 + |b|^2 = 1$. The quantum circuit for teleportation we have three inputs, where A is the input $|\psi\rangle$, B the input $|0\rangle$ and C the input $|0\rangle$. We study what happens when we feed the product state $|\psi 00\rangle$ into the quantum circuit. We have the following eight 8×8 unitary matrices (left to right)

$$U_1 = I_2 \otimes U_H \otimes I_2, \qquad U_2 = I_2 \otimes U_{XOR},$$

$$U_3 = U_{XOR} \otimes I_2, \qquad U_4 = U_H \otimes I_2 \otimes I_2,$$

$$U_5 = I_2 \otimes U_{XOR}, \qquad U_6 = I_2 \otimes I_2 \otimes U_H,$$

$$U_7 = I_4 \oplus U_{NOT} \oplus U_{NOT}, \qquad U_8 = I_2 \otimes I_2 \otimes U_H$$

where $\oplus$ denotes the direct sum of matrices and $\otimes$ denotes the Kronecker product. Now U_H denotes the *Hadamard gate*

$$U_H = \frac{1}{\sqrt{2}} \begin{pmatrix} 1 & 1 \\ 1 & -1 \end{pmatrix}$$

U_{XOR} denotes the *XOR*-gate

$$U_{XOR} = \begin{pmatrix} 1 & 0 & 0 & 0 \\ 0 & 1 & 0 & 0 \\ 0 & 0 & 0 & 1 \\ 0 & 0 & 1 & 0 \end{pmatrix}.$$

and U_{NOT} denotes the NOT-gate

$$U_{NOT} := \begin{pmatrix} 0 & 1 \\ 1 & 0 \end{pmatrix}.$$

With the input state $|\psi 00\rangle \equiv |\psi\rangle \otimes |0\rangle \otimes |0\rangle$ teleportation is then described by

$$U_8 U_7 U_6 U_5 U_4 U_3 U_2 U_1 |\psi 00\rangle.$$

Applying the first four unitary matrices to the input state $|\psi 00\rangle$ we obtain

$$U_4 U_3 U_2 U_1 |\psi 00\rangle = \frac{a}{2}(|000\rangle + |100\rangle + |011\rangle + |111\rangle)$$
$$+ \frac{b}{2}(|010\rangle - |110\rangle + |001\rangle - |101\rangle).$$

This state can be rewritten as

$$U_4 U_3 U_2 U_1 |\psi 00\rangle = \frac{1}{\sqrt{2}}(|0\rangle + |1\rangle) \otimes \left(\frac{a}{\sqrt{2}}(|00\rangle + |11\rangle) \right)$$
$$+ \frac{1}{\sqrt{2}}(|0\rangle - |1\rangle) \otimes \left(\frac{b}{\sqrt{2}}(|01\rangle + |10\rangle) \right)$$

where $|00\rangle \equiv |0\rangle \otimes |0\rangle$. Applying all eight unitary matrices to the input state we obtain

$$U_8 U_7 U_6 U_5 U_4 U_3 U_2 U_1 |\psi 00\rangle = \frac{a}{2}(|000\rangle + |100\rangle + |010\rangle + |110\rangle)$$
$$+ \frac{b}{2}(|011\rangle + |111\rangle + |001\rangle + |101\rangle).$$

This normalized state can be rewritten as

$$\left(\frac{1}{\sqrt{2}}(|0\rangle + |1\rangle) \right) \otimes \left(\frac{1}{\sqrt{2}}(|0\rangle + |1\rangle) \right) \otimes |\psi\rangle.$$

The state $|\psi\rangle$ will be transferred to the lower output, where both other outputs will come out in the state

$$\frac{1}{\sqrt{2}}(|0\rangle + |1\rangle).$$

If the two upper outputs are measured in the standard basis ($|0\rangle$ versus $|1\rangle$), two random classical bits will be obtained in addition to the quantum state $|\psi\rangle$ on the lower output.

```cpp
// teleportion.cpp
#include <iostream>
#include "symbolicc++.h"
using namespace std;

Symbolic Hadamard(const Symbolic &v)
{
Symbolic H("",2,2);
Symbolic sqrt12 = sqrt(1/Symbolic(2));
H(0,0) = sqrt12; H(0,1) =  sqrt12; H(1,0) = sqrt12; H(1,1) = -sqrt12;
return (H*v);
}


Symbolic XOR(const Symbolic &v)
{
Symbolic X("",4,4);
X(0,0) = 1; X(0,1) = 0; X(0,2) = 0; X(0,3) = 0;
X(1,0) = 0; X(1,1) = 1; X(1,2) = 0; X(1,3) = 0;
X(2,0) = 0; X(2,1) = 0; X(2,2) = 0; X(2,3) = 1;
X(3,0) = 0; X(3,1) = 0; X(3,2) = 1; X(3,3) = 0;
return (X*v);
}


Symbolic Bell(const Symbolic &v)
{
Symbolic I("",2,2), H("",2,2), X("",4,4);
Symbolic sqrt12 = sqrt(1/Symbolic(2));
I = I.identity();
H(0,0) = sqrt12; H(0,1) =  sqrt12;
H(1,0) = sqrt12; H(1,1) = -sqrt12;
Symbolic UH = kron(H,I);
X(0,0) = 1; X(0,1) = 0; X(0,2) = 0; X(0,3) = 0;
X(1,0) = 0; X(1,1) = 1; X(1,2) = 0; X(1,3) = 0;
X(2,0) = 0; X(2,1) = 0; X(2,2) = 0; X(2,3) = 1;
X(3,0) = 0; X(3,1) = 0; X(3,2) = 1; X(3,3) = 0;
return (X*(UH*v));
}
```

```
Symbolic Swap(const Symbolic &v)
{
Symbolic S("",4,4);
S(0,0) = 1; S(0,1) = 0; S(0,2) = 0; S(0,3) = 0;
S(1,0) = 0; S(1,1) = 0; S(1,2) = 0; S(1,3) = 1;
S(2,0) = 0; S(2,1) = 0; S(2,2) = 1; S(2,3) = 0;
S(3,0) = 0; S(3,1) = 1; S(3,2) = 0; S(3,3) = 0;
return XOR(S*XOR(v));
}


Symbolic Teleport(const Symbolic &v)
{
Symbolic result;
Symbolic NOT("",2,2),H("",2,2),I("",2,2),X("",4,4);
Symbolic sqrt12 = sqrt(1/Symbolic(2));
NOT(0,0) = 0; NOT(0,1) = 1; NOT(1,0) = 1; NOT(1,1) = 0;
H(0,0) = sqrt12; H(0,1) =  sqrt12;
H(1,0) = sqrt12; H(1,1) = -sqrt12;
I = I.identity();
X(0,0) = 1; X(0,1) = 0; X(0,2) = 0; X(0,3) = 0;
X(1,0) = 0; X(1,1) = 1; X(1,2) = 0; X(1,3) = 0;
X(2,0) = 0; X(2,1) = 0; X(2,2) = 0; X(2,3) = 1;
X(3,0) = 0; X(3,1) = 0; X(3,2) = 1; X(3,3) = 0;
Symbolic U1=kron(I,kron(H,I));
Symbolic U2=kron(I,X);
Symbolic U3=kron(X,I);
Symbolic U4=kron(H,kron(I,I));
Symbolic U5=kron(I,X);
Symbolic U6=kron(I,kron(I,H));
Symbolic U7=dsum(I,dsum(I,dsum(NOT,NOT)));
Symbolic U8=kron(I,kron(I,H));
result=U8*(U7*(U6*(U5*(U4*(U3*(U2*(U1*v)))))));
return result;
}


// The outcome after measuring value for qubit.
// Since the probabilities may be symbolic this function
// cannot simulate a measurement where random outcomes
// have the correct distribution
Symbolic Measure(const Symbolic &v,unsigned int qubit,
                 unsigned int value)
{
int i, len, skip = 1-value;
Symbolic result(v);
Symbolic D;
```

```
len = v.rows()/int(pow(2.0,qubit+1.0));
for(i=0;i<v.rows();i++)
{
if(!(i%len)) skip = 1-skip;
if(skip) result(i) = 0; else D += result(i)*result(i);
}
return result/sqrt(D);
}


// for output clarity
ostream &print(ostream &o,const Symbolic &v)
{
char *b2[2]={"|0>","|1>"};
char *b4[4]={"|00>","|01>","|10>","|11>"};
char *b8[8]={"|000>","|001>","|010>","|011>",
             "|100>","|101>","|110>","|111>"};
char **b;
if(v.rows()==2) b=b2;
if(v.rows()==4) b=b4;
if(v.rows()==8) b=b8;
for(int i=0;i<v.rows();i++)
if(v(i)!=0) o << "+(" << v(i) << ")" << b[i];
return o;
}


int main(void)
{
Symbolic zero("",2),one("",2);
Symbolic zz("",4),zo("",4),oz("",4),oo("",4),qreg;
Symbolic tp00, tp01, tp10, tp11, psiGHZ;
Symbolic a("a"), b("b");
Symbolic sqrt12 = sqrt(1/Symbolic(2));
zero(0) = 1; zero(1) = 0; one(0) = 0; one(1) = 1;
zz = kron(zero,zero); zo = kron(zero,one);
oz = kron(one,zero);  oo = kron(one,one);
cout << "UH|0> = "; print(cout,Hadamard(zero))<< endl;
cout << "UH|1> = "; print(cout,Hadamard(one)) << endl;
cout << endl;
cout << "UXOR|00> = "; print(cout,XOR(zz)) << endl;
cout << "UXOR|01> = "; print(cout,XOR(zo)) << endl;
cout << "UXOR|10> = "; print(cout,XOR(oz)) << endl;
cout << "UXOR|11> = "; print(cout,XOR(oo)) << endl;
cout << endl;
cout << "UBELL|00> = "; print(cout,Bell(zz)) << endl;
cout << "UBELL|01> = "; print(cout,Bell(zo)) << endl;
cout << "UBELL|10> = "; print(cout,Bell(oz)) << endl;
```

```
cout << "UBELL|11> = "; print(cout,Bell(oo)) << endl;
cout << endl;
cout << "USWAP|00> = "; print(cout,Swap(zz)) << endl;
cout << "USWAP|01> = "; print(cout,Swap(zo)) << endl;
cout << "USWAP|10> = "; print(cout,Swap(oz)) << endl;
cout << "USWAP|11> = "; print(cout,Swap(oo)) << endl;
cout << endl;
qreg=kron(a*zero+b*one,kron(zero,zero));
cout << "UTELEPORT("; print(cout,qreg) << ") = ";
print(cout,qreg=Teleport(qreg)) << endl;
cout << "Results after measurement of first 2 qubits:" << endl;
tp00 = Measure(Measure(qreg,0,0),1,0);
tp01 = Measure(Measure(qreg,0,0),1,1);
tp10 = Measure(Measure(qreg,0,1),1,0);
tp11 = Measure(Measure(qreg,0,1),1,1);
Equations simplify = (a*a==1-b*b,1/sqrt(1/Symbolic(4))==2);
tp00 = tp00.subst_all(simplify);
tp01 = tp01.subst_all(simplify);
tp10 = tp10.subst_all(simplify);
tp11 = tp11.subst_all(simplify);
cout << "  |00> : " ; print(cout,tp00) << endl;
cout << "  |01> : " ; print(cout,tp01) << endl;
cout << "  |10> : " ; print(cout,tp10) << endl;
cout << "  |11> : " ; print(cout,tp11) << endl;
cout << endl;
psiGHZ=kron(zz,zero)*sqrt12+kron(oo,one)*sqrt12;
cout << "Greenberger-Horne-Zeilinger state : ";
print(cout,psiGHZ) << endl;
cout << "Measuring qubit 0 as 1 yields : ";
print(cout,Measure(psiGHZ,0,1)) <<endl;
cout << "Measuring qubit 1 as 1 yields : ";
print(cout,Measure(psiGHZ,1,1)) <<endl;
cout << "Measuring qubit 2 as 0 yields : ";
print(cout,Measure(psiGHZ,2,0)) <<endl;
return 0;
}
```

♣

Example 5.17. An application of the Kronecker product for the two-point Ising model with external field is as follows. We consider the two-point Ising model with an external field

$$\hat{H} = J\sigma_{3,1}\sigma_{3,2} + B(\sigma_{1,1} + \sigma_{1,2})$$

where σ_1, σ_2, σ_3 are the Pauli spin matrices and $J \neq 0$ and $B > 0$ are constants. We have

$$\sigma_{3,1} = \sigma_3 \otimes I_2, \quad \sigma_{3,2} = I_2 \otimes \sigma_3, \quad \sigma_{1,1} = \sigma_1 \otimes I_2, \quad \sigma_{1,2} = I_2 \otimes \sigma_1.$$

Thus

$$\hat{H} = J(\sigma_3 \otimes \sigma_3) + B(\sigma_1 \otimes I_2 + I_2 \otimes \sigma_1)$$

is a 4×4 matrix. We find that the 4×4 Hermitian matrix $\hat{H}$ takes the form

$$\hat{H} = \begin{pmatrix} J & B & B & 0 \\ B & -J & 0 & B \\ B & 0 & -J & B \\ 0 & B & B & J \end{pmatrix}.$$

The eigenvalues of $\hat{H}$ are

$$E_{1,2} = \pm\sqrt{J^2 + 4B^2}, \qquad E_3 = -J, \qquad E_4 = J.$$

Thus the eigenvalue of the ground state is given by

$$-\sqrt{J^2 + 4B^2}$$

which is nondegenerate. The four eigenvectors (not yet normalized) are

$$\begin{pmatrix} 1 \\ -(\sqrt{J^2+4B^2}+J)/2B \\ -(\sqrt{J^2+4B^2}+J)/2B \\ 1 \end{pmatrix}, \quad \begin{pmatrix} 1 \\ (\sqrt{J^2+4B^2}-J)/2B \\ (\sqrt{J^2+4B^2}-J)/2B \\ 1 \end{pmatrix},$$

$$\begin{pmatrix} 1 \\ 0 \\ 0 \\ -1 \end{pmatrix}, \quad \begin{pmatrix} 0 \\ 1 \\ -1 \\ 0 \end{pmatrix}.$$

After normalization the eigenvectors form an orthonormal basis in the Hilbert space $\mathbb{C}^4$.

In the Maxima program we evaluate the Hamilton operator $\hat{H}$. Finally we determine the eigenvalues and eigenvectors of $\hat{H}$.

```
/* HeisenbergModel.mac */
I2: matrix([1,0],[0,1]);
sig1: matrix([0,1],[1,0]);
sig2: matrix([0,-%i],[%i,0]);
sig3: matrix([1,0],[0,-1]);
H: J*kronecker_product(sig3,sig3)+
   B*(kronecker_product(sig1,I2)+kronecker_product(I2,sig1));
print("H=",H);
r1: eigenvalues(H); print("r1=",r1);
r2: eigenvectors(H); print("r2=",r2);
```

♣

Example 5.18. An application of the Kronecker product for the two-point Heisenberg model is as follows. The two-point Heisenberg model is given by

$$\hat{H} = J \sum_{j=1}^{2} \mathbf{S}_j \cdot \mathbf{S}_{j+1}$$

where J is the so-called exchange constant ($J > 0$ or $J < 0$) and $\cdot$ denotes the scalar product. We impose cyclic boundary conditions, i.e. $\mathbf{S}_3 = \mathbf{S}_1$. The Hamilton operator $\hat{H}$ is given by the 4×4 symmetric matrix

$$\hat{H} = \frac{J}{2} \begin{pmatrix} 1 & 0 & 0 & 0 \\ 0 & -1 & 2 & 0 \\ 0 & 2 & -1 & 0 \\ 0 & 0 & 0 & 1 \end{pmatrix} \equiv \frac{J}{2} \left((1) \oplus \begin{pmatrix} -1 & 2 \\ 2 & -1 \end{pmatrix} \oplus (1) \right)$$

where $\oplus$ denotes the *direct sum*. In the Maxima program we evaluate the eigenvalues and eigenvectors of $\hat{H}$. The eigenvalues are given by

$$E_1 = \frac{J}{2} \quad \text{three times degenerate,} \qquad E_2 = \frac{-3J}{2}.$$

If J is positive, then E_2 is the ground state energy. If J is negative, then E_1 is the ground state energy.

```
/* HeisenbergModel.mac */
sig1: matrix([0,1],[1,0]);
s1: 1/2*sig1;
sig2: matrix([0,-%i],[%i,0]);
s2: 1/2*sig2;
sig3: matrix([1,0],[0,-1]);
s3: 1/2*sig3;
H: kronecker_product(s1,s1)+kronecker_product(s2,s2)
   +kronecker_product(s3,s3);
H: 2*J*H; print("H=",H);
r1: eigenvalues(H); print("r1=",r1);
r2: eigenvectors(H); print("r2=",r2);
```

This list gives the eigenvalues J (3 times degenerate) and $-3J/2$ with the corresponding eigenvectors. ♣

Example 5.19. Let A, B be 2×2 matrices over $\mathbb{C}$. Show that

$$\det(A \otimes B - B \otimes A) = 0.$$

Let C, D be 3×3 matrices over $\mathbb{C}$. Show that

$$\det(C \otimes D - D \otimes C) = 0.$$

A Maxima implementation is

```
A: matrix([a11,a12],[a21,a22]);
B: matrix([b11,b12],[b21,b22]);
T1: kronecker_product(A,B);
T2: kronecker_product(B,A);
T3: T1-T2;
R: determinant(T3);
R: ratsimp(R);
C: matrix([c11,c12,c13],[c21,c22,c23],[c31,c32,c33]);
D: matrix([d11,d12,d13],[d21,d22,d23],[d31,d32,d33]);
T1: kronecker_product(C,D);
T2: kronecker_product(D,C);
T3: T1-T2;
R: determinant(T3);
R: ratsimp(R);
```
♣

Example 5.20. Let A, B be $m \times n$ matrices. The entrywise product $A \bullet B$ is defined by elementwise multiplication of the entries of the matrices A and B, i.e.

$$(A \bullet B)_{i,j} = a_{ij}b_{ij}, \quad i = 1,\ldots,m; \quad j = 1,\ldots,n.$$

The entrywise product $A \bullet B$ is a principal submatrix of the Kronecker product $A \otimes B$. For $n \times n$ sized matrices A and B, one can write

$$A \bullet B = J^{T}(A \otimes B)J$$

where J is an $n^2 \times n$ sized matrix (called *selection matrix*) with entries 0's and 1's satisfying $J^{T}J = I_n$. In the Maxima program we give an implementation for $n = 2$. Note that in Maxima the entrywise multiplication of matrices is given by $*$, matrix multiplication is given by . and the transpose by transpose().

```
/* selectionmatrix.mac */
A: matrix([a11,a12],[a21,a22]);
B: matrix([b11,b12],[b21,b22]);
EW: A*B;
ABK: kronecker_product(A,B);
SMT: matrix([1,0,0,0],[0,0,0,1]);
SM: transpose(SMT);
F: EW-SMT . ABK . SM;
```
♣

Example 5.21. Consider the Lie group $SU(2)$ which consists of all unitary matrices $U = (u_{k\ell})_{1 \leq k,\ell \leq 2}$ with $\det(U) = 1$. For example, the Pauli spin matrices are not elements of $SU(2)$, but $i\sigma_1$, $i\sigma_2$, $i\sigma_3$ are. Let

$$j \in \{1/2, 1, 3/2, 2, \ldots\}.$$

The matrix $U_H = \frac{1}{\sqrt{2}}(\sigma_1 + \sigma_3)$ is not element of $SU(2)$, but iU_H is. The unitary irreducible representations of the Lie group $SU(2)$ are given by

$$D^j_{m,m'}(U) = \sqrt{(j+m)!(j-m)!(j+m')!(j-m')!}$$

$$\times \sum_k \frac{u_{11}^k u_{12}^{j+m-k} u_{21}^{j+m'-k} u_{22}^{k-m-m'}}{k!(j+m-k)!(j+m'-k)!(k-m-m')!}$$

where m, m' take the values $-j$, $-j+1$, ...,j. The sum k runs over all nonnegative values of k for which all factorials in the denominator are nonnegative. We give an implementation in SymbolicC++ and apply it to the matrices $i\sigma_1$, $i\sigma_2$, $i\sigma_3$,

$$iU_H = \frac{1}{\sqrt{2}} \begin{pmatrix} i & i \\ i & -i \end{pmatrix}.$$

We set $u_{11} =$u11, $u_{12} =$u12, $u_{21} =$u22, $I_2 =$I2.

```cpp
// rep.cpp
#include <iostream>
#include "symbolicc++.h"
using namespace std;

int main(void)
{
Symbolic j, m, mp;
Symbolic u11("u11"), u12("u12"), u21("u21"), u22("u22");
using SymbolicConstant::i;
int k, kl, ku;
for(j=1/Symbolic(2);j!=2;j+=1/Symbolic(2))
{
cout << "j = " << j << endl;
Symbolic Dj("",int(2*j+1),int(2*j+1));

for(m=-j;int(m+j)<=int(2*j);m++)
 for(mp=-j;int(mp+j)<=int(2*j);mp++)
 {
Symbolic sq
  = sqrt(factorial(j+m)*factorial(j-m)*factorial(j+mp)*factorial(j-mp));
Symbolic sum = 0;
kl = (int(m+mp)>0) ? int(m+mp) : 0;
ku = (int(j+m)>int(j+mp)) ? int(j+mp) : int(j+m);
for(k=kl;k<=ku;k++)
sum += (u11^k)*(u12^(j+m-k))*(u21^(j+mp-k))*(u22^(k-m-mp))
    /(factorial(k)*factorial(j+m-k)*factorial(j+mp-k)*factorial(k-m-mp));
Dj(int(j-m),int(j-mp)) = sq*sum;
}
```

```
Dj = Dj[sqrt(Symbolic(4))==2,sqrt(Symbolic(12))==2*sqrt(Symbolic(3)),
        sqrt(Symbolic(36))==6];
cout << Dj << endl;
cout << "For u=I2:" << endl
        << Dj[u11==1,u12==0,u21==0,u22==1] << endl;
cout << "For u=i sigma 1:" << endl
        << Dj[u11==0,u12==i,u21==i,u22==0] << endl;
cout << "For u = i sigma 2:" << endl
        << Dj[u11==0,u12==1,u21==-1,u22==0] << endl;
cout << "For u = i sigma 3:" << endl
        << Dj[u11==i,u12==0,u21==0,u22==-i] << endl;
cout << "For u = i U_H:" << endl
        << Dj[u11==i/sqrt(Symbolic(2)),u12==i/sqrt(Symbolic(2)),
              u21==i/sqrt(Symbolic(2)),u22==-i/sqrt(Symbolic(2))] << endl;
}
return 0;
}
```

For $j = 1/2$ the output is

$$\begin{pmatrix} u11 & u12 \\ u21 & u22 \end{pmatrix}, \quad \begin{pmatrix} 0 & i \\ i & 0 \end{pmatrix}, \quad \begin{pmatrix} 0 & 1 \\ -1 & 0 \end{pmatrix}, \quad \begin{pmatrix} i & 0 \\ 0 & -i \end{pmatrix}$$

and

$$\begin{pmatrix} i*(2)^{(-1/2)} & i*(2)^{(-1/2)} \\ i*(2)^{(-1/2)} & -i*(2)^{(-1/2)} \end{pmatrix}.$$

For $j = 1$ the output is

```
[      u11^(2)          (2)^(1/2)*u11*u12        u12^(2)        ]
[(2)^(1/2)*u11*u21   u12*u21+u11*u22   (2)^(1/2)*u12*u22]
[      u21^(2)          (2)^(1/2)*u21*u22        u22^(2)        ]
```

$$\begin{pmatrix} 1 & 0 & 0 \\ 0 & 1 & 0 \\ 0 & 0 & 1 \end{pmatrix}, \quad \begin{pmatrix} 0 & 0 & -1 \\ 0 & -1 & 0 \\ -1 & 0 & 0 \end{pmatrix}, \quad \begin{pmatrix} 0 & 0 & 1 \\ 0 & -1 & 0 \\ 1 & 0 & 0 \end{pmatrix}, \quad \begin{pmatrix} -1 & 0 & 0 \\ 0 & 1 & 0 \\ 0 & 0 & -1 \end{pmatrix}.$$

For u = i U_H:
```
[    -1/2      -(2)^(-1/2)       -1/2     ]
[-(2)^(-1/2)        0         (2)^(-1/2)]
[    -1/2       (2)^(-1/2)       -1/2     ]
```

For $j = 3/2$ the output is

For u = I2:
```
[1 0 0 0]
[0 1 0 0]
[0 0 1 0]
```

```
[0 0 0 1]

For u = i sigma 1:
[     0           0          0       (-1)^(3/2)]
[     0           0         -i          0      ]
[     0          -i          0          0      ]
[(-1)^(3/2)       0          0          0      ]
For u = i sigma 2:
[ 0  0  0  1]
[ 0  0 -1  0]
[ 0  1  0  0]
[-1  0  0  0]
For u = i sigma 3:
[ (-1)^(3/2)      0          0          0        ]
[     0           i          0          0        ]
[     0           0         -i          0        ]
[     0           0          0     -(-1)^(3/2)]
```

Example 5.22. *Dimino's algorithm* computes a complete enumeration of the elements of a finite group given a generating set

$$\{\, g_0, g_1, \ldots, g_{n-1} \,\}.$$

Dimino's algorithm enumerates, successively, the elements of each of the subgroups

$$G_k \,=\, <\, g_0, g_1, \ldots, g_{k-1}\, >$$

of G which form a chain

$$<\, g_0\, > \,=\, G_0 \leq G_1 \leq \cdots \leq G_{k-1} \leq \cdots \leq G_{n-1} = G.$$

These elements can be enumerated by performing products of the generators g_0, g_1, $\ldots$, g_{n-1} in all possible ways until all the elements of G have been found. We apply Dimino's algorithm to the 4×4 permutation matrices

$$g_0 = \begin{pmatrix} 0 & 1 & 0 & 0 \\ 1 & 0 & 0 & 0 \\ 0 & 0 & 1 & 0 \\ 0 & 0 & 0 & 1 \end{pmatrix}, \quad g_1 = \begin{pmatrix} 1 & 0 & 0 & 0 \\ 0 & 0 & 1 & 0 \\ 0 & 1 & 0 & 0 \\ 0 & 0 & 0 & 1 \end{pmatrix}, \quad g_2 = \begin{pmatrix} 1 & 0 & 0 & 0 \\ 0 & 1 & 0 & 0 \\ 0 & 0 & 0 & 1 \\ 0 & 0 & 1 & 0 \end{pmatrix}.$$

The SymbolicC++ program using the `list` class is

```cpp
// dimino.cpp
#include <iostream>
#include <list>
#include "symbolicc++.h"
```

```
using namespace std;

template<class T>
void unique_add(list<T> &l,const T &t)
{
 typename list<T>::iterator li;
 for(li=l.begin();li!=l.end();++li)
  if(*li==t) return;
 l.push_back(t);
}

list<Symbolic> dimino1(const list<Symbolic> &g)
{
 size_t ln;
 list<Symbolic> l;
 Symbolic g0 = g.front();
 Symbolic g0n = g0;
 l.push_back(g0);
 ln = 0;
 // generate all powers until we repeat a result
 while(ln != l.size())
 { ln = l.size(); g0n *= g0; unique_add(l,g0n); }
 return l;
}

// generate G_n from <G_{n-1}> and gn
list<Symbolic> generate(const list<Symbolic> &g,
    const Symbolic &gn)
{
 size_t ln;
 list<Symbolic> l = g;
 list<Symbolic>::iterator li;
 ln = 0;
 while(ln != l.size())
 {
  ln = l.size();
  for(li=l.begin();li!=l.end();++li)
  { unique_add(l,*li * gn); unique_add(l,gn* *li); }
 }
```

```
  return l;
}

list<Symbolic> dimino(const list<Symbolic> &g)
{
 size_t ln;
 list<Symbolic> l;
 list<Symbolic>::const_iterator li;
 if(g.size()==1) return dimino1(g);
 else
 {
  ln = 0; l.push_back(g.front()); l = dimino1(l);
  while(ln != l.size())
  {
   ln = l.size(); li = g.begin(); ++li;
   for(;li!=g.end();++li) l = generate(l,*li);
  }
 }
 return l;
}

int main(void)
{
 list<Symbolic> s;
 list<Symbolic>::iterator li;
 Symbolic g0 = ((Symbolic(0),1,0,0),(Symbolic(1),0,0,0),
                (Symbolic(0),0,1,0),(Symbolic(0),0,0,1));
 Symbolic g1 = ((Symbolic(1),0,0,0),(Symbolic(0),0,1,0),
                (Symbolic(0),1,0,0),(Symbolic(0),0,0,1));
 Symbolic g2 = ((Symbolic(1),0,0,0),(Symbolic(0),1,0,0),
                (Symbolic(0),0,0,1),(Symbolic(0),0,1,0));
 s.push_back(g0); s.push_back(g1); s.push_back(g2);
 cout << "Generators" << endl << "==============" << endl;
 for(li=s.begin();li!=s.end();++li) cout << *li;
 cout << endl << "Group" << endl << "==============" << endl;
 list<Symbolic> g = dimino(s);
 for(li=g.begin();li!=g.end();++li) cout << *li;
 return 0;
}
```

We find that g_0, g_1, g_2 generates all $4! = 24$ permutation matrices. ♣

Example 5.23. Consider a normal 4×4 matrix V with eigenvalues

$$\lambda_1 = i, \quad \lambda_2 = -1, \quad \lambda_3 = -i, \quad \lambda_4 = 1$$

and the corresponding normalized eigenvectors

$$\mathbf{v}_1 = \frac{1}{\sqrt{2}} \begin{pmatrix} 1 \\ 0 \\ 0 \\ 1 \end{pmatrix}, \quad \mathbf{v}_2 = \frac{1}{\sqrt{2}} \begin{pmatrix} 0 \\ 1 \\ 1 \\ 0 \end{pmatrix},$$

$$\mathbf{v}_3 = \frac{1}{\sqrt{2}} \begin{pmatrix} 0 \\ 1 \\ -1 \\ 0 \end{pmatrix}, \quad \mathbf{v}_4 = \frac{1}{\sqrt{2}} \begin{pmatrix} 1 \\ 0 \\ 0 \\ -1 \end{pmatrix}.$$

Reconstruct the matrix V using the spectral theorem. Show that V is unitary. Show that $V^4 = I_4$.

```
/* spectral.mac */
v1: matrix([1],[0],[0],[1])/sqrt(2);
v2: matrix([0],[1],[1],[0])/sqrt(2);
v3: matrix([0],[1],[-1],[0])/sqrt(2);
v4: matrix([1],[0],[0],[-1])/sqrt(2);
v1T: transpose(v1);
v2T: transpose(v2);
v3T: transpose(v3);
v4T: transpose(v4);
lam1: %i; lam2: -1; lam3: -%i; lam4: 1;
V: lam1*(v1 . v1T)+lam2*(v2 . v2T)+lam3*(v3 . v3T)+lam4*(v4 . v4T);
VT: transpose(V);
VTC: conjugate(VT);
R: V . VTC; R: ratsimp(R);
F: V . V . V. V; F: ratsimp(F);
```
♣

Example 5.24. Consider the three 4×4 matrices

$$g_1 = \begin{pmatrix} 0 & 1 & 0 & 0 \\ 1 & 0 & 0 & 0 \\ 0 & 0 & 1 & 0 \\ 0 & 0 & 0 & 1 \end{pmatrix}, \quad g_2 = \begin{pmatrix} 1 & 0 & 0 & 0 \\ 0 & 0 & 1 & 0 \\ 0 & 1 & 0 & 0 \\ 0 & 0 & 0 & 1 \end{pmatrix}, \quad g_3 = \begin{pmatrix} 1 & 0 & 0 & 0 \\ 0 & 1 & 0 & 0 \\ 0 & 0 & 0 & 1 \\ 0 & 0 & 1 & 0 \end{pmatrix}$$

which are generators of the 24 4×4 permutation matrices. Show that

$$g_1 g_2 g_1 = g_2 g_1 g_2, \quad g_2 g_3 g_2 = g_3 g_2 g_3$$

and $g_1 g_3 g_1 - g_3 g_1 g_3$ is the direct sum of two 2×2 matrices. Let

$$g_{11} = g_1 \otimes g_1, \quad g_{22} = g_2 \otimes g_2, \quad g_{33} = g_3 \otimes g_3.$$

Show that

$$g_{11} g_{22} g_{11} = g_{22} g_{11} g_{22}, \quad g_{22} g_{33} g_{22} = g_{33} g_{22} g_{33}$$

and $g_{11} g_{33} g_{11} - g_{33} g_{11} g_{33}$ is the direct sum of two 8×8 matrices.

```
/* braid0.mac */
g1: matrix([0,1,0,0],[1,0,0,0],[0,0,1,0],[0,0,0,1]);
g2: matrix([1,0,0,0],[0,0,1,0],[0,1,0,0],[0,0,0,1]);
g3: matrix([1,0,0,0],[0,1,0,0],[0,0,0,1],[0,0,1,0]);
R12: g1 . g2 . g1 - g2 . g1 . g2;
R13: g1 . g3 . g1 - g3 . g1 . g3;
R23: g2 . g3 . g2 - g3 . g2 . g3;
g11: kronecker_product(g1,g1);
g22: kronecker_product(g2,g2);
g33: kronecker_product(g3,g3);
R1122: g11 . g22 . g11 - g22 . g11 . g22;
R1133: g11 . g33 . g11 - g33 . g11 . g33;
R2233: g22 . g33 . g22 - g33 . g22 . g33;
```

♣

Example 5.25. Give a C++ implantation for matrices based on the **vector** class of the Standard Template Library. The underlying field is the complex numbers. Implement the transpose, matrix multiplication, Kronecker product and the cyclic matrix. Furthermore implement the unitary matrix

$$U(\phi, \theta) = \begin{pmatrix} \cos(\theta) & e^{i\phi} \sin(\theta) \\ e^{-i\phi} \sin(\theta) & -\cos(\theta) \end{pmatrix}.$$

```cpp
// matrixvector.cpp
// c++ -std=c++11 -o matrixvector matrixvector.cpp
// ./matrixvector

#include <cassert>
#include <complex>
#include <iostream>
#include <vector>
using namespace std;

vector<vector<complex<double> > >
transpose(const vector<vector<complex<double> > > &M)
{
  int i, j;
```

```
vector<vector<complex<double> > > T;
if(M.size()==0) return T;
T.resize(M[0].size());
for(i=0;i<T.size();++i)
{ T[i].resize(M.size()); for(j=0;j<M.size();++j) T[i][j]=M[j][i]; }
return T;
}


vector<complex<double> >
mul(const vector<vector<complex<double> > > &M,
    const vector<complex<double> > &v)
{
 int i, j;
 vector<complex<double> > u(M.size());
 assert(M.size()>0 && M[0].size()==v.size());
 for(i=0;i<M.size();++i)
 {
 u[i] = 0.0;
 for(j=0;j<v.size();++j) u[i] += M[i][j]*v[j];
 }
 return u;
}


vector<complex<double> >
mulT(const vector<complex<double> > &v,
     const vector<vector<complex<double> > > &M)
{
 int i, j;
 assert(M.size()>0 && M[0].size()==v.size());
 vector<complex<double> > u(M[0].size());
 for(i=0;i<M[0].size();++i)
 {
 u[i] = 0.0;
 for(j=0;j<v.size();++j) u[i] += M[j][i]*v[j];
 }
 return u;
}


vector<vector<complex<double> > >
mul(const vector<vector<complex<double> > > &M,
    const vector<vector<complex<double> > > &A)
{
 int i, j, k;
 vector<vector<complex<double> > > B(M.size());
 assert(M.size()>0 && A.size()>0 && M[0].size()==A.size());
 for(i=0;i<M.size();++i)
```

```
{
B[i].resize(A[0].size());
for(j=0;j<A[0].size();++j)
{
B[i][j] = 0.0;
for(k=0;k<M[0].size();++k) B[i][j] += M[i][k]*A[k][j];
}
}
return B;
}


vector<complex<double> >
kron(const vector<complex<double> > &u,
     const vector<complex<double> > &v)
{
 int i, j;
 vector<complex<double> > k(u.size()*v.size());
 for(i=0;i<u.size();++i)
  for(j=0;j<v.size();++j) k[i*v.size()+j] = u[i]*v[j];
 return k;
}


vector<vector<complex<double> > >
kron(const vector<vector<complex<double> > > &A,
     const vector<vector<complex<double> > > &B)
{
 int i, j, k, l;
 vector<vector<complex<double> > > K(A.size()*B.size());
 assert(A.size()>0 && B.size()>0);
 for(i=0;i<A.size();++i)
  for(j=0;j<B.size();++j)
  {
  K[i*B.size()+j].resize(A[0].size()*B[0].size());
  for(k=0;k<A[0].size();++k)
   for(l=0;l<B[0].size();++l)
    K[i*B.size()+j][k*B[0].size()+l] = A[i][k]*B[j][l];
  }
 return K;
}


template<class T>
ostream& operator << (ostream& o,const vector<T>& v)
{
 int i;
 o << "[ ";
 for(i=0;i<v.size();++i) o << v[i] << " ";
```

```
  o << "]";
  return o;
}

vector<vector<complex<double> > >
cyclic(const unsigned int n)
{
 int i, j;
 vector<vector<complex<double> > > C(n);
 for(i=0;i<n;i++) C[i].resize(n);
 for(i=0;i<n;++i)
  for(j=0;j<n;++j) { C[i][j] = 0.0; }
 for(i=0;i<n;i++) C[i][(i+1)%n]=1.0;
 return C;
}

vector<vector<complex<double> > >
U(const double theta,const double phi)
{
 vector<vector<complex<double> > > M(2);
 M[0].resize(2); M[1].resize(2);
 M[0][0] = cos(theta);
 M[0][1] = sin(theta)*(exp(complex<double>(0,phi)));
 M[1][0] = sin(theta)*(exp(complex<double>(0,-phi)));
 M[1][1] = -cos(theta);
 return M;
}

int main(void)
{
  vector<vector<complex<double> > > H;
  vector<complex<double> > v = { 0.0, 1.0, 2.0, 3.0, 4.0, 5.0, 6.0, 7.0 };
  H = cyclic(8);
  cout << "H = " << H << endl;
  cout << "H^T = " << transpose(H) << endl;
  cout << "v = " << v << endl;
  cout << "H v = " << mul(H,v) << endl;
  cout << "v^T H = " << mulT(v,H) << endl;
  vector<vector<complex<double> > > A =
   { { 1.0, 0.0, 0.0, 0.0, 0.0, 0.0, 0.0, 1.0 },
     { 0.0, 1.0, 0.0, 0.0, 0.0, 0.0, 1.0, 0.0 },
     { 0.0, 0.0, 1.0, 0.0, 0.0, 1.0, 0.0, 0.0 },
     { 0.0, 0.0, 0.0, 1.0, 1.0, 0.0, 0.0, 0.0 },
     { 0.0, 0.0, 0.0, 1.0, 1.0, 0.0, 0.0, 0.0 },
     { 0.0, 0.0, 1.0, 0.0, 0.0, 1.0, 0.0, 0.0 },
     { 0.0, 1.0, 0.0, 0.0, 0.0, 0.0, 1.0, 0.0 },
```

```
    { 1.0, 0.0, 0.0, 0.0, 0.0, 0.0, 0.0, 1.0 } };
  cout << "A = " << A << endl;
  cout << "A*H^T = " << mul(A,transpose(H)) << endl;
  cout << "H A H^T = " << mul(H,mul(A,transpose(H))) << endl;
  cout << "kron(v,v) = " << kron(v,v) << endl;
  vector<vector<complex<double> > > B =
    { { 1.0,-1.0 },{ 0.0,2.0 },{ 3.0,0.0 } };
  vector<vector<complex<double> > > C =
    { { 1.0,2.0,0.5 },{ -1.0,1.0,0.0 } };
  cout << "B = " << B << endl;
  cout << "C = " << C << endl;
  cout << "kron(B,C) = " << kron(B,C) << endl;
  double pi = 4.0*atan(1.0);
  cout << "pi = " << pi << endl;
  cout << "U(0.0,0.0) = " << U(0.0,0.0) << endl;
  cout << "U(pi/2.0,0.0) = " << U(pi/2.0,0.0) << endl;
  cout << "U(pi/2.0,-pi/2.0) = " << U(pi/2.0,-pi/2.0) << endl;
  cout << "kronecker(U(0,0),U(0,0)) = " << kron(U(0.0,0.0),U(0.0,0.0)));
  return 0;
}
```

♣

Example 5.26. The *Cayley-Hamilton theorem* tells us that every linear transformation of an n-dimensional vector space to itself ($n \times n$ matrices) is a zero of its characteristic polynomial. In the program we consider the 4×4 matrix

$$A = \begin{pmatrix} 1 & 0 & 0 & 1 \\ 0 & 1 & 1 & 0 \\ 0 & 1 & 1 & 0 \\ 1 & 0 & 0 & 1 \end{pmatrix}.$$

In Maxima the command to find the characteristic polynomial is `charpoly`.

```
/* CayleyHamilton.mac */
I4: matrix([1,0,0,0],[0,1,0,0],[0,0,1,0],[0,0,0,1]);
A: matrix([1,0,0,1],[0,1,1,0],[0,1,1,0],[1,0,0,1]);
expr: expand(charpoly(A,lambda));
t0: coeff(expr,lambda,0);
t1: coeff(expr,lambda,1);
t2: coeff(expr,lambda,2);
t3: coeff(expr,lambda,3);
t4: coeff(expr,lambda,4);
M: t0*I4+t1*A+t2*(A . A)+t3*(A . A . A)+t4*(A . A . A . A);
```

So M is the 4×4 zero matrix.

♣

Example 5.27. The *spectral theorem* is tested for the 2×2 Hermitian matrix

$$A = \begin{pmatrix} 0 & 2 \\ 2 & 1 \end{pmatrix}.$$

The eigenvalues and the eigenvectors are calculated. The eigenvectors are then normalized so that the spectral theorem can be applied. The eigenvalues are given by

$$\lambda_1 = -\frac{1}{2}(\sqrt{17} - 1), \quad \lambda_2 = \frac{1}{2}(\sqrt{17} + 1).$$

```
/* Spectral1.mac */
A: matrix([0,2],[2,1]);
list: eigenvectors(A);
part1: part(list,1);
part11: first(part1);
lam1: first(part11);
lam2: second(part11);
part2: part(list,2);
v1: first(part2);
v2: second(part2);
v1: first(v1);
v2: first(v2);
v1n: v1 . transpose(v1);
v2n: v2 . transpose(v2);
v1: v1/sqrt(v1n);
v2: v2/sqrt(v2n);
B: lam1*(transpose(v1) . v1) + lam2*(transpose(v2) . v2);
B: ratsimp(B);
```

♣

Example 5.28. We consider the compact Lie group $SO(2)$ with the element

$$G = \begin{pmatrix} \cos(x) & -\sin(x) \\ \sin(x) & \cos(x) \end{pmatrix}.$$

Taking the derivative of the entries of the matrix with respect to x and then setting $x = 0$ we obtain the generator of the Lie group $SO(2)$. The eigenvalues of the generator are $-i$ and $+i$ with the normalized eigenvectors

$$\frac{1}{\sqrt{2}} \begin{pmatrix} 1 \\ i \end{pmatrix}, \quad \frac{1}{\sqrt{2}} \begin{pmatrix} 1 \\ -i \end{pmatrix}.$$

Then we consider the Lie group $SO(1,1)$ with the element

$$K = \begin{pmatrix} \cosh(x) & \sinh(x) \\ \cosh(x) & \sinh(x) \end{pmatrix}.$$

Taking the derivative of the entries of the matrix with respect to x and then setting $x = 0$ we obtain the generator of the Lie group SO(1,1). The eigenvalues of the generator are -1 and $+1$ with the normalized eigenvectors

$$\frac{1}{\sqrt{2}}\begin{pmatrix}1\\-1\end{pmatrix}, \quad \frac{1}{\sqrt{2}}\begin{pmatrix}1\\1\end{pmatrix}.$$

```
/* LieGroup.mac */
G: matrix([cos(x),-sin(x)],[sin(x),cos(x)]);
GD: diff(G,x);
GDO: subst(0,x,GD);
R: eigenvectors(GDO);
K: matrix([cosh(x),sinh(x)],[sinh(x),cosh(x)]);
KD: diff(K,x);
KDO: subst(0,x,KD);
S: eigenvectors(KDO);
```

♣

Example 5.29. Consider the *rotation matrix*

$$R(\phi) = \begin{pmatrix}\cos(\phi) & -\sin(\phi)\\\sin(\phi) & \cos(\phi)\end{pmatrix}.$$

Then

$$X = \left.\frac{dR(\phi)}{d\phi}\right|_{\phi=0} = \begin{pmatrix}0 & -1\\1 & 0\end{pmatrix}.$$

We reconstruct $R(\phi) = \exp(\phi X)$ from the eigenvalues $\lambda_1 = i$, $\lambda_2 = -i$ and normalized eigenvectors of X and the *spectral theorem*.

```
/* Spectral2.mac */
lam1: +%i; lam2: -%i;
v1: matrix([1],[-%i])/sqrt(2);
v1T: transpose(v1); v1TC: conjugate(v1T);
v2: matrix([1],[%i])/sqrt(2);
v2T: transpose(v2); v2TC: conjugate(v2T);
R: exp(%i*phi)*(v1 . v1TC) + exp(-%i*phi)*(v2 . v2TC);
R: subst(cos(phi)+%i*sin(phi),exp(%i*phi),R);
R: subst(cos(phi)-%i*sin(phi),exp(-%i*phi),R);
R: trigsimp(R);
```

♣

Example 5.30. Consider the Pauli spin matrices. The Pauli spin matrix σ_1 admits the eigenvalues $\lambda_1 = +1$ and $\lambda_2 = -1$ with the corresponding normalized eigenvectors

$$\mathbf{v}_1 = \frac{1}{\sqrt{2}}\begin{pmatrix}1\\1\end{pmatrix}, \quad \mathbf{v}_2 = \frac{1}{\sqrt{2}}\begin{pmatrix}1\\-1\end{pmatrix}.$$

The Pauli spin matrix σ_3 admits the eigenvalues $\mu_1 = +1$ and $\mu_2 = -1$ with the corresponding normalized eigenvectors

$$\mathbf{w}_1 = \begin{pmatrix} 1 \\ 0 \end{pmatrix}, \quad \mathbf{w}_2 = \begin{pmatrix} 0 \\ 1 \end{pmatrix}.$$

We find $\sigma_1 \otimes \sigma_3$ and $\sigma_3 \otimes \sigma_1$ using the spectral representations of σ_1 and σ_3 and show that $\sigma_1 \otimes \sigma_3 \neq \sigma_3 \otimes \sigma_1$.

```
/* Spectral3.mac */
lam1: +1; lam2: -1;
v1: matrix([1],[1])/sqrt(2); v2: matrix([1],[-1])/sqrt(2);
mu1: +1; mu2: -1;
w1: matrix([1],[0]); w2: matrix([0],[1]);
v1T: transpose(v1); v2T: transpose(v2);
w1T: transpose(w1); w2T: transpose(w2);
Tv: lam1*(v1 . v1T) + lam2*(v2 . v2T);
Tw: mu1*(w1 . w1T) + mu2*(w2 . w2T);
sig13: kronecker_product(Tv,Tw);
sig31: kronecker_product(Tw,Tv);
if equal(sig13,sig31) then print ("equal") else print("inequal");
```
♣

Example 5.31. The complex simple Lie algebras can be classified up to isomorphism. There are five exceptional Lie algebras g_2, f_4, e_6, e_7, e_8 with dimensions 14, 52, 52, 78, 133. We consider the Lie algebra g_2 with dimension 14. Given the Cartan-Weyl basis and the commutation relation we obtain the adjoint representation given by 14×14 matrices. Obviously the representation for the elements $H(0)$ and $H(1)$ are diagonal matrices. Note that the adjoint representation of a semisimple Lie algebra is faithful. The adjoint representation of a simple Lie algebra is irreducible.

```cpp
// adjoint2.cpp
#include <iostream>
#include "symbolicc++.h"
using namespace std;

int main(void)
{
 int n=14, i, j ,k;
 Symbolic V("V",n), H("H",2), X("X",6), Y("Y",6);
 Symbolic C("",n,n); // commutator of the 14 basis elements
 Symbolic adV("adV",n);
 V(0) = H(0); V(1) = H(1);
 V(2) = X(0); V(3) = X(1); V(4) = X(2);
 V(5) = X(3); V(6) = X(4); V(7) = X(5);
```

```
V(8) = Y(0); V(9) = Y(1); V(10) = Y(2);
V(11) = Y(3); V(12) = Y(4); V(13) = Y(5);
C(0,0) = Symbolic(0), C(0,1) = Symbolic(0);
C(0,2) = -2*X(0); C(0,3) = -3*X(1); C(0,4) = -X(2); C(0,5) = X(3);
C(0,6) = 3*X(4); C(0,7) = Symbolic(0);
C(0,8) = -2*Y(0); C(0,9) = -3*Y(1);
C(0,10) = Y(2); C(0,11) = -Y(3);
C(0,12) = -3*Y(4); C(0,13) = Symbolic(0);
C(1,1) = Symbolic(0); C(1,2) = -X(0);
C(1,3) = 2*X(1); C(1,4) = X(2);
C(1,5) = Symbolic(0); C(1,6) = -X(4);
C(1,7) = X(5); C(1,8) = Y(0);
C(1,9) = -2*Y(1); C(1,10) = -Y(2);
C(1,11) = Symbolic(0); C(1,12) = Y(4); C(1,13) = -Y(5);
C(2,2) = Symbolic(0); C(2,3) = X(2);
C(2,4) = 2*X(3); C(2,5) = -3*X(4);
C(2,6) = Symbolic(0); C(2,7) = Symbolic(0);
C(2,8) = H(0); C(2,9) = Symbolic(0);
C(2,10) = -3*Y(1); C(2,11) = -2*Y(2);
C(2,12) = Y(3); C(2,13) = Symbolic(0);
C(3,3) = Symbolic(0); C(3,4) = Symbolic(0); C(3,5) = Symbolic(0);
C(3,6) = -X(5); C(3,7) = Symbolic(0);
C(3,8) = Symbolic(0); C(3,9) = H(1);
C(3,10) = Y(0); C(3,11) = Symbolic(0);
C(3,12) = Symbolic(0); C(3,13) = Y(4);
C(4,4) = Symbolic(0); C(4,5) = -3*X(5); C(4,6) = Symbolic(0);
C(4,7) = Symbolic(0); C(4,8) = -3*X(1);
C(4,9) = X(0); C(4,10) = H(0)+3*H(1);
C(4,11) = 2*Y(0); C(4,12) = Symbolic(0); C(4,13) = Y(3);
C(5,5) = Symbolic(0); C(5,6) = Symbolic(0); C(5,7) = Symbolic(0);
C(5,8) = -2*X(2); C(5,9) = Symbolic(0); C(5,10) = 2*X(0);
C(5,11) = 2*H(0)+3*H(1); C(5,12) = -Y(0); C(5,13) = Y(2);
C(6,6) = Symbolic(0); C(6,7) = Symbolic(0);
C(6,8) = X(3); C(6,9) = Symbolic(0);
C(6,10) = Symbolic(0); C(6,11) = -X(0);
C(6,12) = H(0)+H(1); C(6,13) = -Y(1);
C(7,7) = Symbolic(0); C(7,8) = Symbolic(0);
C(7,9) = X(4); C(7,10) = X(3);
C(7,11) = -X(2); C(7,12) = -X(1); C(7,13) = H(0)+H(1);
C(8,8) = Symbolic(0); C(8,9) = -Y(2);
C(8,10) = -2*Y(3); C(8,11) = 3*Y(4);
C(8,12) = Symbolic(0); C(8,13) = Symbolic(0);
C(9,9) = Symbolic(0); C(9,10) = Symbolic(0); C(9,11) = Symbolic(0);
C(9,12) = Y(5); C(9,13) = Symbolic(0);
C(10,10) = Symbolic(0); C(10,11) = 3*Y(5);
C(10,12) = Symbolic(0); C(10,13) = Symbolic(0);
```

```
C(11,11) = Symbolic(0); C(11,12) = Symbolic(0);
C(11,13) = Symbolic(0); C(12,12) = Symbolic(0);
C(12,13) = Symbolic(0); C(13,13) = Symbolic(0);
for(i=0;i<n;i++)
 for(j=0;j<i;j++) C(i,j) = -C(j,i);
for(i=0;i<n;i++)
{
  Symbolic rep("",n,n);
  for(j=0;j<n;j++)
   for(k=0;k<n;k++) rep(k,j) = C(i,j).coeff(V(k));
  adV(i) = rep;
  cout << "ad" << V(i) << " = " << adV(i) << endl;
}
return 0;
}
```

♣

Example 5.32. We introduce the $n \times n$ Fourier matrices for $n = 2, 3, 4$ and then evaluate FAF^* for an $n \times n$ matrix A.

```
// fourier.cpp
#include <iostream>
#include "symbolicc++.h"

const Symbolic fourier(const int n)
{
 int j, k;
 const Symbolic &i = SymbolicConstant::i;
 const Symbolic &pi = SymbolicConstant::pi;
 Symbolic F("",n,n);
 Symbolic w, wj = 1;
 if(n<1) throw SymbolicError(SymbolicError::NotMatrix);

 switch(n)
 {
  case  1: w = 1; break;
  case  2: w = -1; break;
  case  3: w = (-1+i*sqrt(Symbolic(3)))/2; break;
  case  4: w = i; break;
  case  6: w = (1+i*sqrt(Symbolic(3)))/2; break;
  case  8: w = (1+i)/sqrt(Symbolic(2)); break;
  case 12: w = (i+sqrt(Symbolic(3)))/2; break;
  default: w = exp(2*i*pi/n);
   break;
 }
```

```
 for(j=0;j<n;++j)
 {
  Symbolic wjk = 1;
  for(k=0;k<n;++k) { F(j,k) = wjk; wjk *= wj; }
  wj *= w;
 }
 return F/sqrt(Symbolic(n));
}

Symbolic ctranspose(const Symbolic &A)
{
 const Symbolic &i = SymbolicConstant::i;
 return A.transpose().subst(i==-i);
}

Symbolic FT(const Symbolic &A)
{
 int n = A.rows();
 if(A.columns() != A.rows())
   throw SymbolicError(SymbolicError::NotMatrix);
 Symbolic F = fourier(n);
 return F*A*ctranspose(F);
}

int main(void)
{
 cout << fourier(2) << endl;
 cout << fourier(3) << endl;
 cout << fourier(4) << endl;
 cout << ctranspose(fourier(4)) << endl;
 Symbolic A = ((Symbolic(1),1),(Symbolic(2),2));
 cout << "A = " << A << endl;
 cout << FT(A) << endl;
 Symbolic B = ((Symbolic(0),1,0),(Symbolic(0),0,1),
               (Symbolic(1),0,0));
 cout << "B = " << B << endl;
 cout << FT(B) << endl;
 return 0;
}
```

♣

Example 5.33. We consider mutually unbiased bases. For example starting from the standard basis in $\mathbb{C}^3$ we find the mutually unbiased basis

$$\begin{pmatrix} 1/\sqrt{3} \\ 1/\sqrt{3} \\ 1/\sqrt{3} \end{pmatrix}, \quad \begin{pmatrix} 1/\sqrt{3} \\ -1/(2\sqrt{3}) - i/2 \\ -1/(2\sqrt{3}) + i/2 \end{pmatrix}, \quad \begin{pmatrix} 1/\sqrt{3} \\ -1/(2\sqrt{3}) + i/2 \\ -1/(2\sqrt{3}) - i/2 \end{pmatrix}.$$

```cpp
// mub.cpp
#include <iostream>
#include "symbolicc++.h"
using namespace std;

using SymbolicConstant::i;
using SymbolicConstant::pi;

Symbolic conjugate(const Symbolic &x) { return x[i==-i]; }

Symbolic ip(const Symbolic &x,const Symbolic &y)
{ return conjugate(y.transpose())*x; }

Symbolic ipmag(const Symbolic &x,const Symbolic &y)
{ Symbolic ipxy = ip(x,y); return sqrt(ipxy*conjugate(ipxy)); }

list<list<Symbolic> > mub(const list<Symbolic> &basis)
{
 int d = basis.size();
 int b, m, n;
 list<list<Symbolic> > bases;
 list<Symbolic>::const_iterator bi;
 for(b=0;b<d;++b)
 {
  bases.push_back(list<Symbolic>());
  list<Symbolic> &newbasis = bases.back();
  for(m=0;m<d;++m)
  {
   Symbolic sum = 0;
   for(n=0,bi=basis.begin();n<d;++n,++bi)
    sum += exp((b*n*(n-1)-2*n*m)*i*pi/d)* *bi;
   sum /= sqrt(Symbolic(d));
   newbasis.push_back(sum);
  }
 }
 return bases;
}

void output_mub_information(const Symbolic &b1,const Symbolic &b2,
                            const Symbolic &b3)
```

```
{
 list<Symbolic> b;
 list<Symbolic>::iterator li, lj, lk;
 list<list<Symbolic> > bases;
 list<list<Symbolic> >::iterator lb, lb2;
 Symbolic sqrt3 = sqrt(Symbolic(3));
 Symbolic x("x");
 Equations simprules = (
  exp(1*i*pi/3)==1/Symbolic(2)+i*sqrt3/2,
  exp(2*i*pi/3)==-1/Symbolic(2)+i*sqrt3/2,
  exp(3*i*pi/3)==-1,
  exp(4*i*pi/3)==-1/Symbolic(2)-i*sqrt3/2,
  exp(5*i*pi/3)==1/Symbolic(2)-i*sqrt3/2,
  exp(6*i*pi/3)==1,
  exp(7*i*pi/3)==1/Symbolic(2)+i*sqrt3/2,
  exp(8*i*pi/3)==-1/Symbolic(2)+i*sqrt3/2,
  exp(9*i*pi/3)==-1,
  exp(-1*i*pi/3)==1/Symbolic(2)-i*sqrt3/2,
  exp(-2*i*pi/3)==-1/Symbolic(2)-i*sqrt3/2,
  exp(-3*i*pi/3)==-1,
  exp(-4*i*pi/3)==-1/Symbolic(2)+i*sqrt3/2,
  exp(-5*i*pi/3)==1/Symbolic(2)+i*sqrt3/2,
  exp(-6*i*pi/3)==1,
  exp(-7*i*pi/3)==1/Symbolic(2)-i*sqrt3/2,
  exp(-8*i*pi/3)==-1/Symbolic(2)-i*sqrt3/2,
  exp(-9*i*pi/3)==1);
 cout << "Basis: " << endl; cout << b1 << endl;
 cout << b2 << endl; cout << b3 << endl;
 b.push_back(b1); b.push_back(b2); b.push_back(b3);
 cout << "Test orthonormality: " << endl << endl;
 for(li=b.begin();li!=b.end();++li,cout<<endl)
  for(lj=b.begin();lj!=b.end();++lj)
   cout << ip(*li,*lj) << " ";
 cout << endl;
 bases = mub(b);
 cout << "Mutually unbiased bases: " << endl << endl;
 for(lb=bases.begin();lb!=bases.end();++lb)
  for(li=lb->begin();li!=lb->end();++li)
   *li = (*li)[simprules];

 for(lb=bases.begin();lb!=bases.end();++lb)
 {
  cout << "MUB: " << endl;
  for(li=lb->begin();li!=lb->end();++li) cout << *li << endl;
  cout << "Test orthonormality: " << endl << endl;
  for(li=lb->begin();li!=lb->end();++li,cout << endl)
```

```
  for(lj=lb->begin();lj!=lb->end();++lj) cout << ip(*li,*lj) << " ";
 cout << endl;
 cout << "Inner product magnitudes with original basis:"
      << endl << endl;
 for(li=lb->begin();li!=lb->end();++li)
  for(lj=b.begin();lj!=b.end();++lj) cout << ipmag(*li,*lj) << endl;
 cout << endl;
}

cout << "Inner product magnitudes between MUB basis:"
     << endl << endl;
for(lb=bases.begin();lb!=bases.end();++lb)
{
 for(lb2=bases.begin();lb2!=bases.end();++lb2)
  if(lb != lb2)
   for(li=lb->begin();li!=lb->end();++li)
    for(lj=lb2->begin();lj!=lb2->end();++lj)
     cout << ipmag(*li,*lj) << endl;
}
 cout << endl;
}

int main(void)
{
 Symbolic sqrt2 = sqrt(Symbolic(2)); Symbolic sqrt3 = sqrt(Symbolic(3));
 Symbolic sqrt6 = sqrt2*sqrt3;
 Symbolic b1("",3), b2("",3), b3("",3);
 b1(0) = 1; b1(1) = 0; b1(2) = 0;
 b2(0) = 0; b2(1) = 1; b2(2) = 0;
 b3(0) = 0; b3(1) = 0; b3(2) = 1;
 output_mub_information(b1,b2,b3);
 cout << endl << "=========================" << endl << endl;
 b1(0) = 1; b1(1) = 0; b1(2) = 0;
 b2(0) = 0; b2(1) = 1/sqrt2; b2(2) = 1/sqrt2;
 b3(0) = 0; b3(1) = 1/sqrt2; b3(2) = -1/sqrt2;
 output_mub_information(b1,b2,b3);
 cout << endl << "=========================" << endl << endl;
 b1(0) = 1/sqrt3; b1(1) = 1/sqrt3; b1(2) = 1/sqrt3;
 b2(0) = -2/sqrt6; b2(1) = 1/sqrt6; b2(2) = 1/sqrt6;
 b3(0) = 0; b3(1) = 1/sqrt2; b3(2) = -1/sqrt2;
 output_mub_information(b1,b2,b3);
 return 0;
}
```

♣

Example 5.34. Consider the 2×2 matrix

$$F(s) = \begin{pmatrix} \cosh(s) & \sinh(s) \\ \sinh(s) & \cosh(s) \end{pmatrix}.$$

We integrate entrywise from 0 to $t > 0$ and then evaluate the commutator of the two matrices. The result is the 2×2 zero matrix.

```
/* integration.mac */
F: matrix([cosh(s),sinh(s)],[sinh(s),cosh(s)]);
assume(t > 0);
g11(t) := ''(integrate(F[1,1],s,0,t));
g12(t) := ''(integrate(F[1,2],s,0,t));
g21(t) := ''(integrate(F[2,1],s,0,t));
g22(t) := ''(integrate(F[2,2],s,0,t));
G: matrix([g11(t),g12(t)],[g21(t),g22(t)]);
C: F . G - G . F;
C: expand(C);
```

♣

Example 5.35. Consider the matrices

$$A = \begin{pmatrix} 0 & 1 \\ 1 & 0 \end{pmatrix}, \quad B = \begin{pmatrix} 1 & 0 \\ 0 & -1 \end{pmatrix}, \quad \rho = \begin{pmatrix} 1/2 & 0 \\ 0 & 1/2 \end{pmatrix}$$

where ρ is a density matrix, $A = \sigma_1$ and $B = \sigma_3$. We show that (*uncertainty relation*)

$$(\text{tr}(A^2\rho) - (\text{tr}(A\rho))^2)(\text{tr}(B^2\rho) - (\text{tr}(B\rho))^2)$$

$$\geq \left(\frac{1}{2}(\text{tr}(AB\rho) + \text{tr}(BA\rho)) - \text{tr}(A\rho)\text{tr}(B\rho) \right)^2 - \frac{1}{4}(\text{tr}(AB\rho) - \text{tr}(BA\rho))^2.$$

```
/* uncertainty.mac */
load("nchrpl");
A: matrix([1,0],[0,-1]);
B: matrix([0,1],[1,0]);
rho: matrix([1/2,0],[0,1/2]);
/* left-hand side */
T1: mattrace(A . rho);
T2: mattrace(A . A . rho);
T3: mattrace(B . rho);
T4: mattrace(B . B . rho);
LHS: (T2-T1*T1)*(T4-T3*T3);
/* right-hand side */
V1: mattrace(A . B . rho);
V2: mattrace(B . A . rho);
RHS: (1/2*(V1+V2)-T1*T3)^2 - 1/4*(V1-V2)^2;
```

♣

Example 5.36. We provide a C++ implementation of the *Cartesian product* of two sets.

```cpp
// cproduct.cpp
// c++ -std=c++11 -o cproduct cproduct.cpp

#include <iostream>
#include <set>
#include <string>
#include <utility>
using namespace std;

template <class T1,class T2>
set<pair<T1,T2>>
cartesian_product(const set<T1> &X,const set<T2> &Y)
{
  set<pair<T1,T2>> CP;
  for(auto &x : X)
   for(auto &y : Y) CP.insert(make_pair(x,y));
  return CP;
}

template <class T>
ostream& operator << (ostream& o,const set<T> &X)
{
  string comma = " ";
  o << "{";
  for(auto &x : X) { o << comma << x; comma = ","; }
  o << " }";
  return o;
}

template <class T1,class T2>
ostream& operator << (ostream& o,const pair<T1,T2>& p)
{ return o << "(" << p.first << "," << p.second << ")"; }

int main(void)
{
 set<string> A = { "x", "y", "z" }; set<int> B = { 3, 5 };
 cout << "A = " << A << endl;
 cout << "B = " << B << endl;
 cout << "A x B = " << cartesian_product(A,B) << endl;
 return 0;
}
```

♣

Example 5.37. In Java the `AffineTransform` class represents a two dimensional affine transform that performs a linear map from two dimensional coordinates (x_1, x_2) to other two dimensional coordinates (x_1', x_2') that preserves the straightness and parallelness of lines. The transform is given by

$$\begin{pmatrix} x_1' \\ x_2' \\ 1 \end{pmatrix} = \begin{pmatrix} t_{00} & t_{01} & t_{02} \\ t_{10} & t_{11} & t_{12} \\ 0 & 0 & 1 \end{pmatrix} \begin{pmatrix} x_1 \\ x_2 \\ 1 \end{pmatrix}.$$

```java
// Graph2D2.java

import java.awt.Frame;
import java.awt.event.*;
import java.awt.Graphics;
import java.awt.Graphics2D;
import java.awt.Color;
import java.awt.geom.AffineTransform;
import java.awt.geom.GeneralPath;
import java.awt.Font;
import java.awt.RenderingHints;

public class Graph2D2 extends Frame
{
 public Graph2D2()
 {
 addWindowListener(new WindowAdapter()
 { public void windowClosing(WindowEvent event)
 { System.exit(0); }});
 setTitle("Graphics2D Application");
 setSize(400,400);
 }

 public void paint(Graphics g)
 {
 Graphics2D g2d = (Graphics2D) g;  // type conersion
 g2d.setRenderingHint(RenderingHints.KEY_ANTIALIASING,
                      RenderingHints.VALUE_ANTIALIAS_ON);
 GeneralPath path = new GeneralPath(GeneralPath.WIND_EVEN_ODD);
 path.moveTo(0.0f,0.0f);
 path.lineTo(0.0f,125.0f);
 path.quadTo(100.0f,100.0f,225.0f,125.0f);
 path.curveTo(260.0f,100.0f,130.0f,50.0f,225.0f,0.0f);
 path.closePath();
 AffineTransform at = new AffineTransform();
 at.setToRotation(-Math.PI/8.0);
```

```
g2d.transform(at);
at.setToTranslation(0.0f,150.0f);
g2d.transform(at);
g2d.setColor(Color.red);
g2d.fill(path);
Font fo = new Font("TimesRoman",Font.PLAIN,40);
g2d.setFont(fo);
g2d.setColor(Color.blue);
g2d.drawString("ISSC",0.0f,0.0f);
} // end paint

public static void main(String[] args)
{ Frame f = new Graph2D2(); f.setVisible(true); }
}
```

♣

Exercise. (1) What is the output of the following Maxima program?

```
/* CayleyInverse.mac */
I2: matrix([1,0],[0,1]);
sig2: matrix([0,-%i],[%i,0]);
A: a*sig2;
Ap: A + %i*I2;
Am: A - %i*I2;
Api: invert(Ap);
Api: ratsimp(Api);
UC: Am . Api; UC: ratsimp(UC);
UCT: transpose(UC); UCT: ratsimp(UCT);
F: UC . UCT;
F: ratsimp(F);
```

(2) The six 2×2 matrices

$$I = \begin{pmatrix} 1 & 0 \\ 0 & 1 \end{pmatrix}, \quad A = \begin{pmatrix} -1 & -\sqrt{3} \\ \sqrt{3} & -1 \end{pmatrix}, \quad B = \begin{pmatrix} -1 & \sqrt{3} \\ -\sqrt{3} & -1 \end{pmatrix},$$

$$C = \begin{pmatrix} 1 & \sqrt{3} \\ \sqrt{3} & -1 \end{pmatrix}, \quad D = \begin{pmatrix} -1 & 0 \\ 0 & 1 \end{pmatrix}, \quad E = \begin{pmatrix} 1 & -\sqrt{3} \\ -\sqrt{3} & -1 \end{pmatrix}$$

form a group (*dihedral group*) under matrix multiplication. The first part of the following Maxima program provides the group table for the group. What does the second part provide?

```
/* grouptable.mac */
I: matrix([1,0],[0,1])$
```

```
A: matrix([-1,-sqrt(3)],[sqrt(3),-1])/2$
B: matrix([-1,sqrt(3)],[-sqrt(3),-1])/2$
D: matrix([-1,0],[0,1])$
C: matrix([ 1,sqrt(3)],[sqrt(3),-1])/2$
E: matrix([ 1,-sqrt(3)],[-sqrt(3),-1])/2$

for i in [I,A,B,C,D,E] do block(
 [x],
 x: [],
 for j in [I,A,B,C,D,E] do
  x: endcons(subst([I='I,A='A,B='B,C='C,D='D,E='E],i . j),x),
 print(x)
)$


p(x,y):= c20*x^2+c11*x*y+c02*y^2+c10*x+c01*y+c00$

for i in [I,A,B,C,D,E] do block(
 [x,y,xp,yp,z,v,pp,k,R],
 R: genmatrix(R,6,6),
 v: matrix([x],[y]), v: invert(i) . v,
 xp: matrix([1,0]). v, yp: matrix([0,1]). v,
 pp: expand(p(xp,yp)),
 print("p -> ",pp),
 k: expand(subst([x=0,y=0],diff(pp,x,2)/2)),
 R[1,1]: coeff(k,c20),R[1,2]: coeff(k,c11),R[1,3]: coeff(k,c02),
 R[1,4]: coeff(k,c10),R[1,5]: coeff(k,c01),R[1,6]: coeff(k,c00),
 k: subst([x=0, y=0], diff(pp,x,1,y,1)),
 R[2,1]: coeff(k,c20),R[2,2]: coeff(k,c11),R[2,3]: coeff(k,c02),
 R[2,4]: coeff(k,c10),R[2,5]: coeff(k,c01),R[2,6]: coeff(k,c00),
 k: expand(subst([x=0,y=0],diff(pp,y,2)/2)),
 R[3,1]: coeff(k,c20),R[3,2]: coeff(k,c11),R[3,3]: coeff(k,c02),
 R[3,4]: coeff(k,c10),R[3,5]: coeff(k,c01),R[3,6]: coeff(k,c00),
 k: subst([x=0,y=0],diff(pp,x,1)),
 R[4,1]: coeff(k,c20),R[4,2]: coeff(k,c11),R[4,3]: coeff(k,c02),
 R[4,4]: coeff(k,c10),R[4,5]: coeff(k,c01),R[4,6]: coeff(k,c00),
 k: subst([x=0,y=0],diff(pp,y,1)),
 R[5,1]: coeff(k,c20),R[5,2]: coeff(k,c11),R[5,3]: coeff(k,c02),
 R[5,4]: coeff(k,c10),R[5,5]: coeff(k,c01),R[5,6]: coeff(k,c00),
 k: subst([x=0,y=0],pp),
 R[6,1]: coeff(k,c20),R[6,2]: coeff(k,c11),R[6,3]: coeff(k,c02),
 R[6,4]: coeff(k,c10),R[6,5]: coeff(k,c01),R[6,6]: coeff(k,c00),
 print(R)
)$
```

Bibliography

[1] Abe E., *Hopf Algebras*, Cambridge University Press, 1977.

[2] Axler S. J., *Linear Algebra Done Right*, Springer, 1997.

[3] Barnett S., "Matrix Differential Equations And Kronecker Products", *SIAM J. Appl. Math.*, 1973, **24**, 1–5.

[4] Barnett S., "Inversion of partitioned matrices with patterned blocks", *Int. J. Systems Sci.*, 1983, **14**, 235–237.

[5] Barouch E., "Lax Pair for the Free Fermion Eight-Vertex Model", *Stud. Appl. Math.*, 1982, **70**, 151–162.

[6] Baumslag B. and Chandler B., *Group Theory*, Schaum's Outline Series, McGraw-Hill, 1968.

[7] Bäuerle G. G. A. and de Kerf E. A., *Lie Algebras*, North-Holland, 1990.

[8] Baxter R. J., *Exactly Solved Models in Statistical Mechanics*, Academic Press, 1982.

[9] Berberian S. K., *Introduction to Hilbert Space*, AMS, 1999.

[10] Biedenharn L. C. and Louck J. D., *Angular Momentum in Quantum Physics*, Cambridge University Press, 2009.

[11] Bohm A., *Quantum Mechanics*, Springer, 1994.

[12] Brewer J. W., "Kronecker Products and Matrix Calculus in System Theory", *IEEE Trans. Circ. Syst.*, 1978, CAS. 25, 772–781.

[13] Cartwright D. and Sturmfels B., "The Number of Eigenvalues of a Tensor", arXiv:1004.4953v2

[14] Ciarlet P. G., *Introduction to numerical linear algebra and optimisation*, Cambridge University Press, 1989.

[15] Choquet-Bruhat Y. and DeWitt-Morette C., *Analysis, Manifolds and Physics*, North Holland, Amsterdam (19??).

[16] Dăscălescu S., Năstăsescu C. and Raianu Ş., *Hopf Algebras: An Introduction*, Chapman & Hall/CRC Pure and Applied Mathematics, 2000.

[17] Davis P. J., *Circulant Matrices*, Wiley, 1979.

[18] Deif A. S., *Advanced Matrix Theory for Scientists and Engineers*, Abacus Press-Halsted Press, 1982.

[19] Elby A. and Bub J., "Triorthogonal uniqueness theorem and its relevance to the interpretation of quantum mechanics", *Phys. Rev. A*, 1994, **49**, 4213–4216.

[20] Elliott D. F. and Rao K. R., *Fast Fourier Transforms*, Academic Press, 1982.

[21] Fetter A. and Walecka J., *Quantum Theory of Many Particle Systems*, Dover Publications, 2003.

[22] Fletcher J. P., "Symbolic Processing of Clifford Numbers in C++", Paper 25, AGACSE, 2001.

[23] Fulton W. and Harris J., *Representation Theory*, Springer, 1991.

[24] Gantmacher F. R., *The Theory of Matrices*, Chelsea Publishing Company, 1960.

[25] Graham A., *Kronecker Product and Matrix Calculus with Applications*, Ellis Horwood Limited, 1981.

[26] Gröbner W., *Matrizenrechnung*, Bibliographisches Institut, 1966.

[27] Halmos P. R., *Introduction to Hilbert space and the theory of spectral multiplicity*, AMS, 2000.

[28] Hardy Y., Tan K. S. and Steeb W.-H., *Computer Algebra with SymbolicC++*, World Scientific, 2008.

[29] Hardy Y. and Steeb W.-H., "Vec-Operator, Kronecker Product and Entanglement", *Int. J. Alg. Comp.*, 2010, **20**, 71–76.

[30] Henderson H. V. and Searle S. R., "The vec-permutation matrix, the vec operator and Kronecker products: A Review", *Lin. Mult. Alg.*, 1981, **9**, 271–288.

[31] Henderson H. V., Pukelsheim F. and Searle S. R., "On the History of the Kronecker Product", *Lin. Mult. Alg.*, 1983, **14**, 113–120.

[32] Horn R. A. and Johnson C. R., *Topics in Matrix Analysis*, Cambridge University Press, 1991.

[33] Huang K., *Statistical Mechanics*, Second Edition, Wiley, 1987.

[34] Humphreys J. E., *Introduction to Lie Algebras and Representation Theory*, Springer, 1980.

[35] Jacobson N., *Lie Algebras*, Interscience Publisher, 1962.

[36] Kassel C. , *Quantum Groups*, Springer, 1995.

[37] Kaufman B., "Crystal Statistic. II. Partition Function Evaluated by Spinor Analysis", *Phys. Rev.*, 1949, **76**, 1232–1243.

[38] Kent A., "A note on Schmidt states and consistency", *Phys. Lett. A*, 1995, **196**, 313–317.

[39] Kulish P. P. and Sklyanin E. K., "Quantum Inverse Scattering Method and the Heisenberg Ferromagnet", *Phys. Lett.*, 1979, **70A**, 461–463.

[40] Lancaster P., *Theory of Matrices*, Academic Press, 1969.

[41] Laub A. J., *Matrix Analysis for Scientists and Engineers*, SIAM, 2005.

[42] Lieb E. H., "Solution of the Dimer Problem by the Transfer Matrix Method", *Journ. Math. Phys.*, 1967, **8**, 2339–2343.

[43] Ludwig W. and Falter C., *Symmetries in Physics*, Springer, 1988.

[44] Miller W., *Symmetry Groups and Their Applications*, Academic Press, 1972.

[45] Neudecker H., "Some Theorems on matrix differentiation with special reference to Kronecker matrix products", *J. Amer. Stat. Ass.*, 1969, **64**, 953–963.

[46] Onsager L., "Zero Field Properties of the Plane Square Lattice", *Phys. Rev.*, 1944, **65**, 117–149.

[47] Percus J. K., *Combinatorical Methods*, Springer, 1971.

[48] Prugovečki E., *Quantum Mechanics in Hilbert Space*, Second Edition, Academic Press, 1981.

[49] Regalia P. A. and Mitra S. K., "Kronecker products, unitary matrices and signal processing applications", *SIAM Rev.*, 1989, 586–613.

[50] Searle S. R., *Matrix Algebra Useful For Statistics*, Wiley, 1982.

[51] Sogo K. and Wadati M., "Boost Operator and Its Application to Quantum Gelfand-Levitan Equation for Heisenberg-Ising Chain with Spin One-Half", *Prog. Theor. Phys.*, 1983, **69**, 431–450.

[52] Steeb W.-H., "A Comment on Trace Calculations for Fermi Systems", *Acta Phys. Hung.*, 1977, **42**, 171–177.

[53] Steeb W.-H., "The Relation between the Kronecker Product, the Trace Calculation, and the One-dimensional Ising Model", *Lin. Alg. Appl.*, 1981, **37**, 261–265.

[54] Steeb W.-H. and Wilhelm F., "Exponential Functions of Kronecker Products and Trace Calculation", *Lin. Mult. Alg.*, 1981, **9**, 345–346.

[55] Steeb W.-H., *Hilbert Spaces, Wavelets, Generalized Functions and Modern Quantum Mechanics*, Kluwer, 1998.

[56] Steeb W.-H., *Continuous Symmetries, Lie Algebras, Differential Equations and Computer Algebra*, second edition, World Scientific, 2007.

[57] Steeb W.-H., *Problems and Solutions in Introductory and Advanced Matrix Calculus*, World Scientific, 2006.

[58] Steeb W.-H., "Diabolic Points and Entanglement", *Phys. Scr.*, 2010, **81**, 025012.

[59] Steeb W.-H. and Hardy Y., *Problems and Solutions in Quantum Computing and Quantum Information 2nd ed.*, World Scientific, 2006.

[60] Steeb W.-H., *Theoretical and Mathematical Physics: Problems and Solutions*, World Scientific, 2018.

[61] Steeb W.-H. and Lai Choy Heng, "Lax Representation and Kronecker Product", *Int. J. Theor. Phys.*, 1996, **35**, 475–479.

[62] Steeb W.-H. and Hardy, Y., "Entangled Quantum States and the Kronecker Product", *Z. Naturforschung*, 2002, **57a**, 689–691.

[63] Steeb W.-H, Hardy Y. and Stoop R., "Lax Representation and Kronecker Product", *Phys. Scrip.*, 2003, **67**, 464–465.

[64] Steiner P. A. J., "Real Clifford algebras and their representations over the reals", *J. Phys. A: Math. Gen.*, 1987, **20**, 3095–3098.

[65] Stewart G. W., "Computing the CS Decomposition of a Partitioned Orthonormal Matrix", *Numer. Math.*, 1982, **40**, 297–306.

[66] Sweedler M. E., *Hopf Algebras*, Benjamin, 1969.

[67] Takhtadzhyan L. A. and Faddeev L. D., "The Quantum Method of the Inverse Problem and the Heisenberg XYZ-Model", *Russ. Math. Surv.*, 1979, **34**, 11–68.

[68] Takhtajan L. A, *Introduction to Quantum Groups*, Springer, 1990.

[69] Van Loan C. F., "Computing the CS and the Generalized Singular Value Decompositions", *Numer. Math.*, 1985, **46**, 479–491.

[70] Van Loan C. F., "The ubiquitous Kronecker product", *J. Comp. Appl. Math.*, 2000 **123**, 85–100.

[71] Van Loan C.F. and Pitsianis N., "Approximation with Kronecker Products", *Linear Algebra for Large Scale and Real-Time Applications*, M.S. Moonen and G.H. Golub (eds), Kluwer Publications, 293–314, 1993.

[72] Villet C.M. and Steeb W.-H., "Kronecker Products of Matrices and an Application to Fermi Systems", *International Journal of Theoretical Physics*, 1986, **25**, 67–72.

[73] Weidmann J., *Linear Operators in Hilbert Space*, Springer-Verlag, New York 1980.

[74] Young N., *An Introduction to Hilbert space*, Cambridge Mathematical Textbooks, 1988.

[75] Zachos C., "Elementary Paradigms of Quantum Algebras", *Contemporary Mathematics*, 1992, **134**, 351–377.

[76] Maxima, a Computer Algebra System
http://maxima.sourceforge.net/

Index

369

Printed in the United States
By Bookmasters